FLORIDA STATE
UNIVERSITY LIBRARIES

OCT 3 0 2000

TALLAHASSEE, FLORIDA

Manufacturing
Montreal

Creating the
North American
Landscape

Gregory Conniff
Edward K. Muller
David Schuyler
Consulting Editors

George F. Thompson
Series Founder and Director

Published in cooperation
with the Center for American Places,
Santa Fe, New Mexico, and Harrisonburg, Virginia.

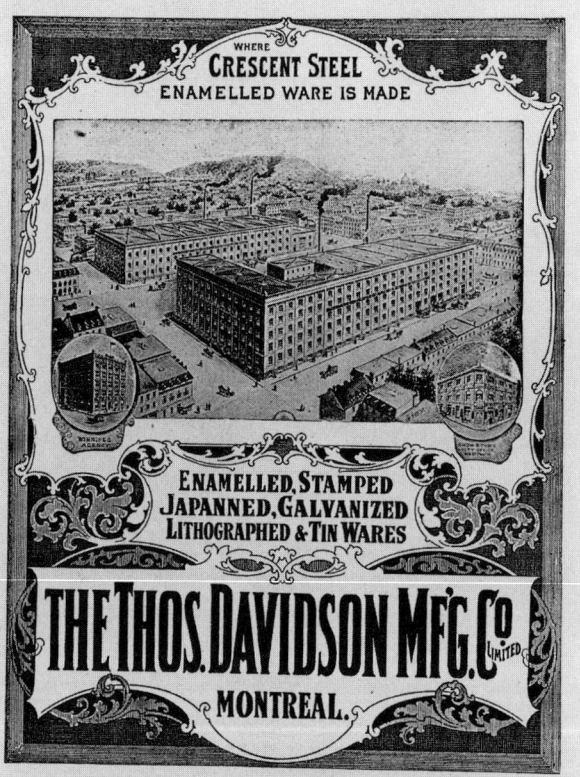

Manufacturing Montreal

The Making of an Industrial Landscape, 1850 to 1930

Robert Lewis

The Johns Hopkins University Press
BALTIMORE & LONDON

© 2000 The Johns Hopkins University Press
All rights reserved. Published 2000
Printed in the United States of America on acid-free paper
9 8 7 6 5 4 3 2 1

The Johns Hopkins University Press
2715 North Charles Street
Baltimore, Maryland 21218-4363
www.press.jhu.edu

Library of Congress Cataloging-in-Publication Data will be found at the end of this book.
A catalog record for this book is available from the British Library.

ISBN 0-8018-6349-X

Frontispiece: The Thomas Davidson Manufacturing Company. (Ernest Chambers, *The Book of Canada* [Montreal: Book of Canada Co., 1905], 333.)

HC
118
.M6
L49
2000

Contents

List of Illustrations   ix
List of Tables   xiii
Acknowledgments   xv

CHAPTER 1   "Living Tendrils"
North American Urban Manufacturing Districts   1

PART I: MONTREAL'S MANUFACTURING DISTRICTS, 1850 TO 1890

CHAPTER 2   "Marvellous Rapidity"
Montreal's Industrial Expansion   25

CHAPTER 3   "One Vast Block"
The Making of the Central Manufacturing Districts   49

CHAPTER 4   "Factories and Industrial Establishments of Various Kinds" in the Eastern Manufacturing Districts   78

CHAPTER 5   "The Whirr of Machinery and the Booming Noise of a Thousand Hammers" in the Western Manufacturing Districts   100

PART II: MONTREAL'S MANUFACTURING DISTRICTS, 1890 TO 1929

CHAPTER 6 "One of the Most Magnificent Cities"
Montreal's Economy, Planning, and Housing, 1890 to 1929   131

CHAPTER 7 "Pierced by Another Giant Skyscraper"
The Changing Fortunes of the Central Manufacturing Districts   157

CHAPTER 8 "Busy Hives of Industry" of the East and North Ends   185

CHAPTER 9 "Expanded in All Directions"
The Western Manufacturing Districts   221

Conclusion
Montreal's Manufacturing Districts, 1850 to 1929   255

Appendix: Sources, Case Studies, and Scale   273
Notes   279
Index   327

Illustrations

Figures

3.1	Owen McGarvey Store and Factory, 1890	54
3.2	Robert Mitchell Foundry, Montreal, QC, 1869	56
3.3	Harbor from the Canadian Pacific Railway Elevator, Montreal, QC, c. 1885	62
3.4	H. Shorey Clothing Factory, 1890	73
4.1	Michel Lefebvre's Vinegar Works, 1890	84
4.2	Some Views of Canadian Rubber, 1889	86
4.3	The Hudon Cotton Mill, 1882	89
5.1	Thomas Robertson's Lead Pipe Works, 1890	110
5.2	Foundries and Warerooms of Ives and Company, 1890	112
5.3	Frothingham and Workman's Côte Saint-Paul Factory Complex, 1880	120
5.4	The Montreal Rolling Mills, 1868	127
6.1	The Royal Electric Company, 1897	132
6.2	Belding, Paul, 1905	137
7.1	Herald Building, 1915	168
7.2	Landau and Cormack Cigarette Factory, 1907	175
7.3	L. O. Grothé Cigar Factory, 1906	177
7.4	Samuel Davis Cigar Factory, 1898	178
7.5	Caron Building, 1924	181
7.6	Amherst Building, 1925	182
7.7	John Peck Clothing Factory, 1915	184
8.1	The Canadian Rubber Factory, 1915	188
8.2	The Hochelaga Textile Mills, 1915	190

8.3	Dominion Oil Cloth Company, 1906	192
8.4	Montreal Locomotive and Machine Company, 1905	197
8.5	Canadian Steel Foundries, 1912	198
8.6	United Shoe Machinery, 1915	202
8.7	Dufresne and Locke, 1912	204
8.8	Brandram-Henderson Factory, 1915	209
8.9	Amherst Park, 1905	214
8.10	Workers' Cottages at Pointe-aux-Trembles, 1918	217
8.11	St. Lawrence Sugar Company, 1900	218
9.1	Montreal from Street Railway Power House Chimney, QC, 1896	224
9.2	Montreal from Street Railway Power House Chimney, QC, 1896	225
9.3	The Valley of the Lachine Canal, QC, c. 1910	226
9.4	The Clendinneng Foundry, 1892	231
9.5	Northern Electric, 1914	233
9.6	R. C. Jamieson Paint Factory, 1915	234
9.7	Plan of Canada Paint, 1896	236
9.8	Wellington Street, Pointe St. Charles, Montreal, QC, c. 1910	249
9.9	Merchant Cotton Advertisement, 1905	251

Maps

2.1	Montreal and Its Manufacturing Districts, 1890	43
3.1	Old Montreal: The Financial, Commercial, and Retailing Center, 1860	51
3.2	Redevelopment of the Hôtel-Dieu and Grey Nun Lands, 1852 and 1881	61
3.3	The Geography of the Printing Industry, 1890	68
3.4	The Geography of the Carriage Making Industry, 1890	70
3.5	The Geography of the Clothing Industry, 1890	75
4.1	Sainte-Marie and Hochelaga Manufacturing Districts, 1890	82
5.1	Expansion of the Redpath Sugar Refinery, 1856 to 1881	104
5.2	Canal and Griffintown Manufacturing Districts, 1881	108
5.3	The Western Suburban Manufacturing Districts, 1890	124
6.1	Montreal and Its Manufacturing Districts, 1929	134
7.1	Montreal's Clothing and Printing Districts, 1929	162
7.2	Central Montreal's Retail, Financial, and Legal Functions, 1909	164
8.1	Sainte-Marie and Hochelaga Manufacturing Districts, 1929	187
8.2	The New East End Manufacturing Districts, 1929	196
8.3	The Mile End Manufacturing District, 1929	207
9.1	Griffintown and Canal Manufacturing Districts, 1929	228
9.2	Mill Street, 1874 and 1918	229
9.3	The Older Western Suburban Manufacturing District, 1929	238
9.4	The Lachine Manufacturing District, 1929	239

Tables

1.1	Montreal Manufacturing Districts by Period of Establishment before 1929	8
1.2	Rent Structure of Montreal's Manufacturing Districts, 1861, 1890, and 1929	9
2.1	Rent Characteristics of Selected Montreal Manufacturing Sectors, 1861 and 1890	33
2.2	Population of Montreal and Suburbs, 1851 to 1891	41
2.3	Manufacturing Districts in Montreal, 1861 and 1890	47
3.1	Manufacturing Structure of Old Montreal and the Outer Core, 1861 and 1890	50
3.2	Occupational Structure of the Central Manufacturing Districts, 1861 and 1901	64
4.1	Manufacturing Structure of Sainte-Marie and Hochelaga, 1861 and 1890	80
4.2	Population, Place of Origin, and Religion in Sainte-Marie and Hochelaga, 1851 to 1901	93
4.3	Occupational Structure of Sainte-Marie and Hochelaga, 1861 to 1901	94
5.1	Manufacturing Structure of Griffintown and Canal, 1861 and 1890	107
5.2	Population, Place of Origin, and Religion in Saint-Ann and Saint-Gabriel, 1851 to 1901	113
5.3	Occupational Structure of Saint-Ann and Saint-Gabriel, 1861 and 1901	114
5.4	Population, Place of Origin, and Religion in Selected Western Suburbs, 1871 to 1901	118

6.1	Manufacturing Districts in Montreal, 1890 and 1929	133
6.2	Rent Characteristics of Selected Montreal Manufacturing Sectors, 1890 and 1929	140
6.3	Population of Montreal and Suburbs, 1891 to 1931	144
7.1	Manufacturing Structure of Old Montreal and the Outer Core, 1890 and 1929	159
8.1	Manufacturing Structure of Sainte-Marie and Hochelagas, 1890 and 1929	186
8.2	Manufacturing Structure of Maisonneuve, Mercier, and Montréal Est, 1929	195
8.3	Manufacturing Structure of Mile End and Plateau, 1929	206
9.1	Manufacturing Structure of Griffintown and Canal, 1890 and 1929	227
9.2	Manufacturing Structure of the Western Suburbs, 1890 and 1929	237

Acknowledgments

This book, like urban manufacturing districts, has a long and rocky history. It had its early beginnings as a dissertation at McGill University. While in Montreal, I received support from several people. Damaris Rose provided both encouragement and excellent comments in the early stages of the writing. Not only did I enjoy the many trips I made with David Hanna to the suburban city halls in search of assessment rolls and other municipal sources, but my understanding of the relationship between manufacturing and working-class housing in Montreal has benefited from his detailed and original work. Thanks also go to the Concordia team headed by Pat Thornton and Brian Slack, who provided an early forum for my ideas and logistic support for some of the data collection.

After its initial start at McGill, the manuscript for this book went through various stages of retooling at McMaster University and the University of Toronto. I have been fortunate to have received financial support and encouragement from the Departments of Geography at McMaster and the University of Toronto. In particular, their chairs, Martin Taylor (now at the University of Victoria), Carl Amrhein (now dean of the Faculty of Arts and Science at Toronto), and Joe Desloges provided me with an environment conducive to new research and writing.

Toronto's "smoking seminar" colleagues, especially the regulars, Rick Difrancesco, Larry Bourne, Alan Waterhouse, and Jim Lemon, provided me with a congenial environment from which I could draw some fresh air after the stuffiness of my smoke-free office. Gunter Gad has been both a friend and a great professional support. His encouragement over the years has been greatly appreciated. I have also been very fortunate to have had the aid of two excellent research assistants. Willie Jenkins was ever able and performed the many tricky tasks that I asked him to do with skill and perseverance. As a computer mapping illiterate, I relied on Carl Drouin to make the maps. He provided exemplary assistance, from the actual creation of the maps to his helpful comments on map layout and format.

A book of this kind depends on the support given by various manuscript depositories and libraries. The water tax rolls and various other materials were collected from the City of Montreal archives. My thanks to the staff, who cheerfully tolerated someone whose knowledge of French left a lot to be desired. Even though I could peruse most of the industrial journals at various libraries, the only repository of the *Canadian Cigar and Tobacco Journal* was the Imperial Tobacco Company factory in Saint-Henri (Montreal). I would like to thank the company's library staff for the unlimited access they gave me to the journal and, almost as important, for the privilege to consume the firm's product in the relaxing, if not hallowed, halls of the factory's library. I would also like to extend my gratitude to the staff at the McLennan Rare Books Library (McGill University), the Société Historique de Saint-Henri, and the interlibrary loan offices at the McGill, McMaster, and Toronto Universities for their kindness in helping me hunt down refuge theses, maps, directories, and various other oddities of historical research. The wonderful Notman photographs that grace the pages of this book were made available by the generosity of Brenda Klinkow of the McCord Museum of Canadian History, Montreal.

Writing and research time is also dependent upon money. I would like to thank the Social Science and Humanities Research Council and the Geography Department of McMaster University for postdoctoral scholarships, the University of Toronto for a Connaught Scholarship, and the Fonds pour la Formation de Chercheurs et l'Aide à la Recherche for a doctoral scholarship.

Several people have, one way or another, played a central role in the production of this book. Not only has Ted Muller's exemplary work on nineteenth-century manufacturing districts in Baltimore and Pittsburgh been an inspiration to me over the years, but, as an editor of Creating the North American Landscape series, he also showed me the kindness of asking that my dissertation be turned into a book. The book has benefited in untold ways from the editorial expertise of George F. Thompson and Randy Jones at the Center for American Places and Elizabeth Gratch and Kimberly Johnson at the Johns Hopkins University Press. Richard Harris has been friend, collaborator, and a great support over the years. He is the only person who read the final version of the final manuscript and, with little notice and a request for a quick turnover, made some extremely valuable comments. My greatest intellectual debt is to Sherry Olson. I cannot say too much about her support, patience, and encouragement, despite her own immense workload and responsibilities. Her research on the social geography of nineteenth-century Montreal is an inspiration to me. Finally, my greatest personal debt is to Lisa, Yonah, and Lev. Thank you for everything.

Manufacturing
Montreal

CHAPTER ONE

"Living Tendrils"

North American Urban Manufacturing Districts

> Huge industrial plants are uprooting themselves bodily from the cities. With households, small stores, lodges, churches, schools—clinging to them like living tendrils, they set themselves down ten miles away in the open.
> —G. Ferguson, "Decentralization of Industry,"
> *Journal of the Town Planning Institute of Canada*, April 1924

In a comprehensive review published in 1943 Jean Delage outlined the locational dynamics of metropolitan Montreal's manufacturing. Building his interpretation on an earlier study by Gérald Martin, Delage identified several manufacturing districts: the port, Lower Lachine Canal, East, North, and suburbs. There are several revealing aspects of this picture. Apart from their geographic coordinates Delage presents no logic binding the internal structure of these districts together. The port district, for example, consists of the area running from Old Montreal through Sainte-Marie, Hochelaga, Maisonneuve, and Longue Pointe. Here he lumps together areas built at different times and under different circumstances. Furthermore, manufacturing decentralization was driven by two factors: large firms seeking transportation facilities (highway, railroad, and port) and large lots of cheap land. Finally, the 1920s are considered to be the beginning of manufacturing diffusion from the central city to the suburbs. The location of Pepsi-Cola, Wellcome Drug, Continental Can, and Canadian Copper Refineries in the suburbs of Saint-Laurent, Lasalle, and Montréal Est are touted as the first indications of substantial industrial decentralization. For Delage the municipal boundary separating the city from the suburbs signified that industrial dynamics functioned in different ways on each side of the boundary. While the emerging industrial suburbs are considered to be the destinations of decentralizing firms, the central city, with its long history of manufacturing districts, social areas, institutions, and annexed suburbs, is seen as a unified and coherent spatial unit with no apparent legacy of suburbanized manufacturing.[1]

Delage and Martin were not alone in believing that industrial decentralization was a new phenomenon beginning in the 1920s. Most writers argue that interwar industrial diffusion paved the way for mass suburbanization after World War II. According to Leo Schnore, one of the most influential postwar commentators on metropolitan development, manufacturing at the turn of the century "was still concentrated in the large city." Before the 1920s and continuing until after World War II, according to Robert Fishman, the rail lines of the central "factory zone . . . gave enterprises located there a significant advantage over those anywhere else in the region." Although there may have been a slow dispersal of firms from the central city to the suburbs between the 1920s and the mid-1940s, Brian Berry and Yehoshua Cohen argue that "the massive decentralization of industry that followed" the war was a new phenomenon. According to Allen Scott, before World War II "the core of the large cities in the United States were still typically given over to a considerable degree of industrial production." David Slater, in an analysis of Canadian metropolitan industrial decentralization, makes a similar point. He argues that a "massive diffusion of manufacturing jobs" from the central city to the suburbs took place between 1939 and 1956. Although some firms had moved to the suburbs before the end of the war, Slater claims that "the change has mainly been concentrated in the period since 1948." Even those more sensitive to the historical nuances of metropolitan change understand the geography of manufacturing in the same terms. In his history of the American city Sam Bass Warner states that "industrial dispersion awaited the building of war plants on the fringes and the postwar manufacturing and housing boom."[2]

These writers also attribute manufacturing decentralization to the same factors identified by Delage and Martin: large firms seeking large amounts of cheap land where they could build single-story factories and the diffusion of the automobile and truck. From the 1920s, as Fishman explains, "trucks made it possible for factory owners to leave the crowded streets of the industrial zone for cheaper land on the periphery." In the 1920s, according to Schnore, "both industry and population were scattering as a response to the development of the motor vehicle." This process was accelerated after the war. Warner dates dispersal to the postwar period, when "the truck and automobile did at last free many firms from traditional central locations." Berry and Cohen argue that "superhighways, long-haul trucks, and piggyback combination of road and rail transport reversed the pull of rail terminals and docks in central area on industrial location." The search for large lots of land and the declining importance of centrally located transportation terminals, they continue, ensured that "suddenly . . . peripheral locations became as (or even more) accessible to the metropolitan region than was the traditional city center." Playing down transportation's role, Scott argues that postwar decentralization was led by large, high-volume firms seeking cheap land on the urban fringe. Finally, in their influential text *Urban America* David Goldfield and Blaine Brownell state that after World War II "industries seeking space, better access to transportation (the motor truck), tax advantages, and proximity to a skilled labor pool moved to the fringe." Regardless of the particular weight given to these factors, few geographers, economists, or historians would disagree with the basic picture of centralized manufacturing before World War I, the beginnings of decentralization in the 1920s, and massive industrial suburbanization after 1945.[3]

Similarly, writers have stressed that residential decentralization before World War I was mainly a middle-class phenomenon. Linking social mobility and home ownership to urban fringe development, scholars have highlighted the emergence and growth of middle-class residential suburbs before 1945. While it is acknowledged that some working-class suburban development occurred in places such as Pullman, these scholars emphasize two quite separate spatial and class-based realms. In particular, the image of a poor, immigrant inner city is contrasted with the wealthy, native or assimilated suburbs. Moreover, this class and ethnic metropolitan geography is directly linked to the location of manufacturing jobs. Reflective of this is Fishman's argument that prewar suburbs "remained an elite enclave, completely dependent on the central city for jobs and essential services," and the central factory zone was "the home of a large skilled and unskilled work force which only those enterprises within the zone could tap." In works of the Chicago School to those of recent urban historians this dualism between middle-class suburbs and the working-class inner-city has dominated the understanding of the social history and geography of American and Canadian cities.[4]

The origins of industrial and working-class suburbanization, however, are not to be found in the twentieth century. The movement of industry and the working class to the urban fringe and suburban municipalities has been taking place since the middle of the nineteenth century and has been documented in a small number of city studies. In New York the larger metalworking firms, with their highly developed division of labor and greater technical sophistication, left central Manhattan for suburban locations as early as the 1840s. In Boston the suburbs of Roxbury and Jamaica Plain contained an assortment of factories producing various products. By the 1850s a large textile manufacturing cluster had developed in the suburbs of Kensington and Germantown outside Philadelphia. In Baltimore several fringe manufacturing districts had formed around shipyards, railroad shops, food processing mills, and metal fabricating factories by 1860.[5]

The development of new urban fringe manufacturing districts continued in each cycle of growth. In the only full-length monograph on industrial suburbs, *Satellite Cities*, Graham Taylor points to the exodus of industry and the working class to satellite cities surrounding St. Louis, Chicago, and Cincinnati from the 1880s. Similarly, longitudinal studies of industrial decentralization employing census data that divides metropolitan areas into central cities and suburbs point to early industrial decentralization. Glenn McLaughlin's survey of the thirteen largest American industrial metropolitan districts, for example, concludes that "the dominance of the major city in manufacturing employment has tended to lessen significantly since 1879. A great expansion in factory employment, particularly since 1899, has occurred in suburban districts either in smaller communities or in previously unsettled districts where satellite manufacturing centers have grown up."[6]

This development is confirmed by several case studies. By the late nineteenth century the incorporation of towns and small cities from New Jersey, Long Island, and southern Connecticut extended and diversified the manufacturing geography of the New York metropolitan area. It also took place in the newer industrial cities in the interior and along the Great Lakes, as industrialization spread across the continent. By the mid-1920s the Milwaukee region had eleven industrial districts, six of them outside the city, while Detroit's automobile and ancillary industries formed an industrial crescent along the major transportation facilities encircling the old nineteenth-century city. On the West Coast a similar pattern of specialized manufacturing districts on the edge of the expanding city has been documented for Los Angeles and San Francisco.[7]

Working-class and immigrant suburbs coalesced around industrial districts. In the South, the least industrialized region of the United States, working-class suburbs grew up around the relatively small cities of Knoxville, Nashville, and Birmingham. In the larger metropolitan areas of the Northeast and the Midwest, such as Pittsburgh, Milwaukee, and Chicago, workers followed industry out to Homestead, West Allis, and Cicero. On the West Coast suburban factories in Los Angeles contributed to the formation of four major working-class suburban concentrations: Whittier-Fullerton, San Pedro–Long Beach, El Segundo–Manhattan Beach, and Vernon–Huntingdon Park. In Philadelphia's mid-nineteenth-century textile suburbs were substantial numbers of English and Irish immigrants, while twelve different Polish neighborhoods developed between 1880 and 1920, many of them on Philadelphia's built-up fringe and all of them reflecting the industrial structure of the area.[8]

This body of evidence points to the formation of noncentral manufacturing districts from the nineteenth century and suggests that the central city's hold on manufacturing was lessening. In other words, the move of manufacturing and the working class to greenfield sites (city fringe districts, industrial suburbs, and satellite towns) was an important component of the American and Canadian metropolis from as early as the 1850s. But this evidence for the process of industrial and working-class decentralization is suggestive, rather than conclusive. The detailed case studies are restricted to one district, one year, or one growth cycle. The longitudinal studies are limited by their reliance on census-based delimitations of central city and suburbs. There are no studies that employ detailed empirical data to trace the changing manufacturing geography of one city over an extended period of time, encompassing several business cycles, large-scale population growth, and extensive urban expansion. Other than fragmentary evidence and a few detailed case studies, little is known about any one city's manufacturing geography between the mid-nineteenth century and the Great Depression.

"The Best Returns for the Money Invested"

The purpose of this study is to examine Montreal's manufacturing districts between 1850 and 1929. The first comprehensive published study of one North American metropolitan area over an extended period of time, it chronicles the historical geography of successive Montreal manufacturing districts over several growth cycles. Building on studies of North American cities, the argument presented here is that manufacturing pathways, property dynamics, and growth politics laid the basis for the development of both central and suburban manufacturing districts in the nineteenth- and early-twentieth-century city. This account of Montreal's manufacturing districts features case studies of several important industries—clothing, tobacco, metal, paint, electrical, printing, food, textile, shoes, locomotive, and carriage making. This wide selection shows how the different strategies deployed by industries, the dynamics underlying firms' locational decision-making process, and a variety of interfirm and interindustry linkages contributed to the formation and specialization of manufacturing districts.[9]

The historical geography of Montreal's manufacturing districts and industries are based on data taken from a wide assortment of primary sources. Industrial journals, newspapers, government reports, city directories, and booster pamphlets provide descriptions ranging from the layout of factory space to interfirm product flows. Forming the basis for the empirical outline of the city and suburban manufacturing districts are data collected from the water tax rolls. Beginning in 1847, the water tax rolls provide the address and annual rent of every business establishment. In the absence of suitable census data, the tax rolls provide excellent spatial and temporal coverage and an important scale indicator of the metropolitan area's manufacturing base.

To capture the two broad periods of Montreal's growth, the entire set of city and suburban manufacturing firms were collected for 1861, 1890, and 1929. The first date represents the transition from craft production to modern industry and the spatial organization of the initial movement to a modern industrial urban complex. The choice of 1890 highlights the city's industrial geography at the point between the demise of the expansion beginning in the 1840s and the transition to the next one. It captures industry before the deep recession of the mid-1890s and the massive capital investment that came with the Second Industrial Revolution, and it identifies the well-established districts associated with fifty years of industrial growth beginning in the 1840s. The 1929 data portrays Montreal's industrial geography associated with the winding down of the great expansion of corporate capitalism occurring between the late 1890s and the 1920s.[10] (For a discussion of sources and method, see the Appendix.)

Four important features of Montreal's manufacturing districts are discussed in this study. First, in each cycle of growth the crosscurrents generated by the uneven development of technology, the labor process, and firm organization established the parameters for the creation of new manufacturing space farther out from the city center. In 1861 the new districts established on the edge of the built-up city during the 1850s accounted for more than half of the city's total manufacturing rent (table 1.1). By 1890 the districts formed during the 1850s and 1860s had more than 50 percent of the city rent, and a new set of districts farther from the city core accounted for more than 20 percent of the city's rent. By 1929, despite the continuing importance of the pre-1870 districts, surges of industrial investment consolidated the role of the districts formed between 1870 and 1900 and created a set of post-1900 manufacturing districts on the metropolitan fringe.

Second, in each round of growth there was greater specialization over urban space, as firms clustered in new manufacturing districts. New districts differed from older districts in terms of their scale, technology, labor force, and industrial type. As early as 1861, seven manufacturing districts can be identified (table 1.2). With the largest median rent, the new Canal district had significant concentrations of large flour, sugar, metal, and locomotive manufacturing firms. In contrast, Saint-Lawrence and Saint-Jacques had much smaller median rents ($48 and $38, respectively) and a different rent distribution and assortment of industries. This division of labor characterized metropolitan Montreal's geography of manufacturing over the next seventy years. By 1890 a new set of districts had emerged with a different array of industries, while the older ones continued to grow. The industrial suburb of Hochelaga, with its large cotton, locomotive, and tobacco manufacturers, for example, differed from Canal, Saint-Lawrence, and Saint-Jacques. In the early twentieth century new manufacturing districts, with their soot-blackened smokestacks, poorly built and overcrowded housing, and miles of streetcar lines, railway tracks, and harbor facilities, pushed out into the sparsely settled areas surrounding the nineteenth-century manufacturing districts. By 1929 Montréal Est, with a small number of extremely large cement, oil refining, and locomotive firms, stood in sharp contrast to Mile End's mass of smaller firms from a variety of industries, Old Montreal and Saint-Antoine's cluster of clothing and printing shops, and Canal's metal and food manufacturers.

Third, in each growth cycle industrial expansion on the urban fringe set into motion a string of multipliers upon transportation facilities, urban services, and working-class housing. It may have been true, as a commentator noted in 1910, that "there are but few kinds of business which demand a particular and restricted location. For this reason it is obvious that in nearly every kind of business a location can be selected which will furnish the best returns for the money invested." Selection, however, depended on the fashioning of urban space to meet manufacturers' needs. Accordingly, new districts (satellite towns, suburbs, or accretions to the city periphery) were created by alliances of local elites, which were responsible for the creation of a set of locational assets, or ensemble of properties—worker housing, infrastructure hardware such as transportation facilities and water systems, and a good "business climate"—which attract firms to settle at a particular site. For manufacturers calculating the best returns on their investment not only involved a simple locational cost analysis but also an intimate understanding of the industrial, building, and infrastructural dynamics of metropolitan districts.[11]

Table 1.1 Montreal Manufacturing Districts by Period of Establishment before 1929

Period When Districts Were Established	1861				1890					1929				
	No. of Firms	City Share (%)		Mean Rent ($)	No. of Firms	City Share (%)		Mean Rent ($)		No. of Firms	City Share (%)		Mean Rent ($)
		Firm	Rent			Firm	Rent				Firm	Rent	
Pre-1850	323	51.2	44.2	203	445	32.6	28.9	503		376	15.4	10.7	1,989
1850–1869	308	48.8	55.8	270	791	57.9	50.6	494		1,232	50.5	40.0	2,271
1870–1899	—	—	—	—	129	9.5	20.5	1,227		277	11.3	25.3	6,403
1900–1929	—	—	—	—	—	—	—	—		502	20.5	22.6	3,159
Total	631	100	100	235	1,365	100	100	567		2,444	100	100	2,866

Source: Water Tax Rolls for the City of Montreal and surrounding municipalities, 1861, 1890, and 1929.
Note: The 1929 totals do not add up because scattered firms were not included in manufacturing districts.

Table 1.2 Rent Structure of Montreal's Manufacturing Districts, 1861, 1890, and 1929

District	Firm Median Rent ($)	Rent Distribution in 1861 (%)			Three Largest Industries ($ Rent)	
		Large >$299	Medium $50–299	Small <$50	Industries	Total (%)
Old Montreal	120	23	57	20	Clothing, leather, printing	63
Saint-Antoine	100	22	50	28	Clothing, beverage, leather	61
Saint-Lawrence	48	9	40	51	Beverage, food, furniture	64
Saint-Jacques	38	4	25	71	Leather, food, blacksmithing	66
Griffintown	120	20	57	23	Metalworking, lumber, leather	75
Canal	550	72	10	18	Food, locomotive, metalworking	87
Sainte-Marie	44	8	32	60	Beverage, rubber, food	75
City Total	100	23	48	29	Clothing, food, metalworking	47

District	Firm Median Rent ($)	Rent Distribution in 1890 (%)			Three Largest Industries ($ Rent)	
		Large >$449	Medium $100–449	Small <$100	Industries	Total (%)
Old Montreal	300	34	48	18	Clothing, printing, metalworking	55
Saint-Antoine	280	34	50	16	Clothing, leather, furniture	52
Saint-Lawrence	150	18	53	29	Clothing, beverage, metalworking	41
Saint-Jacques	100	5	49	46	Clothing, leather, food	59
Canal	310	46	23	31	Locomotive, metalworking, food	71
Griffintown	300	39	45	16	Metalworking, food, lumber	61
Saint-Henri	150	29	44	27	Metalworking, cotton, food	76
Lachine	60	Too few firms			Metalworking, beverage	96
Sainte-Marie	120	21	39	40	Rubber, beverage, nonmetallic	48
Hochelaga	145	36	28	36	Cotton, locomotive, tobacco	82
City Total	200	24	51	25	Metalworking, clothing, food	41

District	Firm Median Rent ($)	Rent Distribution in 1929 (%)			Three Largest Industries ($ Rent)	
		Large >$1,999	Medium $400–1,999	Small <$400	Industries	Total (%)
Old Montreal	900	24	57	19	Printing, food, clothing	63
Saint-Antoine	1,500	36	54	10	Clothing, printing, beverage	61
Saint-Lawrence	650	13	61	25	Clothing, food, beverage	72
Saint-Jacques	450	11	45	44	Leather, food, clothing	53
Griffintown	1,500	41	44	15	Metalworking, food, textile	65
Canal	2,400	53	31	17	Food, locomotive, electrical	67
Saint-Henri	1,400	38	34	28	Cotton, metalworking, food	58
Lachine	1,800	42	28	30	Metalworking, locomotive, glass	82
Sainte-Marie	700	30	44	26	Textile, rubber, beverage	54
Hochelaga	600	37	39	24	Cotton, tobacco, locomotive	67
Maisonneuve	950	32	34	34	Metalworking, food, textile	68
Montréal Est	11,800	94	—	6	Cement, locomotive, petroleum	90
Mile End	700	30	44	27	Food, paper, clothing	45
Plateau	600	21	56	23	Clothing, food, furniture	76
City Total	800	26	48	26	Clothing, metalworking, food	41

Source: Water Tax Rolls for Montreal and surrounding towns, 1861, 1890, and 1929.

Fourth, parallels are drawn between the historical geography of Montreal's manufacturing geography and that of other Canadian and American cities. Even though Montreal had unique ethnic, industrial, and class structures, the basic processes of capital accumulation, manufacturing growth, technological change, and shifting class relations were common to Toronto, New York, Pittsburgh, and Los Angeles. At this interpretative level the changing dynamics of industrial capitalism gave rise to similar urban patterns; the manufacturing districts identified in Montreal were replicated in many other North American cities. But differences mattered; industrial structure, ethnic composition, and site features, among other things, modified the manner and rate of industrial decentralization, manufacturing district formation, and working-class suburban development. Manufacturing geographies different from Montreal would appear, for example, in Pittsburgh, with its huge steel and glass corporations and its particular topography, and in New York, with its multitude of small, labor-intensive consumer industries and its diverse ethnic structure. Despite these differences, Montreal's story told here is representative of the broad forces driving North American industrialization, urban growth, and manufacturing district formation after 1850.

The story of the formation and development of Montreal's manufacturing districts is told in Chapters two to nine. Part One, which covers the period 1850 and 1890, examines the formation of manufacturing districts associated with the half-century of industrial expansion following the late 1840s and the imprinting of the initial specialization of manufacturing space upon the city's landscape. Chapter Two provides an overview of Montreal's rise as Canada's major manufacturing center from its staple-based, mercantile foundations. The next three chapters examine each of the manufacturing districts developing in this period: the diverse array of manufacturing firms clustering in the central manufacturing districts of Old Montreal and the Outer Core; the expansion of the East End districts, from their early origins in eighteenth-century Sainte-Marie to the emergence of the industrial suburb of Hochelaga in the 1870s; and the emergence of the West End districts alongside the Lachine Canal. Part Two examines the changes to and expansion of Montreal's manufacturing districts between 1890 and 1929. Chapter Six considers Montreal's growth dynamics and outlines the geography of its economic, social, and political changes. The next three chapters explore several features of the major manufacturing belts: the changing fortunes of the central manufacturing districts; the expansion of the East End nineteenth-century manufacturing districts into new territory, both farther east and into the North End; and the changes to the older districts and the formation of a new set of manufacturing districts in the West End.

Before turning to the story of Montreal's manufacturing districts, the broader dynamics of manufacturing and geographic change between 1850 and 1929 need to be reviewed. The rest of this chapter outlines the manner in which the manufacturing geography of North American cities is linked to the structure of industry, the building of working-class residential districts, the establishment of infrastructures, and the group activities of local elites.

"To Concentrate or Localize": Manufacturing Pathways and Manufacturing Districts

A central feature used to explain the manufacturing geography of cities is industrial structure. Most interpretations of urban geography as it was transformed from a small commercial center to a large industrial metropolis are based on a polarized view of industry. The classic story tracks the progress of the firm from a small, labor-intensive proprietorial company serving local markets to a large, capital-intensive corporation occupying several sites, having extensive markets, and exploiting economies of scale and scope. By the end of the nineteenth century the larger scale of operations, the specialization of plants, centralization of research facilities, intensification of the speed of materials through the plants, and greater control over distribution enabled companies to increase productivity, reduce unit costs, and acquire a larger market share. Firms such as American Tobacco, Westinghouse, and U.S. Steel were able to overcome the problems of the uncertainty of production and distribution by internalizing the processes that were previously done outside the firm and by creating economies of scope within firms. In contrast to the corporations manufacturing standardized products for a mass market, small firms were labor intensive, produced batches of goods for specialized or niche markets, and relied on fewer innovative methods of production and distribution. Underpinning these two different strategies was the search for higher profits on a larger output by the large, capital-intensive firm and high margins per unit by the small ones.[12]

This classic scenario identifies some genuine new elements in North American industrial history. There is little doubt that factories by the 1920s differed significantly from nineteenth-century factories in their size, amount of mechanization, specialization of function, and degree of managerial control. At midcentury most firms were small proprietorial concerns with a large share of skilled workers and serving local markets. By the 1920s, although numerous small firms continued to exist, a few capital-intensive corporations employing a large, semiskilled workforce, massive mechanical energy and high-volume techniques dominated many sectors.

But North American industrialization cannot be so neatly divided into two contrasting types. Even though it has been difficult to identify the strategic combinations lying between the extremes of batch and mass production, industry was characterized by a diversity of productive strategies. The traditional model is based on evidence taken from examples that illustrate the corporate case at the expense of the wide array of existing industries. But industries varied from the classic model, as demonstrated by Philadelphia's textiles and locomotive makers, Pittsburgh's steel firms, and American pottery makers. Another problem contributing to the polarized picture of the industrial spectrum derives from how industrial organization is measured. The ascendancy of the large corporation is usually demonstrated by figures indicating that a small number of firms controlled a large share of an industry and nation's employment and output. Despite a trend to concentrate employment and output in some sectors, however, these large-scale firms frequently behaved in different ways. They employed their own forms of technology and organization, undertook distinct strategies of marketing, and constructed different forms of labor relations. Moreover, even in those industries closely resembling the classic model there is evidence showing the deployment of a range of productive strategies. Milwaukee's Allis-Chalmers, for example, made a small number of large, made-to-order machines and a large number of smaller, mass-produced goods. Even within a plant or between plants of a single firm there were considerable differences in the organization of production. The sheer size of the firm tells us little about its social relations, its place in the economy, and its manufacturing strategies.[13]

These deficiencies with the traditional approach to industrial change have generated a rethinking of industrial organization. Building on the work of Scranton and others, the perspective taken in this study stresses the multitude of ways that firms respond to the interplay of technology, the labor process, and production organization. To be sure, firms do face common problems of industrial organization. They have to be linked to ensure that the necessary physical connections are made. Groups of workers have to be coordinated to allow for the smooth functioning of production. Materials, workers, machinery, and finances have to be regulated so that the system is economically viable. Within the context of these similarities, however, firms can employ a number of different productive strategies. Firms with common material, social, and technological structures tend to follow similar lines of development, while those with dissimilar structures face different procedures for organizing production. Firms are influenced by the markets in which they operate, from raw material supply markets to their product markets. Adding to this is the uneven introduction of new technologies to accelerate the transformation of raw materials into finished products. Differences also result from social and political difficulties associated with labor process reorganization. Labor is not uniform across and within industries. Industries, composed as they are of differences in terms of their material basis, the legacy of labor relations, and degree of mechanization, require different types of skills.[14]

These differences propel segments of the industrial system down different development pathways. Despite common problems of organization and similar material properties, industries themselves are characterized by a diversity of social and technological properties. Between the metal and the clothing industries, for example, exists a great variation in the ability to transform iron ore and cotton into steel and garments, respectively. Firms from one industry may have more in common with firms from another industry than they do with firms from their own industry. In the nineteenth-century leather industry footwear manufacture had more in common with other mechanized industries than it did with tanning, which was dependent upon the chemical transformation of hides to leather. In the textile and the metalworking sectors important differences existed between cotton, hosiery, steel, and machinery manufacture. Similarly, the tobacco industry was composed of the cigar, cigarette, and tobacco branches, all of which faced unique constraints and opportunities in terms of their labor force composition, input-output relations, and degree of mechanization. The meshing of different industrial properties produce a wide range of possible manufacturing pathways.

The two contrasting interpretations of industrial structure outlined here have different implications for the geography of urban manufacturing. The classic scenario highlights the impact of a polarized industrial structure on the historical geography of North American urban manufacturing. Geographers have taken the split between mass and batch producers and used it to draw a manufacturing geography featuring a dual spatial structure from the early twentieth century.

In the nineteenth and early twentieth centuries all firms are believed to have clustered in the central core. A central location provided handicraft shops, manufactories, and factories with all of their requirements. As most firms depended on local and regional markets, used little machinery, and produced an assortment of goods, firms remained small and vertically disintegrated and geared to a variety of manufacturing strategies. The closely bunched set of suppliers, subcontractors, customers, and producer services in the city center provided advantages for firms seeking strong interfirm linkages. Proximity to a large labor pool and transportation terminals further contributed to the cumulative attractions of the central districts. In addition, frequent refashioning of the built environment by razing and rebuilding buildings, installing new infrastructures, and redeveloping the street system buttressed the central area's attractions. Finally, fixed-capital inertia was a compelling factor behind the continued importance of central districts. At some point the amount of capital invested in a place becomes too large for firms to move freely. Incremental changes and additions over time froze firms to a location not acceptable under other circumstances. Together, this set of factors underpinned the growth of an extensive set of central factory districts surrounding the city's administrative, financial, and commercial activities between 1850 and 1929.[15]

According to the classic interpretation of urban manufacturing geography, however, the need for a central location diminished in the early twentieth century for a select set of firms. The locational ties holding firms to the central districts became polarized; small firms remained rooted to the city core, while the large corporation was free to decentralize. Beginning before World War I but becoming more significant by the 1920s, manufacturing dispersal to the suburbs was led by large capital-intensive, high-volume corporations employing space-consuming technologies and utilizing new transportation innovations. Even though nineteenth-century industrial suburbs and the dispersal of obnoxious industries to the urban periphery are acknowledged, both the description and explanation of noncentral manufacturing district formation focuses on the large, twentieth-century, capital-intensive firm. These firms—with their internal economies of scale and scope, new technologies, and large dispersed markets—were free to take advantage of ample cheap suburban land. In contrast, small, labor-intensive firms, unable to implement these strategies, were forced to remain in the city core. In other words, geographers have typically tagged industrial decentralization as beginning in the interwar period and directly linked this new manufacturing geography to the classic interpretation of industrial structure. Despite the elegance of the theory, however, this was not what happened.

A different interpretation of urban manufacturing geography results from viewing industrial change through the prism of manufacturing pathways. Manufacturing change after the mid-nineteenth century created greater locational flexibility for a wide set of firms than is generally supposed. The simple dualism of two types of industries mirrored in urban structure characterized by nineteenth-century centralization and early-twentieth-century decentralization cannot account for either the diversity of organizational form or the associated geography of manufacturing. This was acknowledged by G. Ferguson in 1924 when he noted that "industries tend to concentrate or localize, forming industrial centres which are devoted to the production of a limited line of products.... An analogy to this general characteristic of industry is seen in the way the industrial activities of a city tend to group themselves.... On the manufacturing side work shops will form one centre, planing mills another, wood-working another, and so on according to the commercial activities of the city." Firms from an early date had a locational flexibility, which, in turn, laid the foundation for the formation of geographically distinct manufacturing districts.[16]

One element underpinning widening locational choice was uneven technological change after 1850. Even though relatively simple by today's standards, technological change realigned the demand for urban manufacturing space by forcing new firm scale thresholds and redefining production organization. The changes taking place after 1850 are legendary: steam-, hydraulic-, and electrical-powered machinery triggered changes throughout the entire production system; the rotary press and stereotyping reorganized and increased the scale of printing and publishing; rolling mill technology refashioned the logic and scale of metalworking; new milling machinery underpinned the development of large-scale, high-volume flour and sugar mills; and early-twentieth-century propulsive industries such as automobile, electrical, and chemical intensified ongoing industrial change. These changes, however, were not spread uniformly across the industrial spectrum. Technological change was introduced at different times in different industries. In some cases mechanization introduced rapid, large-scale changes to an industry, while in others change was incremental and piecemeal. The overall effect was to promote quite different demands on the locational mobility of industry and to establish variation in the ability of industries to uproot themselves from a central location and to settle in greenfield sites.

The intensification of the division of labor also redrew the nature of industry and had important implications for the demand for manufacturing space. From the early nineteenth century both merchants and artisans initiated a more intensive technical division of labor. Firms grew rapidly by breaking down the work process into a series of discrete tasks, diluting skills, and assigning tasks to semiskilled workers. In many cases work specialization was linked to mechanization. The introduction of new machinery and new products in the steel industry in the 1880s transformed work tasks and led to greater work specialization. In other cases, however, a highly segmented division of labor preceded technological change. By the 1850s a specialized set of tasks in meatpacking, clothing, and shoemaking occurred without a great deal of mechanization. The drive to segment the labor force continued throughout the period and came to its disciplinary apotheosis with Taylorism at the end of the century.

The growing number of firms operating new technologies, producing an assortment of products, servicing larger market areas, and deploying highly segmented work places after midcentury produced much larger firms. As early as 1871, Montreal's largest firm—the locomotive shops of the Grand Trunk Railway—employed almost eight hundred workers. By 1881 the city's largest 15 percent of firms employed more than 70 percent of the city's workers. These firms not only employed a large workforce but also required larger factories and new layouts. The increasing scale of work forced firms to seek out new types of buildings to house the new work processes, the growing size of boilers and machinery, and larger workforces. Increasingly, the old artisan workshop, large enough to employ a small number of workers and often attached to a retail establishment or a private residence, was no longer able to house the growing scale of firms. From the shoe and vinegar factories employing a few hundred workers to the huge steel mills and locomotive plants with thousands of workers, firms needed to find a new set of factory spaces. After the mid-nineteenth century the typical multistory mill structure became the dominant building form, especially in the central city, where land was at a premium. In other cases more space-extensive, horizontal buildings were constructed as firms established their new machinery and their divided labor force in a sprawling set of buildings comprising different functions. Space demands varied from industry to industry and within industries. The needs of rolling and flour mills differed from that of shoe or clothing firms, while the space demands of cigarette manufacturers were dramatically different from cigar makers. In other words, there was great variation in the physical scale of factories and type of space internal to the plant. Over each generation these differences accelerated the search for new manufacturing sites; technological and labor process innovations established new locational options.[17]

In combination these changes realigned the basis of metropolitan locational choice and forced firms to implement a new spatial strategy. Firms from propulsive sectors had the greatest ability to initiate a new spatial strategy, although firms of all scales and types would move to the urban fringe. Many of the firms marching out of the city were textile, food processing, and metalworking in the nineteenth century and chemical, automobile, and electrical in the early twentieth century. Bursts of growth resulting from large capital investments, the demand for new production conditions, and the search for more suitable factory space presented propulsive firms with an opportunity to escape the limitations of existing districts and to take advantage of new locational possibilities. In the vanguard of introducing new technologies, intensifying the division of labor, and capturing new markets, propulsive industries created new thresholds of productivity, giving them "the tactical flexibility to undertake radically new spatial strategies." At the metropolitan scale, after 1850, this was frequently achieved on the urban fringe at greenfield sites—vacant city fringe or suburban sites with no legacy of manufacturing production. The early pioneers of Toronto's suburban West Toronto Junction were packing plants and large metalworking factories. South of the border the steel mills and metal-finishing works moved to the Calumet district south of Chicago from the 1870s; steel mills, glassworks, and electrical appliance factories settled along the valleys of Pittsburgh's Monongahela River; the automobile companies opened up the Detroit suburbs of Highland Park, River Rouge, and Hamtramck after 1900.[18]

Relocation to peripheral districts also permitted faster production turnover time by allowing firms to acquire gains from faster distributional methods, more efficient internal layout, and better labor control. New productive relations demanded new forms for the circulation of capital through individual firms and the city. Older central spaces, with their cramped sites, high costs, inadequate plant layout, and labor troubles, were an obstacle to the implementation of many forms of modern manufacturing practice. At greenfield sites manufacturing firms were at liberty to build new factories unimpeded by the restraints of existing industrial districts. One particular advantage was the ability to create the factory conditions allowing for the efficient through-flow of goods from raw materials to finished product. This was evident not only in individual firms but also in particular districts of the metropolitan area. In Chicago, for example, the need to accommodate the growing scale of individual packing houses and to remedy the high costs and inefficiencies of scattered locations forced manufacturers to consolidate the various components of the meatpacking industry into the suburban Union Stock Yards in 1865. In other cases suburban efficiency, scale, and through-flow were achieved by constructing manufacturing facilities around outlying resource deposits. By the end of the nineteenth century the oil fields of Los Angeles, the agricultural areas outside of San Francisco, and the coal fields of Pittsburgh were the homes of suburban manufacturing districts.[19]

Manufactures appeared on the urban fringe after 1850. Despite strong pressures inducing central factory clustering, a stream of firms compelled by the industrial conditions of the day sought greenfield sites on the urban fringe. Driven by the increasingly complex forms of the division of labor, incessant and insistent technological change, new scale thresholds in each cycle of growth, and the imposition of even more sophisticated forms of work organization, manufacturers periodically had to devise a new spatial strategy. The changes that they undertook in their search for profits induced them to respond to the new competitive conditions by seeking new locations and locational assets. Central factory districts continued to house a wide variety of industries, many of which had instituted modern production methods. After 1850, however, the central districts were increasingly unable to provide suitable sites for many manufacturers. Not only were they facing congested conditions and unsuitable buildings; they also had to deal with pressure on central land values as a result of the growing scale of the total manufacturing ensemble and competition from other economic activities for increasingly more expensive central land. Even though central districts underwent extensive redevelopment and had a significant advantage in the form of existing wholesale, communication, and transport networks, they also suffered from other limitations besides rising land costs. An inadequate plant layout of many buildings, increasing congestion, large-scale industrial growth, and intensifying class conflict all combined to make central districts unattractive locations. Moreover, the large sprawling enterprises that developed were unable to find adequate and suitable land for large-scale manufacturing. Under these circumstances after 1850 an increasing number of firms implemented a spatial strategy requiring noncentral locations.

"Without a Guiding Plan": The Construction of Manufacturing Districts

The dynamics of industrial change, nevertheless, only established the potential for firms' locations. The ability to choose a particular site, wherever it was located within the metropolis, depended on a package of locational assets, including, among other things, infrastructure, factory land, and workers' housing. These locational assets inducing firms to locate at a particular site did not just appear on the urban landscape; they had to be created. In the case of Montreal, according to an editorial in a 1927 *Journal of the Town Planning Institute of Canada,* "The city of Montreal has been allowed to grow up in a haphazard manner— without a guiding chart or plan. The result is, its large population is not economically distributed. While the workers are crowded in tenements which make for expensive but not good living conditions, there are hundreds of acres of vacant land within the city limits. Land values in consequence are not stabilized, and both the community and real estate owners suffer. The value of a comprehensive plan for the district of Montreal is obvious." Despite the complaints of early-twentieth-century urban reformers concerning the haphazard distribution of population and employment, the absence of a comprehensive plan, and the apparent anarchy of urban growth, several common elements contributed to the formation of suburban locational assets after the mid-nineteenth century.[20]

New transportation developments anchored industrial clustering throughout the city. Harbor facilities typically grew alongside the city's central district before 1850 and attracted most of the artisan workshops, manufactories, and factories. Transportation innovations after midcentury, however, established new locational assets for manufacturing. On the one hand, railroad passenger and freight terminals constructed outside of the emerging central business district created nuclei for factory districts and intensified industrial clustering in the central district. On the other hand, factory clusters developed on the urban fringe, as railroad companies built locomotive shops, marshaling yards, and larger freight terminals at a distance from the core. Similarly, new harbor, canal, and intra-urban transit facilities after 1850 stretched into unsettled city and suburban territory. These fringe nodes and corridors spurred decentralization and established noncentral manufacturing clusters in several ways. Increasing scale compelled enterprises increasingly reliant on bulky raw material supplies, such as coal, iron ore, and wheat, to seek factory sites in close proximity to (un)loading points.

Firms manufacturing for nonlocal markets highly valued proximity to transit sites with quick access to national and international markets. For large firms the building of suburban piers and spur lines hastened industrial decentralization. Transportation facilities also had a direct spillover effect on manufacturing. The establishment of locomotive manufacturing shops at fringe locations attracted suppliers of metal products, wood, and leather.

The lumpy character of transport investments was associated with property market cycles. The fifteen-to-twenty-year Kuznets' cycle, with its surges of capital investment in building construction, forged ring and wedge development in unsettled parts of the metropolitan area. Land developers, financial institutions, and individual lenders all sought the gains to be made from property development on the urban fringe. Even though most studies point to the development of middle-class suburbs before World War II, speculators and builders constructed working-class housing on the urban fringe after 1850. Sometimes building around surrounding villages, sometimes starting from scratch on farmland, suburban working-class housing formed in each building cycle. This housing took many forms: from the small contractor-built house to company- and owner-built housing. As the experience of cities as different as Los Angeles, Chicago, Toronto, Hamilton, and Philadelphia indicate, the search for better housing conditions, the desire to own their own home, and the need to be close to suburban jobs spurred the dispersal of workers to the fringes and added outward momentum to the property market. Although there have been few studies of suburban industrial site development, there can be little doubt that the same dynamics operating to produce residential suburbs were at work. For developers the profits to be gained from the provision of factory sites could be substantial. Not only did they cream off a tidy profit from factory site subdivisions, but they also generated demand for adjacent working-class housing. The result was the building of fringe-belt working-class housing and industrial districts in each cycle of growth.[21]

Orchestrating the rhythms of manufacturing and working-class district formation were urban elites. These individuals constantly sought ways to capture new investment and to defend capital already embedded in the built environment. They formed class-based local alliances to promote and protect their local interests, which involved installing new and defending existing place-bound investments of land and industrial capital. In this context the locational possibilities and constraints of manufacturing firms were strongly fashioned by the attempts of major decision makers (industrial, finance, and commercial capital; speculators and developers; the construction industry and the state) to maximize profits and to create a viable built environment for capitalist industrial urbanization. Frequently, however, the internal cohesion of these alliances was hotly contested, as members had varying aims and needs. Even though most manufacturers were not directly interested in land development, they were deeply concerned with the externalities associated with the development of manufacturing and housing sites. They were interested, for example, in the provision of factory sites that came equipped with infrastructure, were readily accessible to a labor force, and permitted an efficient layout. Land speculators, builders, and property owners, on the other hand, through their direct dependency on the creation of urban land, had to reap profits by selling or renting land and buildings. In many cases these differing ends were complementary, and, despite some conflict, manufacturers and the real estate industry worked together to build factory districts.[22]

The state was deeply implicated in the production of the locational assets underpinning working-class and manufacturing district formation. Various levels of the state, especially the local one, play an important role as a site of struggle between opposing factions and as a mediator between the growth coalition and the rest of the population. The state's role in the growth process is to coordinate the coalition, to coax business and residents into accepting a common direction of growth, and to mobilize resources for growth. It thus plays both a direct and an indirect role in the creation of urban space. Directly, it functions as a planner, generator of infrastructure, and provider of resources. Indirectly, it acts as an ideological vehicle for legitimizing land transfers, opening up new land, and maintaining land values. From the building of Houston and the industrial suburb of Maisonneuve to the assistance provided for the electrification of Chicago and the land development of Los Angeles, the state has played a central role in local alliance politics, the shaping of urban form, and the production of working-class and factory districts. Adding to local government's interest in the scale and manner of urban expansion is its own dependence on the viability of local conditions. Its reliance upon a local tax base forces it to take a special interest in the maintenance of the financial and industrial health of a place, which involves the creation of a viable built environment, often in competition with other localities, which will accelerate and expand the circulation of capital.[23]

Despite different aims and ongoing conflict, shifting sets of local alliances worked to promote continued urban and economic expansion and pocketed the profits from it. Even though the ability to do so varied from issue to issue and from city to city, local alliances invariably established the conditions for the construction of new factory and working-class residential areas. These conditions were not uniformly spread over the metropolis. Topography, landholding patterns, the role of governments, the location of transportation corridors, the presence of middle-class residential districts, and other factors all contributed to the location and character of factory and working-class residential spaces. Nor was this a coherently or systematically planned process. Manufacturing districts, for the most part, formed without any overall sense of the direction that metropolitan growth should take. Planning was piecemeal, focusing on individual portions of the metropolitan area and subject to constant readjustment. Despite this haphazard growth, a spatially differentiated set of locational assets emerged and shaped the direction, character, and timing of urban factory districts.

These locational properties, coupled with the drive to find new production spaces, led an assortment of firms to congregate around propulsive firms that were already established on the urban fringe and around existing suburban manufacturing nodes. Once decentralization had been initiated, the dynamics of noncentral agglomeration economies accelerated fringe district formation. Early firms settling on greenfield sites attracted other firms, as the economies of the central district were replicated in suburban manufacturing areas. Contrary to the prevailing view, large suburban firms were not isolated or autonomous entities. In many cases the cumulative effects of the locational assets attracting them to the suburban district in the first place, combined with the necessity of other firms to locate in close proximity to the large ones, provided for the development of an extensive manufacturing district. In Chicago's Calumet district and the Detroit automobile suburbs large firms continued to rely on close ties to suppliers and customers. Even in the relatively small nineteenth-century city the advantages of small firms clustering around a large suburban one were obvious to many observers. While greenfield sites populated by propulsive firms were one form of manufacturing growth on the urban fringe, existing noncentral centers with small-scale manufacturing activities often became nodes of manufacturing. Typically, small villages or rural settlements lying just outside the built-up area became the recipients of manufacturing capital and in the process established the nucleus for extensive fixed-capital formation. These areas were enveloped in the expanding urban fringe and became a magnet for industrial activity.[24]

Part I

Montreal's Manufacturing Districts, 1850 to 1890

Chapter Two

"Marvellous Rapidity"

Montreal's Industrial Expansion

> Montreal has within the past few years displayed rapid progress and grown with marvellous rapidity. . . . Building operation, during the season, were on a larger scale than heretofore. Warehouses, manufactories, shops and residences have been erected on all sides and in vast numbers.
> —Department of Agriculture Montreal's Agent Report, *Sessional Papers,* 1888

An 1864 description of the Canada Railway Advertising Company announced Montreal's economic independence: "It is but a short time ago that Montreal was compelled to depend upon the workshops of other countries for nearly every mechanical appliance of which she stood in need. Fortunately, however, this state of things has passed away, and, at present, not only is this city independent of almost all the staples of foreign industry, but is able to make, as well as to export, what she was formerly obliged to import and to consume." Even though this writer was dealing in the everyday practice of nineteenth-century boosterism, there was a kernel of truth to the inflated praise of Montreal's industrial development by the early 1860s. As a small colonial town on the fringe of European settlement and lodged within the imperatives of an imperial, staple-based economy before the 1820s, Montreal had been a minor player in the network of transatlantic trade relations. Over the next thirty years it sloughed off its frontier status while retaining its role as an exporter of primary goods and importer of British and American manufactured goods. From the late 1840s Montreal experienced the thrusts of industrial capitalism and underwent the transition from a small, staple-based commercial city to the industrial and financial center of an expanding national economy. Paralleling developments in other eastern seaboard commercial cities, Montreal acquired a dynamic manufacturing base, new technologies, a wage-labor workforce, and a large share of expanding local, regional, and national markets. At the same time, the city maintained its position as British North America's most important entrepôt while consolidating its control over the new staples of wheat and lumber. It became the major transportation center of interior North

America and established a financial sector with well-developed connections to New York, London, and regional centers. Together these changes established the basis for the manufacturing geography of the city. By midcentury a range of manufacturing districts was apparent in the city center and on the urban fringe. Over the course of the second half of the century these districts expanded, and new ones developed.[1]

Montreal's Rise to National Preeminence, 1820 to 1870

Early-nineteenth-century Montreal was a small commercial city servicing the staple-based British North American economy geared to the metropolitan center in England. As an outpost of the British imperial division of labor, staple products furnished one path of the colony's economic growth and settlement. The fur trade generated a set of trade lines bounded by the northern reaches of Rupert's Land, the United States and Canadian border, and the Pacific Ocean. Montreal was the organizational nexus connecting the fur posts located in this vast territory to the markets and financial houses across the Atlantic. Providing a livelihood to only a small share of the colony's population, the fur trade nonetheless drew settlers into the St. Lawrence Valley, opened up a vast hinterland for the city's mercantile community, established embryonic financial, handicraft, and transportation activities, and facilitated the development of a Montreal-centered set of institutional linkages. As a gateway city, it funneled raw materials from the continental fur trade to metropolitan markets, and manufactured goods from Europe to the rest of the colony. With the fur trade's dwindling role from the 1820s, Montreal merchants turned their sights on the profits to be gained from the lumber and agricultural sectors. Montreal reaped increasing gains from the wheat economy extending down the St. Lawrence Valley into Ontario and the agroforestry economy branching out along the lumber frontier to the north. The reframing of the colony within the imperial division of labor lay behind the reorganization of Montreal's economy and the widening of its markets.[2]

To compete successfully in the new staple economy demanded new methods of channeling credit and capital into new investment areas, and Montreal merchants proved more than equal to the task. Peter McGill and George Moffatt, for example, diversified their holdings by moving their investments out of fur into new staples (wheat, flour, and lumber), transportation (shipping and railways), real estate, and utilities. Just as important, McGill and Moffatt, along with other Montreal merchants, established an elaborate banking and insurance system. Beginning with the Bank of Montreal in 1817, the city's bourgeoisie gained control over national credit and commercial facilities. Initially established to provide access to commercial credit, the expanding banking institutions channeled capital from the colonial commercial system into the city's burgeoning industrial economy. The repeal of the Corn Laws and the Navigation Act in the 1840s added an extra stimulus to merchants seeking other profit-making avenues outside of purely commercial ventures. Montreal's importance was enhanced by the decline in competition from its major regional rival, Quebec City. By midcentury, despite the efforts of merchants such as William Price, Quebec City's merchants were having difficulty switching from an old-regime, staple-orientated economy to one based on industrial production. By the 1840s, as it extended its hold over its regional competitors and a much larger hinterland, Montreal had become firmly and successfully ensconced in a new set of international economic relations. The city's business and political elite grasped the opportunities thrown up by the dismantling of the Atlantic colonial system, gained control over financial institutions and new avenues of investment, and beat out waning Quebec City competition.[3]

Agricultural change provided an important foundation for Montreal's transition from a small colonial entrepôt to a large commercial-industrial city. Although the majority of people in the nineteenth century lived and worked within rural areas, the increasing linkages between the urban and rural spheres and the modernization of agriculture contributed to Montreal's industrial growth. For some of Quebec's rural inhabitants at the beginning of the nineteenth century, life was relatively autonomous; a traditional preindustrial peasantry labored on the seigneury and the farm, divorced from the transatlantic relations connecting colonial merchants, *voyageurs,* native Canadians, and urban inhabitants to the metropolitan center in London. For many, however, there were strong pressures for change from the late eighteenth century, and over the course of the early nineteenth century agriculture was increasingly integrated into a commercial, market economy. It was not the international market but the regional one that underlay the increasing productivity of the agricultural sector. Only with the development of a British market for wheat in the 1840s would the integration of the agricultural sector into world trade lines take place. Along with the development of small-scale rural industry and an agroforestry economy, agricultural modernization and the growing size of the local and regional market fueled Montreal's growth through the provision of raw materials for processing and food for the growing urban population. Montreal financiers solidified control of regional markets and underwrote agricultural change through their extension of capital, credit, and influence into rural areas.[4]

The creation of regional and national markets added to Montreal's growth. Within close distance of the city, the growing productivity of farms in the Montreal plain accelerated demand for the products of the city's importers and industrial producers. City banks, manufacturers, and merchants extended their grip over a widening regional market. Combining industrial production with financial acumen, Charles-Séraphin Rodier sold threshing machines produced at his Montreal factory to local farmers with mortgages as security. At the same time, a continent-wide market was slowly emerging. If Montreal was to reap the rewards of the new economic order, it had to gain control over the small yet expanding national market that would stretch from the Maritimes to British Columbia. A growing and more urbanized population, along with rising incomes, contributed to the emergence of a capitalist market from coast to coast. Even though the nineteenth-century domestic market was splintered along class and income lines, growing demand for machine-made machines and basic consumer goods generated multiplier effects through many sectors of the Canadian economy. Building on its initial advantages, aggressive Montreal entrepreneurs seeking new sources of profit were strategically placed to capture a large share of the increasing demand for manufactures, financial services, imported goods, and information.[5]

Greater flows of capital into transportation and communications permitted Montreal's capitalists to penetrate the expanding markets of the rural hinterland, Upper Canada, and the United States. The combination of foreign and local investment into railways and the telegraph accelerated Montreal's industrial expansion, commercial supremacy, and financial dominance. From the early nineteenth century, the city's entrepreneurs involved in shipping and shipbuilding contributed to the industrial, financial, and distributive networks, which spurred expansion. Government-funded construction of the Lachine Canal in the 1820s and its refitting in the 1840s underwrote Montreal's growing commercial dominance, integrated the city into the St. Lawrence River canal system, and deepened its connections to markets outside its immediate region. After midcentury the growth of the railway network helped break down local markets and create a national one for a range of products. Two railroad lines, both with their head offices and major depots in Montreal, formed the base of the Canadian transportation network. By the 1870s the Grand Trunk Railway operated lines along the St.

Lawrence and the Great Lakes, south to Portland, Maine, and east to Atlantic Canada. A transcontinental line connecting Montreal with Vancouver and Halifax was completed by 1886. Similarly, after 1847 the Montreal Telegraph Company tied the city's commercial and industrial sectors into a wider market. As early as 1856, the firm's lines ranged more than two thousand miles and linked the city to points in Canada and the United States. By its almost instantaneous speed, the telegraph connected Montreal to a growing intercity network.[6]

Underpinning these changes was the development of a multiskilled labor force. Fed by British, Irish, and American immigrants, Montreal's population grew slowly over the first half of the century. By 1851 the city and suburbs, with a population of more than sixty thousand, formed a substantial population base for industrial growth. The city's labor force underwent a qualitative change. The commercial city's occupational structure, dominated by merchant, artisan, and laborer, was replaced by a new one consisting of a large wage labor force, a growing entrepreneurial and professional middle class, and a small industrial and financial bourgeoisie. Three major groups constituted Montreal's first proletariat. Unskilled famine Irish and skilled British immigrants made up the first two. The third were French-Canadian rural migrants who, dislocated from agriculture and the closure of rural industrial enterprises in the surrounding parishes, streamed into Montreal. These groups formed the basis of Canada's first urban-industrial labor market. Although men accounted for the majority of the city's labor force, women and children were integrated into the industrial paid workforce. In 1871, for example, one-third of Montreal's labor force were women, and one-quarter of all boys between the ages of eleven and fourteen were working.[7]

A principal component of Montreal's rise to preeminence was the creation of an extensive and modern built environment. In some cases the sources and motivation for investment were private and highly speculative, as with the railroads and real estate. In other cases the state, through direct involvement in the city's affairs, such as the funding of the Lachine Canal and harbor improvements, created locational assets. Moreover, as an 1885 federal report made clear, the state should create policies, such as the series of tariff barriers known as the National Policy and railroad freight rates, allowing capital to compete more strenuously with foreign competition. Governments also constructed the legal and ideological context for industrial-urban growth. From the writing of Montreal's municipal code in 1834 to the protection of private capital and property ownership, they were a constant element behind the development of a strong municipal presence. It certainly did no harm to Montreal's case that the colonial government before Confederation was dominated by Montreal mercantile-industrial capitalists. After Confederation the city's elite continued to orchestrate the interventions of national, provincial, and local governments to ensure Montreal's success.[8]

Montreal's Industry and Manufacturing Districts, 1840 to 1890

By the 1850s Montreal's national preeminence centered on its changing position within the international economy, its capture of expanding markets, increasing integration with the rural economy, massive investments in infrastructure built into the landscape, and the development of a wage-labor force. As elsewhere on the eastern seaboard, the transition from a commercial city to an industrial one involved the expansion of industrial production and the emergence of a range of manufacturing pathways and uneven development from handicraft to large-scale industry. Montreal's industrial transformation relied on the ability of local industrialists and merchants to introduce new technologies, divide the labor process, and tap the growth of local and nonlocal markets. Wage labor, firm specialization, a technical division of labor, and increasing capital investment was under way in several industries by the 1820s. Nonetheless, by the mid-1840s manufacturing continued to be based on handicraft production in small workshops, operated by artisans employing a small number of journeymen and apprentices, featuring few machines, little motive power, and dependence on a local market. By the end of the century, however, handicraft production had been largely superseded by a range of new manufacturing pathways.

Montreal's manufacturing growth rested on the demands of regional commercial and staple sectors. From the 1820s the city's entrepreneurs built a small but dynamic manufacturing base to serve local markets. Over the next few decades this base would grow, laying the foundations for the tremendous expansion that would occur in the second half of the nineteenth century. By then Montreal had forged important links to the regional rural resource economy, created the basis of a capitalist labor market, and established viable transportation and distribution networks. Although detailed evidence for Montreal's industrial expansion before the 1860s is fragmentary, there is little doubt that there was a dramatic increase after the late 1840s in the number and scale of firms. As early as 1851, Montreal's manufacturing value-added share of a gross national product reached 18 percent and thereafter grew steadily, reaching 24 percent in 1890. According to the manufacturing census, from a handful of factories in the 1840s Montreal had 1,097 firms in 1871, with capital of $11 million, more than 21,000 workers, and output valued at almost $31 million. Twenty years later more than 1,600 firms employed almost 36,000 workers and more than $45 million in capital. Montreal's industrial base is much larger if firms located in the surrounding Jacques Cartier and Hochelaga Counties are included. Between 1871 and 1891 the metropolitan share of industrial capital and employees found in industrial suburbs and satellite towns such as Saint-Henri, Sainte-Cunégonde, and Lachine increased from 9 to 26 percent and from 8 to 18 percent, respectively.[9]

The transition from commercial to industrial capitalism involved people from various backgrounds and the active collaboration of several forms of capital. In Montreal the line between a mercantile commercial elite and an industrial bourgeoisie was not clear. In many cases the two forms of capital nourished each other and were often deployed by the same individuals. In some cases the local mercantile bourgeoisie channeled capital accumulated from colonial staples into banking, transportation, and industry. Local merchants such as John Frothingham and William Workman switched capital accumulated from their wholesale hardware businesses into rolling mills, nail factories, and edge tool manufacture. Merchants were also instrumental in the creation and growth of banking and credit institutions. Usually working at a distance from actual work floor operations and combining economic rectitude with political savvy, these merchants forged a powerful presence in local, provincial, and national political arenas. In the process they shaped the nature of the local business climate to suit their particular needs. Clothed in the rhetoric of progress, this climate allowed for the merging of local political and business institutions in the hands of new classes. These shifting alliances of capital, credit, and support among a small mercantile-industrial bourgeoisie underpinned the transition from a commercial staple economy to a commercial manufacturing one.[10]

Large merchants were not alone in creating Montreal's manufacturing. In some cases the shift to manufacturing was undertaken by artisans and immigrant entrepreneurs. Heavily involved in the day-to-day running of their establishments, these budding industrialists funneled capital into refineries, mills, and factories. Artisans melded ruthless underconsumption with new work regimes, mechanical ingenuity, and a division of labor to produce small but gradual increments of capital investment. In some industries, such as shoe, bread, and garment, small entrepreneurs started out with little capital but ploughed profits back into the firm to increase its scale. Not only did these worker-industrialists increase their investments in the city's industrial base; they also brought with them a specific set of skills and technological know-how. Some master painters, for example, shifted from an artisanal to an industrial mode by switching their skills at grinding materials for their own consumption into the manufacture of paints, colors, and varnishes for the growing market. Artisanal painters such as John McArthur and Alexander Ramsey laid the foundation for an important local paint manufacturing industry linked to other local industries such as house construction and carriage making. Alongside that of merchants, the ability of skilled artisans to shift from workshop to factory-based manufacture added to the industrial ferment of industrializing Montreal after the early 1840s.[11]

Small, labor-intensive, consumer industries provided a substantial share of Montreal's production output and employment (table 2.1). In 1861 the clothing, leather, beverage, and food sectors accounted for 46 percent of the city's establishments and 53 percent of its manufacturing rent. Over the next thirty years, even though they grew enormously, their share of firms and rent declined. New light industries developed; however, between 1861 and 1890 the tobacco, textile, and printing industries' share of rent rose from 5 percent to almost 18 percent. Montreal's industrial growth also rested on a wider range of industries, with metalworking, chemicals, and transportation equipment playing an important role. Inducing important multiplier effects, these industries accounted for more than a quarter of the city's manufacturing rent in 1861, and they retained this share for the rest of the century.

A range of scales also characterized local industry. In both 1861 and 1890 several sectors, notably blacksmithing and furniture, had low median rents. In contrast, the metal and beverage sectors had high median rents. As early as 1861, a significant share of the city's manufacturing rent was concentrated in a small number of firms: the 110 largest firms (17.4 percent of the total) accounted for 63 percent of Montreal's total rent. By 1890 this concentration had increased: the 197 largest firms (14.4 percent of the city total) accounted for 64 percent of total rent. Montreal's industrial structure, however, was not a polarized one in which various sectors contained either large or small firms. While some sectors, such as transportation equipment, beverage, metal, and food, had a relatively polarized rent distribution, others, including blacksmithing, clothing, chemicals, and leather, did not. There were also differences with respect to the concentration of rent in small and large firms. The beverage,

Table 2.1 Rent Characteristics of Selected Montreal Manufacturing Sectors, 1861 and 1890

	1861							1890						
	Share of	Rent ($)		Percentage of Rent		No. of		Share of	Rent ($)		Percentage of Rent		No. of
Industry	City Rent (%)	Mean	Median	<$144	>$349	Firms		City Rent (%)	Mean	Median	<$300	>$1000	Firms
Clothing	18.8	345	280	3.9	69.2	81		14.2	393	200	18.7	41.5	281
Food	14.5	256	60	17.9	74.9	84		11.4	551	160	14.1	65.8	160
Metalworking	13.4	389	280	5.7	75.1	51		15.5	815	400	6.9	69.7	147
Leather	12.2	177	100	21.0	38.0	102		7.3	511	200	15.2	63.0	110
Transport equipment	9.5	372	66	12.8	79.9	38		7.8	801	150	11.1	78.3	75
Beverage	7.2	534	125	5.0	86.7	20		3.9	852	200	10.1	79.0	35
Furniture	4.2	153	84	32.5	49.9	41		4.7	483	200	14.9	58.0	76
Wood	4.1	159	80	18.8	49.6	38		3.9	564	300	8.0	56.1	53
Printing	3.5	224	140	19.4	58.9	23		6.1	487	200	15.7	56.5	97
Chemical	3.0	211	160	16.2	40.6	21		3.1	675	380	11.4	61.8	35
Blacksmithing	1.9	44	40	94.9	0	64		0.9	71	60	100.0	0	102
Nonmetallic	1.5	108	48	49.6	0	21		1.3	492	315	15.5	40.3	20
Textile	1.3	Too few firms		Too few firms		9		8.3	2,222	450	1.3	92.4	29
Tobacco	0.6	Too few firms		Too few firms		3		3.3	799	280	8.3	71.0	32
All firms	100.0	235	100	15.7	63.0	631		100.0	567	200	12.9	63.6	1,365

Source: Water Tax Rolls for Montreal and surrounding municipalities, 1861 and 1890.

transportation equipment, metalworking, and food industries, for instance, had a substantial share of their rent in large firms, while others, such as blacksmithing, furniture, and leather, had significant amounts of rent in small firms.

The introduction of a range of mechanical technologies paralleled the growing assortment of industries and scales. The Hungarian roller process, for example, underpinned the growing scale and concentration of the domestic flour industry. Numerous small rural mills disappeared from the 1840s, to be replaced by the large, high-volume, mechanized mills located in a new manufacturing district alongside the Lachine Canal. Similarly, the introduction of packing technology from the United States in the 1850s added a strong incentive to the growth of the meatpacking industry. The introduction of vulcanization, machinery, and new products aided the formation of large-scale rubber manufacture. But technologically sophisticated manufacturing was not the only type of production taking place in nineteenth-century Montreal. Many sectors continued to rely heavily on labor power. Baking and blacksmithing experienced little technological change throughout the period; in the brewing industry, machinery, equipment, and techniques remained in place for decades. In other sectors, such as metalworking and printing, technological changes were frequently hitched to existing divisions of labor and work settings without the wholesale revamping of work processes. The variability of the introduction of machinery underpinned the types of manufacturing pathways that developed in nineteenth-century Montreal.[12]

Markets fractured by size and composition constrained the application of new machinery, products, and work methods. Manufacturers' expectations and estimates of the risks involved limited firms' ability to take advantage of markets. Some food processing industries, such as flour and sugar, experiencing increased demand could introduce machinery with little hindrance because of the simple character of production and the ease of transforming raw materials into finished products. In other cases, however, the relatively small extent of the Canadian market hampered industrial growth. Robert Mitchell told an 1876 Parliamentary Committee that if it were possible his brass foundry "would make articles by the hundreds where we now make them by only tens or twenties." The problem was the small size of the domestic market. For similar reasons the machine makers Gardner and Son did not limit their operation to a few types but made a variety of steam engines, lathes, saw mill machinery, printing presses, and machine tools.[13]

The inability of Montreal firms to compete with large American firms featuring new machinery, continuous-processing methods, and great demand added to these difficulties. The smaller Canadian market set limits on manufacturers' options. The cost benefits accruing from well-developed internal economies of scale open to many American firms eluded most local companies. Even firms technically able to implement large-scale, standardized production faced problems. In the case of the textile industry Canadian companies relied on a number of lines rather than one. G. Nye of Hochelaga Cotton expressed the dilemma well in 1876: "In the United States there are 875 cotton mills. Some of the mills from the commencement have been running on one style of goods. . . . Here we have run on different styles, and therefore cannot manufacture so cheaply." Forced to adopt some version of batch production, many industries faced several ongoing problems, including frequent machinery setup and a wide selection of lines. This was especially the case where the small extent of the Canadian market was compounded by the fragmentation of demand. Industries catering to a luxury class of consumers, such as jewelry, or servicing a small number of firms whose demand was made up of a limited time horizon, such as locomotive wheels, were unable to create economies significant enough to allow firms to reap the rewards of technological innovation.[14]

Even in those situations in which mechanization was an important component of a productive strategy, it cannot be assumed that production was trouble free. By investing in machines, Montreal firms faced a set of risks, which in turn affected the options open to manufacturers. The risks involved reliability and the ease of adapting to various product lines. A cotton manufacturer named Peter Wood wrote to his brother John in December 1869: "I now have the wadding machine working well and the grey cottons never were so well made as now. . . . But it seems as tho there is no end of trouble. This week I have had my 'mules' [spinning machines] break down so as to stop the mill." Even when machines were not breaking down, there was the problem of the frequent changes that had to be made in order for the machines to function effectively. In shoe manufacture, as one employer told an 1876 parliamentary committee, "[If] a man is making one kind of work he sets the machine to suit it, but if he has to change it a dozen times a day to do the other kinds of work, it cannot be so regular." At the same time, the expense of changing even simple machinery and equipment in response to changing social conditions could be prohibitive. To compete in the British market, remarked another shoemaker, "I must change my lasts and ties and some other things." Even though he was interested in exporting his goods, he did not "consider that it would be worth [his] while to incur that expense for the present."[15]

A segmented division of labor complemented the technological and market imperatives of Montreal's manufacturing pathways. In the early nineteenth century, despite the small scale of firms and the reliance on traditional tools, several industries broke up work into separate yet interlinked components. In the clothing industry after 1820 work increasingly became subdivided by task and gender. After midcentury the earlier division of labor was intensified, and the industry became mechanized. The changing gender division of labor found in garment manufacture was paralleled in other industries as cheap female labor replaced male labor. At Frederick Harris's cotton factory, denim cloth was manufactured in a highly task-divided workplace by women and children working on machinery such as willows, pickers, carding and drawing machines, spindles, and looms. By 1881 women and children accounted for two-thirds of the Montreal rubber industry, more than 80 percent in clothing, 60 percent in tobacco, and 40 percent in shoemaking. Even in male-dominated industries there was gender and age segmentation within firms. In the Dominion Type Foundry, for example, a few men undertook the difficult tasks of font mattrice making, while some boys and fifty girls performed mundane and unskilled jobs such as breaking off the "jets" and smoothing the surfaces.[16]

Ethnic differentiation existed alongside the gender division of labor. Irish and French Canadians constituted the bulk of workers in the unskilled and semiskilled sectors of industry, while the British and Americans dominated the skilled sectors. In 1859 the English-speaking Protestant group accounted for a fifth of the city's population but half of the bourgeoisie, petite bourgeoisie, and white-collar workers. In contrast, two-thirds of French Canadians were skilled or semiskilled workers, while half of the Irish Catholics were laborers. These divisions were replicated in individual workplaces. In the Grand Trunk Railway shops, for example, Anglophones were concentrated in the skilled trades, with the exception of woodworking; French Canadians were the majority in low-skilled work and in the more skilled woodworking trades. Those in the "other" category were, almost without exception, to be found in low-paid, low-skilled positions. Along with differences between ethnic groups by skill levels, the various departments of the shops had concentrations of different ethnic groups.[17]

All of these elements of Montreal's uneven industrialization contributed to the emergence of a series of pathways. Large flour mills and sugar refineries relied on large capital inputs, a small labor force, and new technologies. The ability of these "milling" manufacturers to capture a large market depended on volume output made from nonlocal raw materials for a relatively large and standardized market. In the "technical" industries—metalworking, printing, and glass—the combination of intricate technical processes, a well-developed set of labor conditions, and differentiated markets produced manufacturing strategies different from that of milling. In the "mechanical" industries—clothing, shoes, tobacco, and biscuit—firms featured a different set of characteristics, including the shift from skilled male workers to female machine operators, the introduction of simple forms of mechanization, a differentiated and seasonal market, and easy entry into the trade. Finally, "craft" traditions continued, even though they were under attack from merchants investing in manufacturing and from artisans seeking to shift their role to manufacturers. By the end of the century bakers and blacksmiths still labored under similar conditions to those of fifty years earlier.

Capitalist industrialization produced new industries, competitive pressures, and opportunities for the manufacturer after midcentury. While the period featured the growth of large and mechanized firms in Montreal, large-scale factories did not dominate the industrial landscape. A sectoral array spanning the gamut from large, mechanized firms specializing in high-volume standardized products to the small artisan workshop and the sweatshop was evident over the course of the century. This complex industrial structure showed up in considerable differences in capital outlay, scale, labor process, markets, and mechanization. Sectors undergoing rapid expansion did not necessarily follow one vector of growth. The technological and social character of a sector could be splintered in such a way that there were many different paths to success, among them increasing firm scale and the introduction of new industries, technologies, skill opportunities, and production systems. One impact of the development of these pathways was the creation of a new manufacturing geography. The commercial city's centralization of work was broken as new industries sought out untapped locations. As a result, after 1850 specialized manufacturing districts appeared in the city and the urban fringes.

Fashioning "So Large and Complete a City"

A visitor to Montreal in 1868 described the tremendous change in the city's physical infrastructure accompanying industrial growth: "I was much struck with the continued rapid growth of this now great northern city. Built as it is almost wholly of stone along the extensive and massive quays which line the bank of the river, and in the business portions, Montreal makes a dignified, indeed, an imposing effect. The beholder for the first time, unless marvellously well up in his geography, is surprised to find so large and so complete a city." The writer's comments described a new phenomenon. Although manufacturing is not mentioned, the description hints at the tremendous change in the city's physical infrastructure accompanying industrial growth. Similar changes were taking place in a number of other North American cities; flows of investment into the urban built environment, orchestrated by the city's elite, were instrumental in creating conditions leading to manufacturing districts. In Montreal the major features of the period involved reorganizing local government, increasing state intervention in the creation of facilities to accommodate large-scale urban growth, and participation in municipal affairs of a new French-Canadian bourgeoisie alongside the decline of the old Tory Anglophone aristocracy. Alliances of the state and business fashioned capitalist property relations, housing markets, and infrastructures and in the process produced a new set of locational assets. These in turn allowed for the emergence of a greater number of locational options, making possible the establishment of industries, technologies, and living habitats in new parts of the expanding city.[18]

From the granting of its first charter in 1831 to the end of the wave of industrial growth in the 1870s, important changes occurred to Montreal's political, economic, and social institutions. The abolition of the seigneurial system in 1854, the rise of a capitalist property market, the development of a basic educational system, and the revision of commercial and civil law all contributed to the laying down of a social foundation based on bourgeois values. Before the 1880s there was little change to local government, despite the development of an industrial capitalist class from the 1840s. The new, mainly British, industrialists and financiers were comfortable with the old British Tory framework and ideology. Marrying into the old elite, the up-and-coming new bourgeoisie was content to use the trappings of the old order to acquire control over the handles of local power. The defining modus operandi of local politics before the 1880s was reducing government costs and allowing entrepreneurs free reign.[19]

By the end of the century, however, the contradictions of unbridled industrial growth and the growing power of new social classes forced changes to the city's political regime. The tactics of the old elite no longer proved responsive to the increasing range and complexity of problems, such as poor water supply, inadequate harbor facilities, high rents, and overcrowding. New social classes, especially French-Canadian land developers and small businesses, placed new demands on the expectations of local politicians. The channeling of workers' discontent into unions and other forms of political dissent added to the search for new local practices. The creation of a new civil code, the changing character of church-state relations, and the demands for an adequate social service delivery system further shook the foundations of the old hegemony. By the 1880s a shift had taken place: the Anglophone Tories was replaced by French-Canadian liberals. Despite the changing nature of local political leadership, Montreal continued to be run by a group of elites who, however much they disagreed on specific issues, were motivated by a similar belief in the city as a growth machine. If Montreal was to remain competitive and an arena in which to elicit private gains, the city had to be collectively managed. One area that came under increasing collective control was the city's infrastructures, notably the harbor.[20]

Under intense pressure from Montreal's merchants, the Lower Canada government in 1850 established the Harbour Commission of Montreal to finance and supervise new port facilities. The broad objective was to undermine the advantages of their competitors in other cities along the St. Lawrence and Hudson Rivers. In 1842, with less than one mile of piers and a very narrow channel to Quebec City, the city's harbor facilities were hardly adequate. Private investment was not forthcoming, and public finances were necessary if the city was to achieve its destiny as the major continental transportation center. The Harbour Commission proved very effective in obtaining public monies for the construction of an extensive set of port facilities that would outrival, with the exception of New York, all other cities in North America.

A small number of Anglophone entrepreneurs representing the city's most important business organizations and heavily involved in all spheres of Montreal business controlled the commission. John Young, for example, a major mover in the formation of the Harbour Commission, had extensive interests in railways, telegraph companies, shipping, insurance, and banking and represented the western section of the city at city hall. Young and his coterie, using provincial and federal funds and with the support of the local government, constructed extensive port facilities in Montreal. By the end of the 1870s there were four miles of piers handling 1.5 million tons of cargo, and the ship channel had been greatly extended. Even though the initial, and successful, intent had been to increase Montreal's share of the continent-wide flow of raw materials and manufactured goods, there were unintended results. The extension of the waterfront into suburban territory, the installation of modern port facilities in existing manufacturing districts, and the more efficient loading and unloading of raw materials and manufactured products contributed to the growth of the city's industrial base and to the increasing specialization of manufacturing space.[21]

The Harbour Commission example illustrates the importance of alliances of local business interests and the state in reshaping Montreal's landscape. As the city grew, there was increasing demand for the creation of heavier-duty infrastructures, which in turn relied upon the investment of large lumps of capital. As the private sphere was unable or unwilling to finance many of these infrastructures, various levels of the state, with their access to large amounts of capital and their concern with social reproduction, became the major provider of critical infrastructures. Just as important, the state coordinated this development through its mobilization of resources and supervision of the ideological content of the debates surrounding infrastructure provision. One unintended effect of the state's involvement was the creation of locational assets, which manufacturing firms would treat as agglomeration economies. Expansion of the harbor was a necessary prerequisite for the establishment of firms reliant on the supply of distant raw materials and long-distance markets. The several rounds of investment in port facilities were directed to the western, mainly English-speaking, part of the city. The construction of the new piers in the west end and large investments in the Lachine Canal attracted firms to that part of the city. A similar set of events was also taking place elsewhere in Montreal. The municipalization and extension of the water system after 1845, for example, provided manufacturers with better and cheaper water. Even though they were not always able to agree on what was to be built, state-supported alliances of merchants, industrialists, and land developers were responsible for the extension and updating of the city's infrastructure. This, in turn, forged the basis of the changes to Montreal's social geography.[22]

After midcentury the combination of growing class and ethnic inequalities, stark divisions within the housing and labor markets, and wage polarization produced a divided social geography. The building of working-class districts and the development of localized labor markets throughout the city and the suburbs offered manufacturers a greater range of locational choices. Adding to this was large-scale population

Table 2.2   Population of Montreal and Suburbs, 1851 to 1891

Place	Year of Annexation	1851	1861	1871	1881	1891
Island of Montreal	—	77,467	116,992	143,209	191,840	277,525
Montreal and suburbs	—	61,507	98,651	124,581	175,882	271,285
Montreal at the census date	—	57,715	90,323	107,225	140,247	216,650
Montreal at its 1882 limits	—	57,715	90,323	107,225	140,247	182,695
Suburbs outside 1882 limits	—	3,792	8,328	17,356	35,635	88,590
Saint-Jean-Baptiste	1886	—	2,269	4,408	5,874	15,523
Saint-Henri	1905	600	1,943	2,467	6,415	13,413
Saint-Gabriel	1887	—	—	—	4,506	9,986
Sainte-Cunégonde	1905	See Saint-Henri		3,656	4,849	9,291
Hochelaga	1883	—	—	1,061	4,111	8,540
Lachine	—	1,089	1,315	1,696	2,406	3,761
Saint-Louis de Mile End	1910	—	—	—	—	3,537
Westmount	—	—	—	—	884	3,076
Côte Saint-Louis	1893	1,089	1,746	2,215	1,571	2,972
Longue Pointe	1910	1,014	1,055	1,011	1,114	2,445
Nôtre Dame de Grace	1910	—	—	—	1,524	2,305

Sources: Yvan Lamonde, *La culture ouvrière à Montréal, 1880–1920* (Québec City: Institute Québécois de Recherche sur la Culture, 1982), 26; Paul-André Linteau, *Montréal* (Montréal: Boréal, 1992), 40, 160, 314.

growth after midcentury on the city fringe and in industrial suburbs. Between 1851 and 1891 the city, excluding annexed areas, grew by 125,000, while the suburban areas, including those annexed to the city during this period, increased by 85,000 (table 2.2). The city reached its population limits in the 1880s, and thereafter suburban growth surpassed that of the city. In the 1880s and early 1890s a large share of Montreal's growth occurred by annexing working-class suburbs such as Hochelaga, Saint-Jean-Baptiste, Saint-Gabriel, and Côte Saint-Louis. Along with the growing population was the changing balance of Montreal's ethnic structure. Between 1842 and 1861 Irish Catholics fleeing the famine made up the majority of newcomers to Montreal, although the Scottish and English formed a sizable proportion. As Irish immigration declined, population growth was fueled by French-Canadian rural migrants. Although about 475,000 people left Quebec between 1840 and 1890, the vast majority of them French Canadians seeking industrial employment in the United States, a large number moved to Montreal and its suburbs.[23]

A growing population, allied with an active land development industry, pushed the boundaries of urban settlement. Rapid growth after midcentury created opportunities for developers and builders, and the prime target for speculators and developers was land lying beyond the built-up area. An expanding population meant a jump in land values, and a speculator, through correct timing, could cream handsome profits as the increase in land prices and demand burst across the urban landscape. By taking advantage of the opportunities open to them, land speculators and local politicians, working in tandem, focused their activities on producing the greatest differentials and pocketing the greatest profit. Evidence of speculators' ability to take advantage of industrial and urban growth has been documented in numerous city and suburban areas: the construction of the wealthy suburban New Town in the 1850s and middle-class Côte à Baron after 1860, the opening of the working-class suburb of Saint-Jean-Baptiste in the 1870s, and the growing involvement of large-scale real estate companies in the financing and construction of housing throughout the city (map 2.1).[24]

The expanding housing market, however, did not benefit everyone. Montreal's working class suffered from shortages resulting from housing market inequalities, labor market conditions, rapid population growth, and devastating fires. Investment in housing tended to lag behind other sectors because house builders could not compete with commercial and industrial investors in money markets. Moreover, when the opportunities for housing construction were present, most builders, from the small-scale, speculative builder to the large-scale developer, sought out the more profitable arena of middle-class housing. In the second half of the nineteenth century distinct differences between areas of the city in terms of median rent, percentage of blue-collar workers, and the concentration of occupational classes became a central feature of Montreal's social geography. The fine-grained segregation that had characterized the early-nineteenth-century city gave way to starker geographic divisions and a more divided housing market and with them a growing mismatch between home and work. The combination of inadequate housing, the growing scale of the city, increasing dependency on more than one income, long hours of work, and the ever-increasing dispersal of jobs throughout the city presented workers with difficult residential choices.[25]

Many working-class families lived in the belt of cheap housing surrounding the city core. These districts were desirable to many because of their proximity to a wide range of jobs and the lack of a suitable mass transportation network. Yet, as the Royal Commission on the Relations of Labour and Capital (RCRLC) revealed, broad swathes of inner-city housing were in poor condition. The high cost of city land and building operations contributed to a situation in which homes of a significant segment of the working class were "scarcely fit for human beings to live

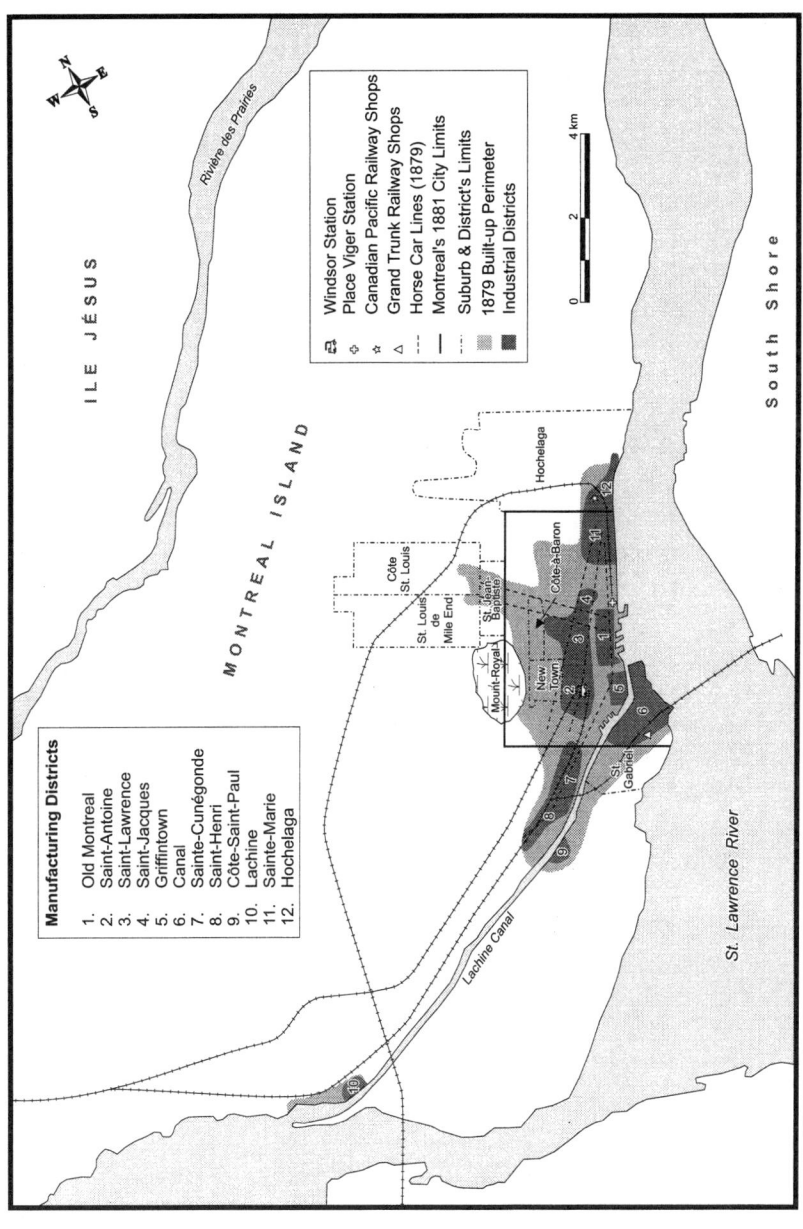

MAP 2.1 MONTREAL AND ITS MANUFACTURING DISTRICTS, 1890. A set of manufacturing districts running along transportation corridors and dispersed throughout the expanding metropolitan area were firmly in place by the last decades of the nineteenth century.

in," according to the testimony of William Costigan, the organizer of the Artizans' Dwelling House Company. He went on to state that "under the present state of things overcrowding is inevitable, and only the cheapest and most inferior class of rookeries, can be paid for out of the current state of wages." According to Dr. Douglass Decrow, a working family typically occupied two- to four-room dwellings, while two families occupying one house was common. Louis Laberge, the city's medical officer, admitted that "a great number" of houses "are badly off" and that workers suffered from "much crowding" and "lack of means." Montreal's working class also found the option of buying a house to be extremely difficult. Home ownership in Montreal was among the lowest in North America, running at less than 20 percent and, in the central parts of the city, even lower. Few workers, a molder told the RCRLC, "get enough money to be able to rent a house, without owning one."[26]

A common response to the problem of central-city housing conditions was to seek out cheap housing elsewhere. Housing in the more distant parts of the city tended to be cheaper than that found more centrally. According to the manager of the Montreal Street Railway in 1888, respectable housing available at a rent affordable to most families was only to be found "in outside localities." One area attractive to an increasing number of workers was the city periphery, close to manufacturing districts and the newly emerging industrial suburbs. As early as 1861, large numbers of workers lived on the city fringe in the industrial wards of Saint-Ann and Sainte-Marie. Over the course of the century this number increased, and working-class suburbs (e.g., Hochelaga and Saint-Henri) appeared, as workers clustered around industrial nodes in the east and west ends. But not all workers hitched their residence so closely to their workplace. Many of them moved to working-class residential suburbs such as Saint-Louis and Saint-Jean-Baptiste, where there was little employment.[27]

Despite the fact that housing on the city fringe and in the suburbs was frequently cheaper and in better condition than central housing, the influx of the working class to the suburbs generated problems. One was the increase in rents and property values. George Muir, the city assessor, told the RCRLC hearings that "property has been rising where there are manufactories established, and railway workshops; the rentals have increased, and, after a while, the values have increased." High property and rental values compounded the difficulties of low-income families, whether in the new suburbs or in the city. According to the RCRLC testimony of the laborer Thomas Gratorex, workers faced high rents in the working-class eastern fringe belt, and, given local wages, "it is a matter of impossibility for any man to pay." There was a seesaw effect, however, as high rents and property values were associated with industrial expansion. Industrial growth produced property booms, while down times led to bad business conditions and falling rents. Of course, employment and wages also declined during depressions, thus providing few workers with the opportunity to take advantage of falling rents. Furthermore, the city's working class did not benefit from rising land values compared to the city's well-off population. Workers were victims of a typical feature of the housing market: as the size of the dwelling increases, the cost per square foot declines. The result, as Muir pointed out, was that "on high priced property the rents are not so high, in proportion, according to the value of the property, . . . the poor man gets much less."[28]

After midcentury the city's social geography became increasingly divided along the lines of class and ethnicity. Wealthy Anglophones moved to the mansions and single-family dwellings of the newly created bourgeois district of upper Saint-Antoine encircling the southern slopes of Mont Royal. To the east rich French Canadians lived in little pockets north of Old Montreal. Separating the elite districts from working-class areas were several middle-class and mixed-class neighborhoods. Even though the inner city became increasingly working class as the bourgeoisie moved to the city periphery, a large number of workers lived in new duplexes adjacent to the city fringe and suburban clusters of manufacturing. Clustered within these broad working-class residential patterns were ethnic groupings. The Irish concentrated in the western districts of Griffintown and Goose Village; French Canadians settled in large numbers on the city's east side; and the British working class resided in the western industrial districts. Suburbanization was common to all classes. The middle class sought delightful views and healthy areas in the new districts on the city fringe or in the suburbs encircling Mont Royal. The working class also moved to the suburbs. Along with looking for better housing conditions than were found in the central city, they also sought cheap land, inexpensive housing, and jobs in the burgeoning industrial districts on the city fringe.[29]

Montreal's Locational Assets and Manufacturing Districts

The changes to nineteenth-century Montreal generated a characteristic common to all expanding North American cities: the coupling of several circuits of capital embedded in land, infrastructure, machinery, and housing. Together, these circuits created spatial packages, or bundles of locational assets. The coupling of manufacturing pathways with the building of infrastructures and fringe working-class housing had profound implications for Montreal's manufacturing geography. Manufacturing pathways established the basis for different sets of locational demands and, in the process, created a constellation of locational assets composed of new forms of transportation, power, labor, and work space. The emergence of wide production cost variations, labor force and power needs, and work space requirements contributed to industrial clustering throughout the city from the 1840s. These changing industrial locational coordinates rested on the creation of a new environment and a new political framework. The remodeling of infrastructures functioned side by side with the changing nature of urban politics. At the same time, a segregated housing market combined with the dispersal of manufacturing to form the basis for the development of noncentral working-class districts. All of these factors came together to provide the opportunity to place factories and workers' housing close together, both in the central city and in the suburbs.

A range of manufacturing districts was a major feature of the city's geography after the late 1840s (table 2.3 and map 2.1). As demonstrated in Chapter Three, the city core (Old Montreal) remained an important magnet for firms seeking the benefits of localized external economies and contained a large share of the city's manufacturing. It did, however, experience changes, notably a decline relative to other city locations. The dynamics producing Old Montreal's manufacturing base also created an extensive set of manufacturing districts in the adjacent area of the Outer Core. Forming a crescent around the city core, the inner-city districts of Saint-Antoine, Saint-Lawrence, and Saint-Jacques took advantage of agglomeration economies derived from Old Montreal.

The extensive industrial complex consisting of Old Montreal and the Outer Core, however, was only one feature of the city's industrial geography. Equally important, manufacturing districts formed on the city fringe and in the suburbs. By the 1890s a more elaborate geography was evident, as new districts emerged and existing ones expanded. Industrialization forced firms from many industries to streamline existing work methods and to introduce new space-extensive machinery. In many cases the restructuring of work in situ or moving to a central location did meet the needs of a firm; in other cases, however, firms had to find other options. One option involved searching for greenfield sites on the city fringe and in new suburban districts, where many firms could in-

Table 2.3 Manufacturing Districts in Montreal, 1861 and 1890

Manufacturing Districts	Date of Emergence	Number of Firms 1861	Number of Firms 1890	Share of Rent (%) 1861	Share of Rent (%) 1890	Median Rent ($) 1861	Median Rent ($) 1890
Old Montreal	pre-1840	270	378	37	24	120	300
OUTER CORE		227	600	24	31	60	200
Saint-Antoine	1840s	144	355	20	25	100	280
Saint-Lawrence	1850s	55	123	4	4	48	150
Saint-Jacques	1850s	28	119	1	2	38	100
WEST END		94	239	34	30	220	250
Griffintown	pre-1840	53	67	7	5	120	300
Canal	1840s	40	68	26	12	550	310
Saint-Henri	1870s	—	88	—	11	—	150
Lachine	1880s	—	14	—	2	—	60
EAST END		40	148	6	15	44	120
Sainte-Marie	pre-1840	40	123	6	7	44	120
Hochelaga	1870s	—	25	—	8	—	145
City Total		631	1,365	100	100	100	200

Source: Water Tax Rolls for Montreal and surrounding municipalities, 1861 and 1890.

troduce their restructured work methods. One suburban manufacturing district, and the concern of Chapter Four, emerged in the East End areas of Sainte-Marie and Hochelaga. Although these towns were small and centered on a narrow range of industries at midcentury, rapid growth occurred over the following decades, and new propulsive industries appeared. Several high-volume, mechanized firms coexisted with small, proprietary ones serving local markets and operating with simple production methods.

As Chapter Five illustrates, the most important fringe manufacturing nucleus emerging in Montreal after midcentury was the Canal district. Located on the unsettled western fringe, it was the locus of a technologically advanced, hydraulically based, energy-intensive form of production consisting of firms from propulsive industries which featured a variety of scales and introduced changes to work organization. These large, capital-intensive firms were looking for some combination of cheap land, access to better transportation facilities, and sites to install new forms of machinery and production rationalization. By the 1860s, the Canal district, along with the adjacent and older district of

Griffintown, constituted a large West End industrial complex of diverse yet linked firms. Although the relative importance of the Canal district diminished in the last quarter of the century, the size and scope of the complex continued to grow. The emergence of industrial suburbs (Saint-Henri and Sainte-Cunégonde) by the 1880s heralded the continuation of existing industrial sectors (food and metal) and the establishment of new ones (textiles and chemicals).

Chapter Three

"One Vast Block"

The Making of the Central Manufacturing Districts

> *The great business warehouses and stores and many of the light industrial establishments are situated in one vast block in the centre of the city to the south.*
> —Anonymous, *Canadian Journal of Commerce,* September 1885

In 1848 the preindustrial walled city core, Old Montreal, contained half of the city's businesses and two-thirds of total business rents. Along with nearly all of the banks, warehouses, offices, and shops, Old Montreal was also the city's most important manufacturing district. Almost without exception, manufacturing firms were small handicraft shops producing goods for the local market.[1] With the great surge of industrialization over the next few decades, Old Montreal's manufacturing base experienced tremendous growth and change. At the same time, adjacent districts developed and, merging with Old Montreal, formed the central manufacturing districts. The districts of Saint-Antoine, Saint-Lawrence, and Saint-Jacques—together making up the Outer Core—formed a new manufacturing belt encircling the older central core and mirroring its industrial base. Throughout the second half of the nineteenth century the central manufacturing districts contained the greatest mass of manufacturing firms in the city. In 1861 their 497 firms accounted for almost 79 percent of the city's firms and 61 percent of total manufacturing rent. Although there was relative decline over the next thirty years, they made up 72 percent and 55 per cent of the city's firms and rent, respectively, in 1890. A common range of dynamics held these different central districts together in one vast manufacturing block. Despite differences in terms of scale, industrial profile, combination of productive strategies, and timing of development, the central manufacturing districts contained a range of industries featuring specific chains of production.

CHAINS OF PRODUCTION AND MANUFACTURING PATHWAYS

With 270 firms and more than a third of the city's total manufacturing rent, a large industrial base had developed in Old Montreal by 1861. The central core contained very high shares of certain sectors, most notably the classic centralizing ones of printing, jewelry, clothing, and leather (table 3.1). Montreal's shoe trade, for example, one of the first industries to undergo the transition from petty commodity handwork to capitalist industrial production, was heavily concentrated in Old Montreal. As one commentator noted in 1888, "The boot and shoe trade, as at present conducted, is one of the most prominent of any of the branches of commercial industry. Mammoth establishments have been established for the manufacture of these most useful articles of wearing apparel, and employment is furnished to hundreds of thousands. Machinery has worked wonders in this line of manufacture, and the cost of boot and shoes is now almost 50 per cent less than it was 25 or 30 years

Table 3.1   Manufacturing Structure of Old Montreal and the Outer Core, 1861 and 1890

	Old Montreal				Outer Core				Central Manufacturing District	
	1861		1890		1861		1890		1861	1890
Industry	No. of Firms	Rent Share (%)	No. of Firms	Rent Share (%)	No. of Firms	Rent Share (%)	No. of Firms	Rent Share (%)	Share of City Rent (%)	
Clothing	56	31.6	89	25.2	24	28.7	160	24.8	98.6	96.4
Leather	57	21.8	23	8.5	32	11.8	59	14.1	89.3	88.4
Printing	22	9.4	64	19.0	1	0.1	29	5.0	100.0	99.0
Metalworking	16	8.0	33	9.9	7	5.1	48	8.9	31.2	34.7
Food	23	6.6	29	8.7	39	24.1	73	15.6	38.2	45.5
Furniture	17	4.8	19	5.5	23	9.1	44	9.9	93.6	92.4
Chemical	6	2.5	5	1.9	5	4.7	11	3.3	68.3	48.4
Jewelry	12	2.3	21	5.2	4	1.1	12	2.8	100.0	100.0
Blacksmithing	19	2.0	20	0.9	32	3.5	39	1.1	82.9	58.3
Tobacco	2	1.6	15	3.4	—	—	9	1.7	89.5	42.0
Carriages	6	1.6	10	1.6	23	4.3	40	4.7	88.8	81.3
Textile	3	1.3	7	1.3	1	0.1	5	1.3	39.2	5.2
Wood	9	1.1	8	3.1	11	1.1	15	2.3	16.0	37.2
Paper	—	—	5	3.2	1	0.2	4	1.0	50.0	61.9
All Firms	270	100.0	378	100.0	227	100.0	600	100.0	61.0	54.7

Source: Water Tax Rolls for Montreal and surrounding municipalities, 1861 and 1890.

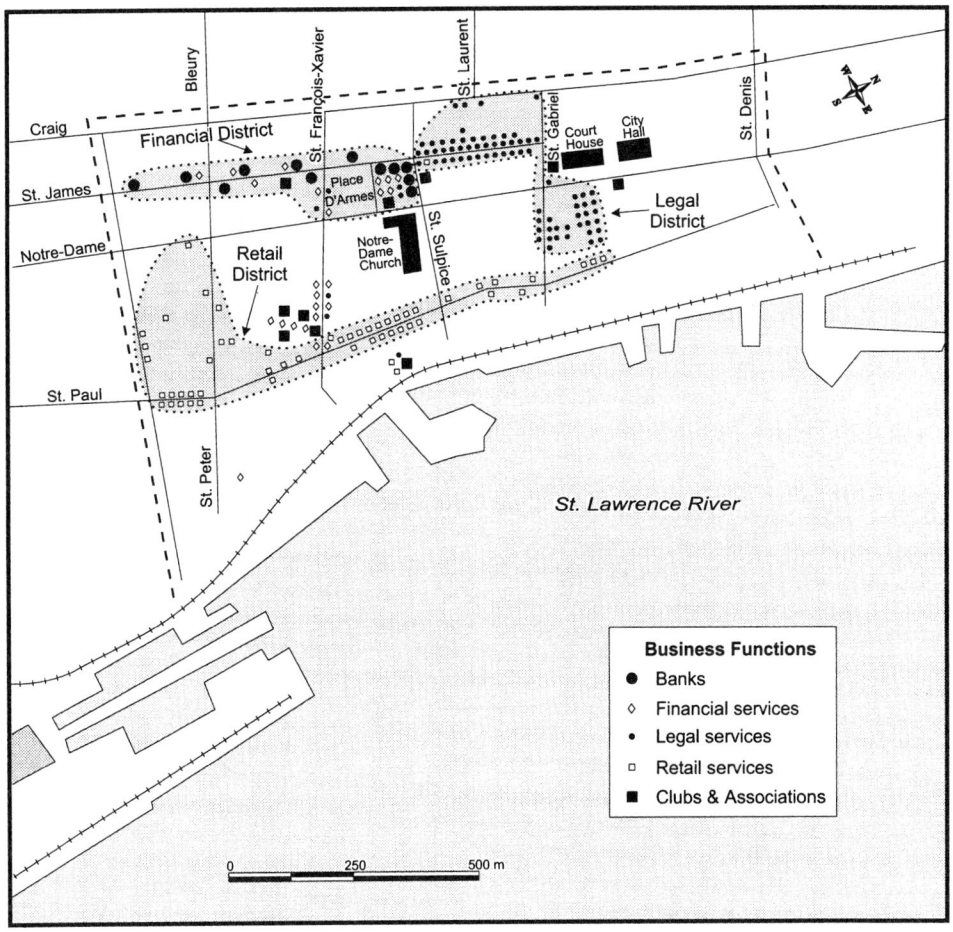

MAP 3.1 OLD MONTREAL: THE FINANCIAL, COMMERCIAL, AND RETAILING CENTER, 1860. Distinct and specialized districts in the city core can be identified from as early as 1860. (Compiled from Lovell, *Montreal City Directory, 1860* [Montreal, 1861].)

ago." As the writer suggests, one feature of the modern shoe industry was large factories with modern machinery and an extensive division of labor. These, however, coexisted with smaller manufactories and workshops employing more traditional methods. Variation in scale, mechanization, and production methods was paralleled in other sectors. A wide assortment of metalworking and chemical firms clustered in the same area as firms from the light consumer good sectors.[2]

The fact that Old Montreal was the original core of the city meant that it had historically been the focus of most business and ancillary functions and presented to manufacturers from an early date a set of locational assets that could not be rivaled elsewhere. As the center of the import-export trade, most of the city's merchants, importers, brokers, and accountants had their businesses along the main arteries, such as St. Paul and Notre Dame (map 3.1). The city's main retailing district lay in close proximity to commercial offices and warehouses. By midcentury Old Montreal had developed as Canada's financial center. The main offices of all of the city's banks and insurance companies ran along St. Jacques Street, as did most of the telegraph, telephone, shipping, and utility firms. Along with the city's communications network, the core also housed the Board of Trade, the Board of Arts and Manufactures, and the prestigious bourgeois social clubs.

Over the next thirty years, despite the increasing number of firms, a relative decline occurred in Old Montreal's share of the city's manufacturing. Even though its total rent share declined, Old Montreal remained the premier district for printing and jewelry and continued to house substantial concentrations of clothing, leather, furniture, and tobacco firms. The inability of manufacturing to compete successfully with financial, administrative, and retailing functions for central land undermined Old Montreal's position as the city's dominant manufacturing center. After 1850, despite the scale of its industrial base and the variety of manufacturing, the conflicting pressures of high central land values and demand from other business functions for a central location posed serious problems for manufacturers. Even though Old Montreal's manufacturing base grew in the second half of the nineteenth century, the supply of central space at a cost manufacturers could bear declined. It became increasingly difficult to find appropriate manufacturing space.[3]

This problem was solved in part by the formation of manufacturing districts immediately outside the original city core. From the 1850s a cluster of firms encircling and adjacent to Old Montreal developed, coalescing in the Outer Core. Offering many of Old Montreal's advantages and similar to the central core in terms of the sectors represented, the size distribution of firms, and the concentration of particular productive strategies, Saint-Antoine, Saint-Lawrence, and Saint-Jacques districts accounted for a third of firms and a quarter of manufacturing rents in 1861 (table 3.1). This was a tremendous increase from 1848, when Saint-Antoine, the most important of the three districts, only had twenty-eight small business establishments of all kinds. By 1890 the Outer Core's share of city firms and manufacturing rent increased to 44 percent and

33 percent, respectively. Most of this growth occurred in the southern portion of Saint-Antoine adjacent to Old Montreal, although appreciable growth took place in Saint-Lawrence and Saint-Jacques. Specializing in the manufacture of baked, beverage, clothing, leather, wood, and carriage products, the Outer Core had an industrial structure similar to that of Old Montreal.[4]

The development of Montreal's central manufacturing districts took the same path as other eastern seaboard cities such as Baltimore, Boston, and New York. They all centered upon the consolidation and concentration of several industries highly dependent upon the breaking up of the chain of production into different productive units. As firms became more specialized, they externalized many steps of the production process. These external economies were associated with an expanding division of labor between production units and within individual firms. The inability to standardize production because of market instabilities such as the vagaries of fashion or seasonal changes forced firms to make close links with other firms. The problem of matching different parts of production within the same plant in an optimal manner forced manufacturers to externalize parts of production to other firms. Moreover, mechanization was inhibited by the need for certain tasks to be performed under conditions of specialized knowledge or skill. As a more elaborate division of labor and more specialized firms developed, the vertical and horizontal linkages between them intensified. A host of interlinked small and medium-sized firms settled in the central districts to take advantage of agglomeration factors such as access to information and credit, supplier and buyer firms, a wide selection of labor skills, and an array of work spaces.[5]

This did not mean that central manufacturing district firms remained small and unsophisticated. Firms serviced a multitude of markets, deployed different degrees of mechanization, and were extremely responsive to the changing conditions of the market. The central manufacturing districts contained a range of smaller firms, such as the family-owned, low-capitalized bakery, blacksmith shop, and saddlery serving local markets and employing one or two workers. But not all small and medium-sized firms functioned under the traditional methods practiced by the bread baker or blacksmith. Many proprietary firms deployed quite sophisticated methods and yet found a central location and its external economies to be extremely advantageous.

A case in point is the furniture business of Owen McGarvey. Five years after his arrival from Ireland in 1838, McGarvey opened a paint, oil, and glass store, where he sold unfinished, imported chairs in the local

FIGURE 3.1 OWEN MCGARVEY STORE AND FACTORY, 1890. McGarvey, like many central district manufacturers, combined manufacturing with retail and wholesale functions. The central location provided McGarvey with many of the locational assets necessary for successful production. (*Special Number of the Dominion Illustrated Devoted to Montreal, the Commercial Metropolis of Canada* [Montreal, 1891], 36. Reproduced courtesy of Toronto Reference Library.)

market. To bypass the difficulties associated with importing, he began manufacturing in the 1850s to supply his retailing store. McGarvey's strategy consisted of manufacturing a variety of styles from the plainest to the most expensive and fashionable and selling in large quantities directly to local customers attracted by the expensive and fancy decor of the store. Common and medium-class chairs were machine-made at L'Assomption (a village forty kilometers from Montreal), hand-finished at his Montreal factory, and sold at his retail store attached to the factory. McGarvey made fancy and custom-made lines at Old Montreal, allowing him to take direct advantage of the district's market, networks of fashion information, and access to a full range of skilled and unskilled labor while maintaining a modern mechanized factory in a country town (fig. 3.1).[6]

Many other centrally located industries combined retailing and manufacturing at various scales. Some of them were relatively large factories employing modern machines and a detailed division of labor requiring extensive space and new layouts. Confectioners, for example, extended their range of goods and manufacturing strategies beyond that of the typical bread baker by utilizing the locational assets of the center. With thirty-five workers, three brick ovens, and a mechanical oven in a two-story building, James Aird's strategy combined the baking of a wide assortment of bread with the manufacture of candies and cakes. At Findlay and McWilliams's confectionery and candy factory a "spirit of goaheadativeness" lay behind the "introduction of machinery that simplifies, and with twenty-five rapidity, accomplishes the work." The company's four-story factory had a highly detailed division of labor both in terms of the physical layout of the plant and the gender structure of tasks.

Comparable in its trajectory was James Griffin's small firm founded in 1857. By the late 1880s he manufactured in a three-story plant: the store was located on the ground floor, baking was undertaken in the basement, and the manufacture of candies and confectionery took place on the second floor. The city's largest caterer, Hall and Scott, supplied confectionery, cakes, and pastries for parties and entertainments. The company's four-story premises consisted of a bakery in the basement, with the other floors being used for a store, showrooms, and banquet hall. Only one floor of the building was actually used for manufacturing, the rest for distribution. The large-scale, highly capitalized forms of manufacturing were not as important in the central districts as they would be in the greenfield sites of the urban periphery. Nevertheless, the growing division of labor, the installation of machinery, and the availability of large premises formed part of the central districts' manufacturing milieu.[7]

The central manufacturing districts had other industries employing modern forms of manufacture and producing an assortment of goods. Charles Garth's large, up-to-date foundry manufactured plumbing goods for the local market, and Robert Mitchell's brass foundry produced a range of architectural bronze products (fig. 3.2). Large-scale shoe manufacturers, trunk makers, clothiers, and hatters operating a highly divided work process and utilizing machinery in three- or four-

FIGURE 3.2 ROBERT MITCHELL FOUNDRY, MONTREAL, QC, 1869. Located in a modern mill structure, the Mitchell foundry, which produced an assortment of products for local and national markets, was typical of nineteenth-century central manufacturing district metalworking plants. It was converted into a company warehouse when Mitchell moved the plant to Sainte-Cunégonde in 1889. (Notman Photographic Archives, McCord Museum of Canadian History, Montreal, accession no. II-346495.)

storey buildings found the locational assets of the central districts irresistible. The shoe manufacturing firm Ames, Millard and Company was located in a four-story building "divided into numerous workshops, all of which are supplied with modern [steam-powered] machinery of the most approved patterns." Ames, Millard, and other shoe manufacturers coexisted side by side with smaller manufacturers and custom shoemakers. The result, as one observer noted in 1874, was that the city core was the home of "nearly 200 different kinds of boots and shoes.... 500 sewing machines, 75 pegging machines, 30 sole-sewers, 30 sole-cutters, —besides large numbers of dieing machines, heeling machines—also machines for eyeletting, punching, skiving, rolling, &." Similarly, the Davis and Lawrence pharmaceutical company's four-story building contained "every modern improvement that would facilitate the business..., such as steam elevators, hand railways, machines for bottle washing, bottle filling, bottle corking, etc." George Barrington made two hundred lines of trunks, valises, and satchels with modern equipment in a five-story building, having "every modern improvement for the perfection of their work." Even in Saint-Jacques, where small-scale, artisanal firms predominated, some large industrial firms with machinery and an intensive division of labor had been established. One was Withall and Hood's soap and candle factory, whose three-story building extended one hundred feet. As early as 1856, the company used steam power in the annual production of twenty thousand boxes of soap and ten thousand boxes of candles, all of which were consumed in Montreal and Quebec.[8]

The importance of centrality, even for the most modern of industries, can be illustrated by cracker and biscuit manufacture. Compared to other branches of the baking industry (bread and confectionery), cracker and biscuit firms employed high-volume techniques in large and highly mechanized factories. As one observer noted, while bread making "demands a great amount of muscle and nimbleness" and "very little machine work ... cracker making is a more mechanical and less free-hand process." From midcentury biscuit manufacturers used machinery, subdivided the labor process, and utilized nationwide distribution networks. In the mid-1880s crackers were made on an assembly line, where, even before the crackers reached the oven, machines and other equipment mixed, thinned, lengthened, dusted, and stamped the dough. Walter Paul, a Montreal grocer, told the 1888 Parliamentary Select Committee on Alleged Combinations that, among the biscuit manufacturers, "machinery has been perfected so that they can turn out enormous proportions to what they used to." Despite mechanization, a large labor force, and high-volume production, the industry continued to find a central location important. Viau and Frère, with two hundred employees, the city's largest biscuit firm, remained centrally located. Its Saint-Jacques four-story building manufactured biscuits, confectionery, and self-rising flour. The company employed a forty-horsepower steam engine and patented machinery. It possessed a 700-acre farm on the eastern limits of the city, where thirty cows furnished milk daily for the company's biscuits. Similarly, Lang Manufacturing, with one hundred employees, manufactured plain and fancy biscuits in its Saint-Antoine factory for eastern Canadian markets.[9]

"Contracted Spirit of the Ancients": Adapting the Built Environment

The transition from a commercial to an industrial city required refashioning the urban fabric, notably redeveloping streets, tearing down and replacing older buildings, and constructing newer and more appropriate infrastructures. Commentators from as early as the 1820s pointed to the problems of Montreal's built form and called for extensive redevelopment. John Duncan, for example, noted that "the streets are for the greater part most inconveniently narrow, and the foot-walks in many places encumbered with cellar doors and other projections." But it was not just the streets that were a problem; Duncan also pointed to "the dull effect" and the "gloom" of the city's housing. If the central manufacturing districts were to industrialize successfully, these problems had to be dealt with.[10]

Over the course of the nineteenth century the central manufacturing districts' streets became a major focus of Montreal's redevelopment. This was evident by the 1860s, as one observer noted: "The original property holders of the city were evidently impressed with the value of every foot of mother earth on their island, therefore, to economise space they built upwards and denied the necessary width for street traffic. This contracted spirit of the ancients is now compelling their descendants to remove whole streets of substantial buildings to enable them to widen the thoroughfares." Many new streets were opened in the Outer Core, and most of Old Montreal's main streets were widened. In some cases the city took full financial responsibility, while in others the city and business interests undertook the changes together. In 1864, for example, the city widened Notre Dame Street, Montreal's main thoroughfare, between McGill Street and Dalhousie Square, at the request of the street's proprietors. The city covered half of the cost, while the other half was borne by the proprietors by means of a special assessment.[11]

The problem of readapting the city's built environment was confounded by the difficulty of converting structures inherited from the past to meet the demands of the present. In Montreal, as in most other North American cities, a warehouse-manufacturing district developed after 1850. Increasing demand for a variety of manufacturing spaces led to the demolition of old buildings and the construction of higher, multistory ones. The introduction of new construction technologies and methods raised buildings from three- to four-storys in 1850 to nine- to twelve-storys by the 1880s. Taking up a significant section of Old Montreal and extending along the major thoroughfares into the Outer Core, Montreal's warehouse district provided many advantages to the nineteenth-century firm. Most important, it offered a range of manufacturing spaces from the small garret to the large multistory building and close proximity to other manufacturers, retailing and wholesaling firms, and the major transportation terminals of the city. For merchant-manufacturers such as McGarvey and the confectioners who combined retailing and manufacturing in the same location, the work spaces found in the warehouse district enabled them to produce for a handful of markets while overseeing several functions at a single location.[12]

A range of work spaces for firms found in the central manufacturing districts contributed to the proliferation of external economies and chains of production. In contrast to the older, cramped, low-rise buildings, the new mill structure of the warehouse district's buildings piled new floors upon existing structures and created entirely new buildings, resulting in the expansion of the floor space available to manufacturers of all types. Manufacturers combined the advantages of these in situ locational assets with new machines, a more intensive division of labor, and new ways of organizing production and distribution to produce increased output, greater productivity, and larger workplaces and to reach much wider markets.

Buildings suitable for manufacturers' changing needs, however, often entailed a great deal of reconstruction, and frequently this was difficult. The physical durability of the buildings themselves were major obstacles to refashioning space. The high cost of making improvements allied with the desire of property owners to put immediate gains ahead of longer-range or high-risk benefits frequently inhibited redevelopment. Despite these difficulties, an ongoing process of converting central manufacturing district space to industrial use occurred. The redevelopment of the land of the Hôtel-Dieu and the Grey Nuns' Hospitals illustrates the process of the construction of new manufacturing spaces in the frame of an old built environment (map 3.2). In 1860 the Hôtel-Dieu, which since 1644 had occupied a large site in Old Montreal, moved to the city's northern fringe. Immediately after the move the Hôtel Dieu converted its central property to commercial and manufacturing uses. This took place in two stages: the remodeling of the physical layout of the land—the building of De Bresoles, Le Royer, and St. Dizier Streets in the space vacated by the hospital; and the building of a series of five- to seven-story warehouses to accommodate a variety of economic activities. By 1881 a range of commercial (hardware, groceries, grain) and manufacturing (cigars, rubber, electroplating, mirrors) interests occupied the site.

A similar process occurred on the waterfront to the south after the Grey Nuns moved their convent and hospital to a new west end location in the early 1870s. The Grey Nuns commissioned the construction of several multistoried warehouses, which were rented out to commercial and manufacturing concerns. In contrast to the Hôtel-Dieu land, the major occupants were produce, provision, and coal merchants as well as other commercial activities dependent on the waterfront for their livelihood. Nonetheless, there were nine manufacturing enterprises, accounting for almost half of the rent taken by the Grey Nuns. In 1881 Both the Hôtel-Dieu and the Grey Nun warehouses covered an area of 100,000 square feet of land with buildings valued at almost a half-million dollars and generating more than $21,000 of yearly rent.[13]

Adding to the central manufacturing districts' locational assets were extensive transportation facilities. One of the largest harbors in North America was constructed over the course of the nineteenth century by aggressive local entrepreneurs hitching state funds to the redevelopment of the harbor (fig. 3.3). Merchant-manufacturers such as John Young, seeking to strengthen their control over a continental market, channeled

---

MAP 3.2 REDEVELOPMENT OF THE HÔTEL-DIEU AND GREY NUN LANDS, 1852 AND 1881. The redevelopment of the Hôtel Dieu and Grey Nun sites illustrate the changing and dynamic character of Old Montreal's built environment. (Compiled from W. McKenzie, *Map of the City of Montreal Showing the Latest Improvements* [Montreal, 1852]; Charles Goad, *Atlas of the City of Montreal, from Special Survey and Official Plans Showing All Buildings and Names of Owners* [Montreal, 1881]; City of Montreal, Water Tax Rolls, Centre ward, 1881; Lovell, *Montreal City Directory, 1881* [Montreal, 1882].)

Making the Central Manufacturing Districts

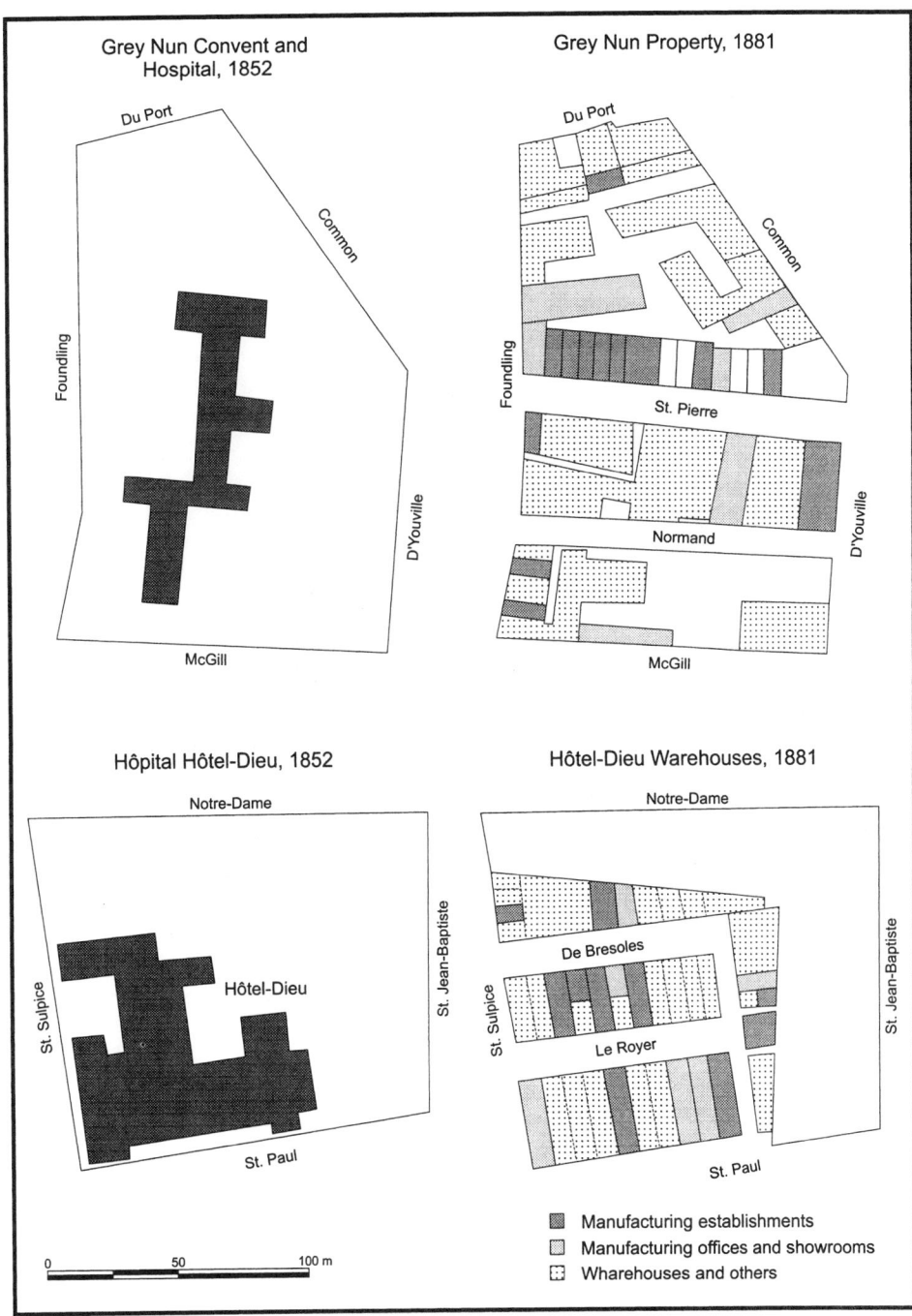

FIGURE 3.3 HARBOR FROM THE CANADIAN PACIFIC RAILWAY ELEVATOR, MONTREAL, QC, c. 1885. Extensive nineteenth-century redevelopment of Old Montreal's harbour was made possible by the ability of the Montreal Harbour Commission to leverage large funds from the federal government. Cycles of investment produced a modern harbor facility producing locational assets to Montreal's manufacturers. (Notman Photographic Archives, McCord Museum of Canadian History, Montreal, accession no. view-1938.)

large chunks of capital into new piers, new handling equipment, and deepening the harbor. In the process they transformed it from a small commercial port in the 1840s to a large, up-to-date facility by the 1870s.

Equally important for manufacturers were the terminals of the national railroads located in Saint-Antoine and Saint-Jacques. The Grand Trunk built the Bonaventure Station in Saint-Antoine, while the creation of the transcontinental railroad by Canadian Pacific in the 1880s led to the construction of two terminals, Windsor station in Saint-Antoine and Viger station in Saint-Jacques. The Canadian Pacific and Grand Trunk central terminals were in turn linked to marshaling and freight yards on the city fringe. The east end Hochelaga yards served the Canadian Pacific until they were shut down in 1903, while the Point Saint-Charles shops in the southwest district were the main Montreal yards of the Grand Trunk. The central districts were also the center of the city's intra-urban transportation system. As the horsecar system expanded after 1861, its lines converged on the financial district and brought in more commuters each year. While the fares of the horsecar were out of reach of most of the working class, skilled workers, such as printers, who commanded high wages and steady employment, as well as central city business owners and white-collar employees rode to work from distant residential districts.[14]

A critical locational asset of the central manufacturing district was the availability of a large labor force. The importance of a large, ethnically diverse, and multiskilled labor force living in the inner city has been a staple of the descriptions of the social geography of the North American city since the writings of the Chicago School in the 1920s. The situation was no different in Montreal. The population of the central manufacturing districts more than doubled in the forty years after 1861, growing from 60,000 to 140,000. Over this period, however, the residential population of Old Montreal fell from 7,000 to 4,000 as commercial, financial, and industrial functions forced residential ones out. In contrast, the population of the three Outer Core districts grew from 54,000 in 1861 to more than 135,000 at the beginning of the twentieth century.[15]

Manufacturing firms were found throughout the central areas, and work and residence were interwoven. There is no evidence in Montreal of a primitive form of Ernest Burgess concentric zones—central business district, "factory district," and "workingmen districts"—upon which North American central cities are believed to have developed. Even though there may have been a fine-grained separation of factories and homes, manufacturing districts did not geographically divide cen-

Table 3.2 Occupational Structure of the Central Manufacturing Districts, 1861 and 1901

	Households (%)									
	Old Montreal		Saint-Antoine		Saint-Lawrence		Saint-Jacques		City Total	
Occupation Class	1861	1901	1861	1901	1861	1901	1861	1901	1861	1901
Business owner	24.2	11.7	23.0	12.7	13.6	6.9	4.3	8.5	10.1	5.4
Professional	22.7	2.8	4.4	7.2	6.0	3.8	3.8	5.5	4.4	3.3
Clerical	16.4	22.8	14.0	32.0	19.2	39.9	10.5	29.4	11.7	22.4
Blue-collar	26.6	15.9	30.3	22.2	37.8	30.2	49.2	29.7	36.2	27.9
Unskilled	10.2	46.9	28.4	26.0	23.4	19.1	32.2	26.9	37.6	40.9
Number of households	256	145	909	3,135	1,454	3,535	844	3,762	5,338	22,530

Source: Compiled from an approximately 50 percent sample of the City of Montreal, Water Tax Rolls, 1861 and 1901.

tral district business functions from housing. Nor did Montreal's central districts uniformly house workers separated into ethnic enclaves; they were, instead, extremely occupationally and ethnically diverse (table 3.2). Unlike the East and West End manufacturing districts, which were almost entirely working class, the residential areas of Old Montreal, Saint-Antoine, Saint-Lawrence, and Saint-Jacques had the full gamut of Montreal's ethnic and class spectrum.[16]

Although each district had its own specific ethnic mix, they were ethnically heterogeneous. French, English, Scottish, and Irish, for example, accounted for approximately a quarter each of Saint-Lawrence's population in 1881. Furthermore, the inner-city's ethnic composition became more diverse with immigration from southern and eastern Europe after 1880. In terms of their occupational class structure the districts tended to be underrepresented in the blue-collar and unskilled classes and overrepresented in the white-collar, professional, and business classes. This picture of the central districts' ecological pattern does not conform to the logic spelled out in Sam Bass Warner's description of 1860 Philadelphia: "a jumble of occupations, classes, homes, immigrants and Native Americans." The social geography of Montreal's central districts was not a jumble; working-class areas existed cheek-by-jowl with factories, boarding houses clustered in Old Montreal, and clumps of high-income residential districts were buffeted from poorer areas by bands of middle-income white-collar and skilled worker areas. The logic of these districts' large and heterogeneous population and varied social composition along with the other magnets of centrality provided a compelling reason for manufacturers to flock to the central manufacturing districts.[17]

Printing, Carriage Making, and Clothing in the Central Manufacturing Districts

In 1868 one observer noted of female clothing workers that "at long rows of wooden frames, their busy fingers were shaping the skirts, while others were engaged in fastening by means of ingeniously constructed machinery, the spangles, clasps, etc., used in joining the tapes and wires; the whirring noise of many sewing machines, also joined in the general chorus." As this description of Cassil and Cameron's crinoline dress factory suggests, the classic North American centralizing sector of clothing was not backward in terms of its degree of mechanization and division of labor. Even though the central manufacturing districts were dominated by small and medium-sized firms, this did not mean that clothing firms were technically or organizationally unsophisticated. An examination of this industry, together with printing and carriage making, illustrates both the forces creating and sustaining Montreal's central manufacturing districts and the variety of manufacturing pathways found in these districts.[18]

Printing

The central district's printing industry grew enormously between 1861 and 1890. Twenty-two of the city's twenty-three printing establishments were located in Old Montreal in 1861. Thirty years later ninety-three out of the city's ninety-seven printing firms were found in Old Montreal and the Outer Core. Even though it remained one of the most centralized of all industries, it was not a backward industry displaying little change or traditional features. Two central features of the nineteenth-century industry were its division into different segments (newspapers, book, and job printing and lithography) and the linking of technological change, new work process and varied markets to different manufacturing strategies. By the end of the century differences in terms of type of work, the nature of the market, the scale of enterprise, and the degree of mechanization within the industry were pronounced.

The most dramatic example of these changes occurred in the newspaper branch, in which the growing market for daily news, growing competition among newspapers, and greater capital availability permitted the introduction of new technologies and a more elaborate division of labor. In the 1850s several technological innovations changed newspaper manufacture: the rotary press and the steam-powered cylinder machine vastly increased productivity and reduced the skills of pressmen; product innovations in the manufacture of paper produced cheaper and better quality newsprint; and the telegraph sent news more quickly from place to place. In the 1870s publishers achieved greater production speed, output, and flexibility when the old process of the blanket sheet and quarto page was replaced by the Bullock press, using the stereotyping process. The linotype machine, installed in all of Montreal's major publishing houses by the end of the century and the smaller ones by 1910, continued the process of skill displacement, faster through-flow, and cheaper costs. By the end of the century the large newspaper houses such as Gazette Printing and Star Printing employed hundreds of workers, had introduced the most modern machinery, and serviced a different type and much larger market than most other printing firms. Taking up several floors of multistoried buildings on the edge of Old Montreal, the Gazette and Star not only printed and published the two leading English-speaking newspapers, the Tory *Montreal Gazette* and the liberal *Montreal Star,* but also undertook significant amounts of contract work.[19]

Uneven development and several productive strategies among its other branches characterized the central manufacturing districts' growing printing complex. In lithography, for example, establishments deployed a range of manufacturing strategies, despite changes to work content, the increasing capitalization of the industry, and growing specialization and segmentation. Clustered within the same small area of the central manufacturing districts were some of Canada's largest lithographing companies. Sabiston Lithographic employed a wide range of skills in its four floors of the Gazette Building. Along with its core set of skilled editors and print composers, it also employed art designers and engravers in its artists' department; photographers, zinc etchers, and plate mounters; eighty-three unskilled binders; and pressmen. Similarly, Canada Bank Note Engraving undertook steel, wood, and stone engraving, chromolithography, and the artotype process to manufacture an assortment of products, including bank notes, debentures, bills of exchange, postage stamps, maps, and books. Its large double building operated steam-powered presses and contained the "finest geometric lathe in America." Similarly, in its six-story building Burland Lithographic, the nation's largest lithographer, operating steam-powered machines, manufactured an assortment of paper and cardboard products, playing cards, tintype mats, and photo mounts. But the large lithographing firm making many products for a range of markets was not the only type to be found in the central manufacturing districts. Numerous small firms, more workshops than factories, clustered close to the large firms in their multistoried mill structure factories. One such small firm was the wood engraving workshop of J. Walker. Catering to "the double demands of the bookmaker and the merchant" and engraving "for publications and business houses," Walker was fixed to a central location, just like the large printing establishments.[20]

In between the extremes of the large newspaper or lithographic house and the small engraver was the mass of printing firms, most of whom were book and job printers. To be successful these firms had to perform a variety of large and small jobs under all types of conditions, as demonstrated by the case of John Lovell. Apprenticed to a Montreal printer in 1823, Lovell gained experience working in several firms and in 1836 opened his own printing house. He quickly became the one of the city's most successful printers, thanks to his drive to mechanize, reorganize the structure of work, and capture several markets. His efforts not only increased the firm's productivity and internal economies of scale but also gave him greater control over the labor process, much to the consternation of printers. Just as important was the spreading of risks through a combination of job and commissioned printing while specializing in lines such as directories, gazetteers, and schoolbooks. With these different market niches, Lovell built up an extensive body of work, expertise, and contacts. Gaining long-term government contracts and rights for foreign books added to his success. With 147 workers and capital of $80,000 in 1871, Lovell had the largest printing establishment in Montreal. Even though most firms did not have Lovell's security, capital, and access to a number of different markets, the ability to respond quickly to the ever-changing demands of the market and technological change was a critical foundation of the printing industry. While Lovell represents the trajectory of a medium-scale and successful firm, the printing segment was characterized by an array of smaller firms. Small book and job printers such as A. Trudeau, with a rent of fifty dollars in 1890, undertook the occasional contract for a large printer or business enterprise and eked out a living using basic printing equipment and employing few, if any, workers. An assortment of manufacturing pathways featuring differences of scale, labor organization, mechanization, and markets characterized the industry over the course of the century.[21]

From the small job printer and engraver to the large publishing and lithographic houses, printing firms of all descriptions clustered in the central districts. Between 1861 and 1890 the industry spread out from two small Old Montreal clusters to encompass a much greater area extending to the Outer Core districts (map 3.3). The industry's central location rested on its provision of products and services to other businesses and

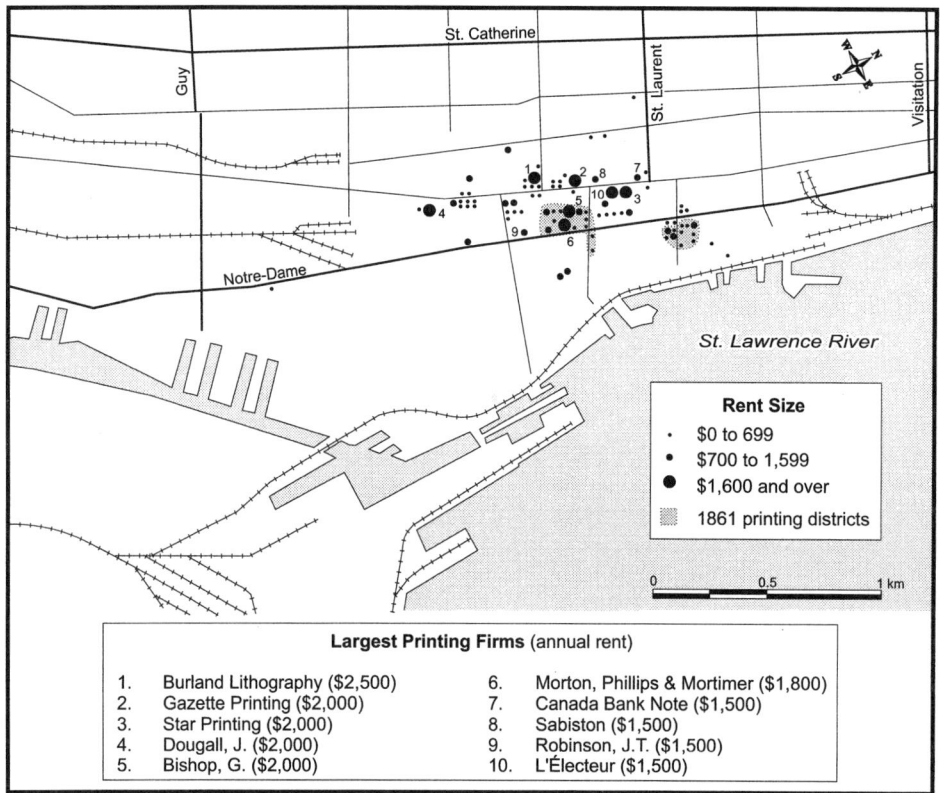

MAP 3.3 THE GEOGRAPHY OF THE PRINTING INDUSTRY, 1890. Despite extensive growth over thirty years, the printing industry remained highly centralized in 1890. (Compiled from City of Montreal, Water Tax Rolls, 1861 and 1890.)

to government offices, rather than to final (private) consumers. Small- and medium-scale printing firms took advantage of the nearness to business clients, face-to-face contacts, and swift messenger services for orders, proofs, and deliveries. Even for those firms with a wider geographic market, the central district provided access to highly valued contacts and networks. Reliance on immediate access to the major printing users of the day—notably lawyers and notaries—merchants, transportation companies, and government offices were critical. Even the newspaper branch, with its elaborate division of labor, mechanization, and high capitalization, sought out a central location, as the core was the most efficient distribution center, having the best access to a market that was shifting from a small commercial elite to one catering to an urban mass market. This highly segmented industry, dependent on elaborate chains of production, formed two specialized clusters in the central core. One was in the west, next to the major retailing and financial activities on Notre Dame and St. James Streets, and the other lay farther east, close to the administrative and legal core of the city.[22]

Carriage Making

Similarly, the structure of carriage making underpinned the industry's clustering in the Outer Core. The initial transformation of the industry was not technologically induced but arose out of a growing division of labor. So advanced was the division of labor that even a small carriage shop had to have at least one blacksmith, painter, woodworker, and trimmer if it were to be successful. As late as the 1870s, firms were small, continued to manufacture carriages mainly by hand, and were geared to producing for the local market: according to the 1871 census, Montreal's thirty-four carriage makers employed little more than $120,000 in capital, while twelve employed only two or three workers each, and only one, Larivière's Canada Coach and Sleigh factory, used steam power. Starting in the 1870s, the handcrafted carriage gave way to the machine-made one. By the end of the century mechanization dominated and was so successful that the cost of a carriage was almost one-sixth the price of a similar product thirty years earlier.[23]

Despite these changes, the Montreal carriage industry remained small in scope, in both its total output and its scale, compared to other North American centers such as Cincinnati and Toronto. In 1891 the city's forty-six establishments employed only 582 workers and produced carriages valued at $733,000. Although the number of workers and amount of capital invested in firms grew in this period, the city's firms remained small. Larivière, the largest firm in 1891, only employed sixty workers. The small scale of the industry, however, did not impede the ability to manufacture a wide range of carriages. At the 1892 World exhibition E. N. Heney, a prominent local manufacturer, displayed nineteen types of carriages, including Victorias, landaus, T carts, phaetons, spiders, and clarence traps, all manufactured at his Saint-Lawrence factory.[24]

Montreal's carriage and ancillary firms formed a crescent in the Outer Core encircling Old Montreal. Within this crescent carriage makers were rooted to two locations, accounting for three-quarters of all

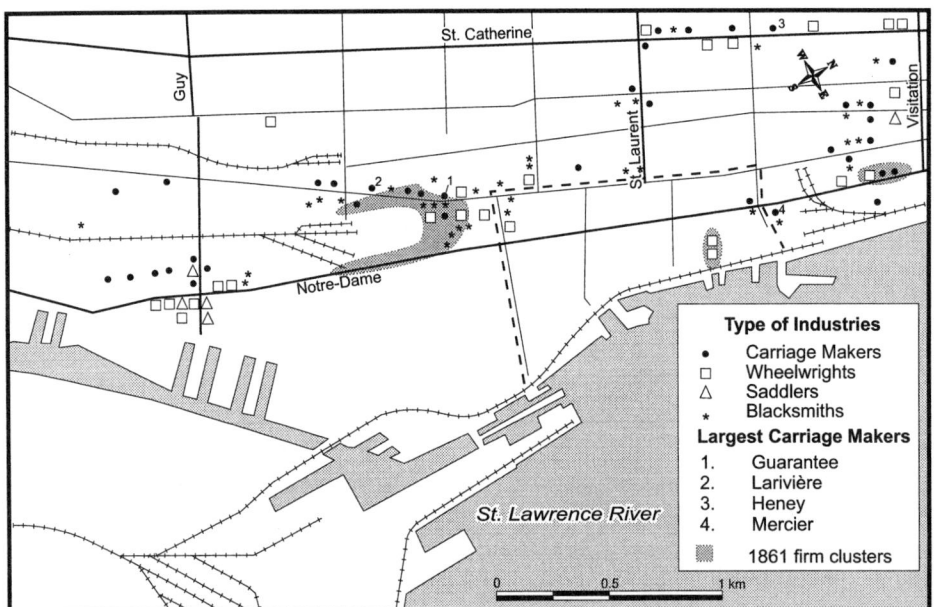

MAP 3.4 THE GEOGRAPHY OF THE CARRIAGE MAKING INDUSTRY, 1890. The assortment of trades involved in carriage making formed nodes around the major railway terminals surrounding Old Montreal. (Compiled from City of Montreal, Water Tax Rolls, 1861 and 1890.)

firms (map 3.4). The most important, centered in Saint-Antoine, clustered along the major avenues leading out of Old Montreal and next to the Grand Trunk Railway terminal. A second group located in a corridor in Saint-Jacques leading from the Canadian Pacific Viger Station. As in the case of the printing industry, carriage makers by 1890 had spread out from their midcentury locations, as financial and commercial functions pushed up land values. Operating a highly competitive industry, manufacturers found in these locations excellent access to off-the-street customers, wealthy clients who worked in Old Montreal and lived in close proximity to the clusters, the large number of carters who frequented the railway terminals, and the knowledge center, where the latest fashions, news, and prices could be best determined. A central location enabled manufacturers such as Martin Gravelle, who faced "great competition, which often renders the profits very low," to keep up with the "constant and improved changes" of the industry. The cluster of suppliers (wood, leather, and paint) and subcontractors (blacksmiths and wheelwrights) found in Saint-Antoine and Saint-Jacques, the metal shops of Saint-Ann and Sainte-Marie, and access to Old Montreal's agents and wholesalers ensured that materials, labor, and distribution networks were readily accessible.[25]

The building requirements of the industry reinforced the central district's advantage. Most large firms operated in three- to four-story warehouse buildings, where gravity-flow production moved the carriage from the top floor down and different departments occupied different floors. Felix Mercier's four-story carriage factory functioned in this way: on the top floor was the wood shop, where two separate processes took place—preparing pieces of the carriage (panels, framework, gears) and assembling the body. After being assembled, the body was ironed in the blacksmith shop. Next, the carriages were painted, varnished, and trimmed in the paint and trimming shop. The last step, the carriage's final assembly, took place on the first floor. The manufacturing pathway of the carriage industry—small-scale firms, splintered markets, tight interindustry linkages, and the demands of competition and fashion—militated against it seeking more peripheral locations.[26]

Clothing

Clothing manufactures also illustrates the overlapping of different productive strategies and the attraction of the central manufacturing districts. Studies of Baltimore, New York, and Toronto, among other places, have shown garment manufacture to be the classic center-seeking industry. Montreal's position as Canada's premier clothing center reflected not only the city's national preeminence but also the industry's ability to fuse technological change, new labor practices, cheaper operating costs, and new work space with the locational assets of the core. The location of the city's clothing firms remained greatly centralized over the course of the nineteenth century even with the tremendous changes taking place to the industry. Production changes started as early as the 1820s, when, despite the small scale of firms and the reliance on traditional tools, clothing was reorganized through an increasing division of labor. Masters and mistresses began to separate themselves from production, taking on a supervisory and coordinating role, while workers were divided by degrees of skill. The intensification of the division of labor led to the employment of large numbers of women and children, as employers reduced costs by lowering wages and equating gender with skill. Of the two main cost items of clothing manufacture—materials and wages—the most successful way to reduce costs was to reorganize labor, as the possibilities of reducing material costs were small and largely out of the clothing manufacturer's control.[27]

After 1850 machinery was superimposed upon the early-nineteenth-century changes to distribution and the division of labor. A variety of innovations, such as steam-powered cutting machines, were introduced, but the most important was the sewing machine, which accelerated the replacement of skilled workers by less skilled ones and, in the process, further spurred the intensification of the division of labor. The advantages of the machines to contemporaries can be discerned from a description of the sewing machine department of McFarlane and Baird. "In one large room thirty sewing machines are in operation humming their lively rattle incessantly, weariness to their metal sinews is a thing unknown, their needle never lags, no romantic musings surround them." The popularity of the sewing machine in Montreal, besides its obvious cost and labor advantages, was related to the availability of spare parts, convenient local machine servicing, the massive advertising campaign of the sewing machine makers, and the establishment of several local sewing machine factories after 1861.[28]

Three different manufacturing pathways characterized Montreal's clothing firms, despite the growing application of more elaborate forms of labor division and more sophisticated machinery after 1850. The first was the small shop employing a few workers and little machinery and servicing the local market. Some were custom shops catering to a local elite market, while others made ready-made clothing for the working class. The small shop was greatly represented among men's tailoring (suits, overcoats, pants, and jackets) and women's clothing: in 1871 firms with less than fifteen workers accounted for 139 of the 179 firms in these two clothing branches. A second type consisted of firms performing in-house production, featuring extensive mechanization, a highly developed division of labor, and a sophisticated system of distribution. In-house work was common in the men's furnishing branch (shirt, collar, cuffs, and ties). Examples include John Aitken's shirt factory, with more than three hundred female employees, and Cassils and Cameron, which employed more than one hundred women in the making of hoopskirts in its four-story building.[29]

A third strategy was the organization of outwork by wholesale manufacturers. As one contemporary noted, "true factory production was almost non-existent," as many Montreal manufacturers "had the bulk of their goods made up in the rural villages beyond the city." According to another observer, "the garments were cut on the premises of the wholesale clothing house, tied into bundles with the linings and trimmings,

FIGURE 3.4 H. SHOREY CLOTHING FACTORY, 1890. One of the largest clothing firms in the city, Shorey employed a large number of rural outworkers and skilled inhouse workers, deployed modern machinery, and served national markets from a central location. (*Souvenir Number of the* Montreal Daily Star *Reviewing the Various Financial and Commercial Interests Represented in the City of Montreal* [Montreal, 1890], between 48–49. Reproduced courtesy of Toronto Reference Library.)

and sent out into the country to be made up. Farmers for miles around the populous centers would drive into the towns, carrying homes the bundles of cut garments and these would be put together at home, being brought back a week or so later when the payment would be made on the basis of so much per garment." According to H. Shorey, one of the city's largest manufacturing wholesalers, most of his seven hundred employees worked at home. He employed seventy skilled inside workers who prepared work to go out. At William Muir's plant, with a similar ratio of inside and outside workers, cutting was done by "a knife which cuts the cloth by steam, so that four cutters will do the work of from twelve to fifteen." The division of labor here was very pronounced: men did the cutting in the factory, while women and children performed the sewing tasks at home. Many of these firms employed a large workforce; in 1871 the six largest firms employed 40 percent of the city's clothing workers and undertook 43 percent of clothing production. Growth continued over the period. In 1883, for example, Shorey added an additional story to his plant, bought an adjacent building, and employed up to four thousand workers, many of whom were "engaged in the many small factories of the firm [i.e., homes] in the neighboring towns and villages, and even as far as the city of Quebec" (fig. 3.4). The dynamic driving outwork was the combination of cheap outside labor, skilled inside cutters, and machinery.[30]

All varieties of clothing firms were concentrated in Old Montreal and the Outer Core (map 3.5). This was to be expected because, as other writers have pointed out, the clothing industry operated under a vertically disintegrated system of production, dependent upon a highly evolved system of external economies. Information flows in terms of price signals, the latest design and materials, proximity to retailing outlets, and access to a large and diverse labor force were critical factors for an industry relying on the reduction of labor costs and subject to the vagaries of seasonal fashion and price changes. As one writer put it, "fashion is a stern dictator and those who faithfully follow her many vagaries require a somewhat plethoric purse to stand the pressure." Manufacturers not only had to change their lines four times a year, but they also had to make constant adjustments to stay abreast of changing fashions. In the 1860s, for example, firms such as Cassils and Cameron turned to the production of crinoline dresses because they had become a "staple of trade," while others switched lines in response to the "discardment of the weighty and multitudinous skirts" that had previously reigned supreme. In addition, intense competition resulted from easy entry into the industry, small initial capital needs, and low barriers to knowledge. The

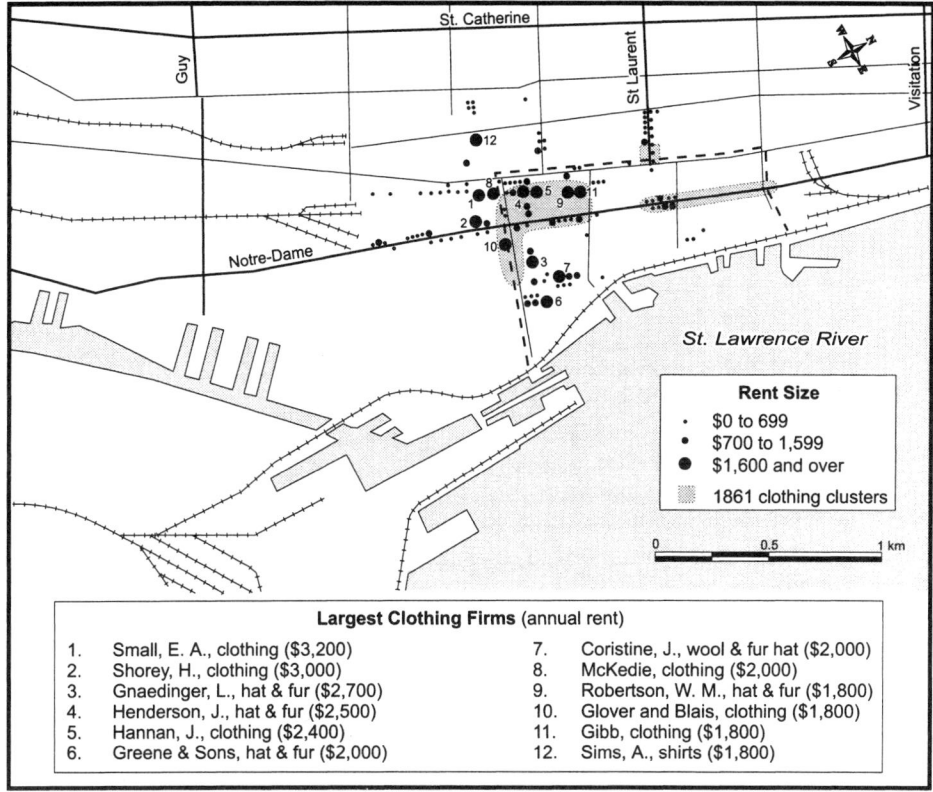

Map 3.5 The Geography of the Clothing Industry, 1890. Between 1861 and 1890 the center of the clothing industry slowly moved northward from Old Montreal into Saint-Antoine and Saint-Lawrence Districts. (Compiled from City of Montreal, Water Tax Rolls, 1861 and 1890.)

intensification of external economies in the clothing industry continued to occur, as it did in the Paterson silk industry, despite the technical and economic changes wrought by mechanization and an expanding market.[31]

The important factors behind the clustering of the clothing industry were those of the market, labor, and work space. The importance of the market lies not in the cost of getting clothing to market but in the numerous linkages between producers and consumers, driven by unpredictable style changes and frequent contact with the market. Makers of custom-made clothing needed to be in close proximity to the city's elite and middle class, most of whom worked in or had easy access to the city core. The market for wholesale ready-made goods was not the final consumer but the wholesale houses, all of which were located in Old Montreal, as were most of the stores selling retail ready-made clothing. The elasticity of the spatial boundary of the labor market—reaching out to the surrounding rural districts—reduced the importance of the core as a location for firms, especially in light of the fact that labor constituted the largest single cost. Although it could have been equally rational to have set up shop close to the low-skilled labor source, manufacturers remained in the core to be close to skilled workers (cutters and pressers) and because centrality proved the most workable option, given the large labor input and the geographical dispersion of outworkers. Moreover, the costs of transporting the material and sewn goods to and from the rural households were either undertaken by the sewing families themselves or constituted only a fraction of a firm's total production costs. Adding to the ability of firms to remain in the core was their ability to pass on the costs of overhead and machinery to workers, thus allowing firms to bear the higher costs of a central location. Equally important, most firms required small working areas, and a range of flexible work spaces were available in the core. Little space was needed to house machinery, and the through-flow of materials, despite an advanced division of labor, was relatively unsophisticated. Moreover, as stocks of raw materials and finished goods were kept at a minimum, little storage space was needed. The core contained a mixture of different work spaces from which different clothing manufacturers could choose: stores along the major retail streets for the custom shops and the merchant tailors and multistoried warehouse structures for the larger in-house and putting-out manufacturers.

As the examples of printing, carriage making, and clothing illustrate, Montreal's central manufacturing districts throughout the nineteenth century, like those of many other North American cities, were the home to innovative and technically sophisticated industries functioning within highly regulated chains of production. These industries contributed to the districts' tremendous manufacturing growth and their consolidation of the locational assets. As early as the 1850s, the central manufacturing districts were characterized by the city's most diversified manufacturing base and a variety of manufacturing pathways: from small- and medium-scale firms operating under handicraft and manufactory conditions to modern firms deploying high-volume production of standardized products. Clustering around harbor and railway facilities and seeking out the large adjacent labor pool, firms both sought and created central agglomeration economies.

Paralleling the specialization of central space with respect to other districts in the city was the specialization to be found within Old Montreal and the Outer Core. A distinct micro-geography developed within the central districts. On the one hand, the financial, government, and commercial functions of the emerging central business district clustered in and adjacent to the western and central areas of Old Montreal. On the other hand, manufacturing industries under constant pressure from these functions were forced out to the surrounding areas and in the process pushed residential settlement further out.

Chapter Four

"Factories and Industrial Establishments of Various Kinds" in the Eastern Manufacturing Districts

> *Factories and industrial establishments of various kinds have been erected in the city and its immediate vicinity, with surprising rapidity. And other manufacturing establishments are being continually added thereto, in which a large number of persons of both sexes are employed.*
> —Department of Agriculture, "Montreal's Agent's Report,"
> *Sessional Papers*, 1888

The large number and variety of firms clustering in the central manufacturing districts was only one feature of Montreal's industrial landscape before 1890. Building on a legacy of petty commodity production stretching back to the eighteenth century, manufacturing districts with factories and an assortment of industrial establishments developed in the sparsely settled eastern city periphery and an adjacent suburb in the second half of the nineteenth century. After 1850 an industrial district physically and socially distant from the central districts emerged. The East End cluster, containing only a few small petty commodity producers centered on the Molson brewery in Sainte-Marie before midcentury, grew dramatically in the second half of the century. A generation after the industrialization of Sainte-Marie the industrial suburb of Hochelaga developed to the east of Sainte-Marie. Very quickly, the suburb, a barely populated area before the 1870s, became an important addition to the city's manufacturing base. Several interlinked features were behind the formation of these peripheral manufacturing districts: the laying down of new technologies, factories, and industries on greenfield sites, the building of working-class housing and community institutions, and the creation of infrastructures by local elites. By the end of the century Sainte-Marie and Hochelaga formed a distinct and substantial industrial and working-class residential belt covering the eastern portions of the emerging metropolis.

Sainte-Marie: Growth and Consolidation of a Nineteenth-Century Manufacturing District

The beginnings of manufacturing in Sainte-Marie can be traced back to 1782, when Thomas Loid built a brewery on the banks of the St. Lawrence River. Four years later John Molson bought out Loid. Over the ensuing years the Molsons, through its ability to control manufacturing and land development, were the most influential family in the eastern part of the city and strongly shaped the district's early development. During the first decades of the nineteenth century the Molsons established several other manufacturing enterprises, including a distillery, shipyard, and foundry, and were involved in transportation and financial concerns. Despite the family's activities, Sainte-Marie's growth was slow. According to the 1842 census, there were only three industrial establishments of any substantial size in Sainte-Marie: two breweries operated by William and Thomas Molson and the William Parkyn foundry. As long as the area's growth occurred within a locally orientated petty commodity mode under the control of the Molsons, its economic and population growth remained restricted. By midcentury, however, a small yet robust manufacturing base had been built upon the embryonic cluster of petty commodity producers and manufactories established over the preceding half-century. The seventeen mostly small handicraft shops in 1851 clustered around the brewery in the southern part of the ward along the roads close to the river. Another cluster, consisting of brickyards, lime kilns, and a tannery, located outside the city's settled area on account of their noxious character and the availability of raw materials. In the following decades these nodes, formed at the junction of water and land, experienced the transition from a small artisanal enclave into a capitalist manufacturing district.[1]

In Sainte-Marie, as in other northeastern North American cities after midcentury, workshops, manufactories, and factories of all scales became important elements of the fringe business landscape. Increasing investment by Montreal's merchant-manufacturers and artisans led to greater amounts of fixed capital, more factory workers, and reorganized work methods. This growth is reflected in an almost sevenfold increase in the number of the district's firms and large increases in scale between 1851 and 1890. Over this period Sainte-Marie's number of firms rose from 17 to 123, its median rent grew from $40 to $120, and its share of the city rent increased to almost 8 percent (table 4.1). Sainte-Marie's manufacturing growth throughout the 1850s resulted from the industrial surge affecting other parts of the city. Two sectors dominated this phase of the

Table 4.1 Manufacturing Structure of Sainte-Marie and Hochelaga, 1861 and 1890

	Sainte-Marie				Hochelaga		Sainte-Marie and Hochelaga	
	1861		1890		1890		1861	1890
Industry	No. of Firms	Rent Share (%)	No. of Firms	Rent Share (%)	No. of Firms	Rent Share (%)	Share of City Rent (%)	
Food	15	60.6	31	21.3	3	10.3	15.5	15.4
Rubber	1	14.5	1	24.5	—	—	60.0	100.0
Chemical	2	5.8	2	1.9	3	0.9	10.8	7.0
Leather	6	3.8	10	4.2	2	0.2	1.7	4.4
Metalworking	2	3.6	6	6.8	3	2.6	1.5	4.5
Textile	2	3.3	4	6.9	2	41.8	14.2	45.4
Nonmetallic	3	2.5	9	9.8	—	—	12.3	55.6
Wood	3	2.3	5	4.4	2	2.5	3.1	13.4
Blacksmithing	3	1.1	12	1.9	5	0.5	3.2	18.8
Transport	2	0.9	8	1.6	1	24.7	0.6	26.5
Tobacco	—	—	3	4.2	1	15.7	—	46.5
Paper	—	—	1	2.7	1	0.3	—	12.7
Clothing	—	—	12	2.5	2	0.3	—	1.5
Furniture	—	—	7	1.6	—	—	—	2.4
Printing	—	—	2	0.2	—	—	—	0.3
Total	40	100.0	123	100.0	25	100.0	5.6	15.1

Source: Water Tax Rolls for Montreal and surrounding municipalities, 1861 and 1890.

district's growth: rubber and food accounted for more than three-quarters of the district's rent. Sainte-Marie was also overrepresented in the textile, chemical, and nonmetallic sectors, although their contribution to its industrial base was minimal. Despite this concentration, an assortment of products was manufactured, including rubber shoes, beer, liquor, bricks, soap, brooms, and crackers. With this sectoral profile Sainte-Marie stands in sharp contrast to the central manufacturing districts with their concentrations of clothing, printing, and leather. Moreover, the eastern and central districts had quite different rent structures. While the latter tended to have small to medium-sized firms, Sainte-Marie's was polarized between a large number of small handicraft shops and manufactories and a few mechanized factories of national stature. Thirty-one of Sainte-Marie's firms in 1861 were small (with rents less than $150), employed few workers, had not subdivided the production process to any substantial degree, and served local markets. In contrast, three firms—Molson's brewery and distillery and Canadian Rubber—were among the largest in the city.[2]

Sainte-Marie firms implemented far-reaching technological and organizational change during the spurt of industrialization in the 1850s. Molson replaced its old preindustrial brewing methods, in place since the eighteenth century, with new equipment and techniques. By the end of the decade the brewery's "grey stone walls might have looked the same" as a generation earlier, "but inside were fanning mills and patent kilns, elevators to raise the grain, steam to operate dependable machinery." Over the following decades the Molson family's primary aim was to ensure the renewal and improvement of equipment and processes. These changes were not in themselves sudden or dramatic. But the accretion of new techniques and the associated reorganization of work, the diversion of capital from commercial and transportation pursuits into manufacture, and a sense of larger, nonlocal markets allowed budding manufacturers to build up a legacy that formed the basis of even more powerful thrusts of capitalist industrialization. Accordingly, Molson was not alone in carrying out changes contributing to the formation of a peripheral manufacturing district. Some smaller firms also installed machinery and expanded their works. In the early 1850s John Converse's two- and three-story rope works extended 1,200 feet in length; in 1856 the firm constructed a new four-story building. In these buildings up to fifty hands operating steam machinery manufactured 150 to 250 tons of Russian hemp annually into a variety of ropes. With the technological change and the growth taking place to Molson, Converse, and other firms such as Canadian Rubber, the Sainte-Marie cluster, which traced its origin back to the eighteenth century, emerged as a large-scale industrial district centered on a variety of firms and manufacturing pathways. Despite this growth, the district's enterprises and most of its housing remained clustered around the brewery in 1861.[3]

Over the next thirty years the East End developed into a major Montreal industrial district. The expansion of older firms such as Molson Brewery and Canadian Rubber and the establishment of several large-scale mechanized firms with elaborate divisions of labor in Sainte-Marie and further east in the new industrial suburb of Hochelaga contributed to dramatic growth (map 4.1). The number of manufacturing firms in Sainte-Marie and Hochelaga more than tripled, reaching 148 in 1890, while their share of the city's firms and rent in this period rose to 10 and 15 percent, respectively (table 4.1). Rent structure remained highly polarized; in 1890 a high proportion of the districts' firms had rents of either less than $300 or more than $2,000. The older pattern continued, with the ubiquitous small handicraft and family firm servicing local needs coexisting with a few very large firms directly integrated into re-

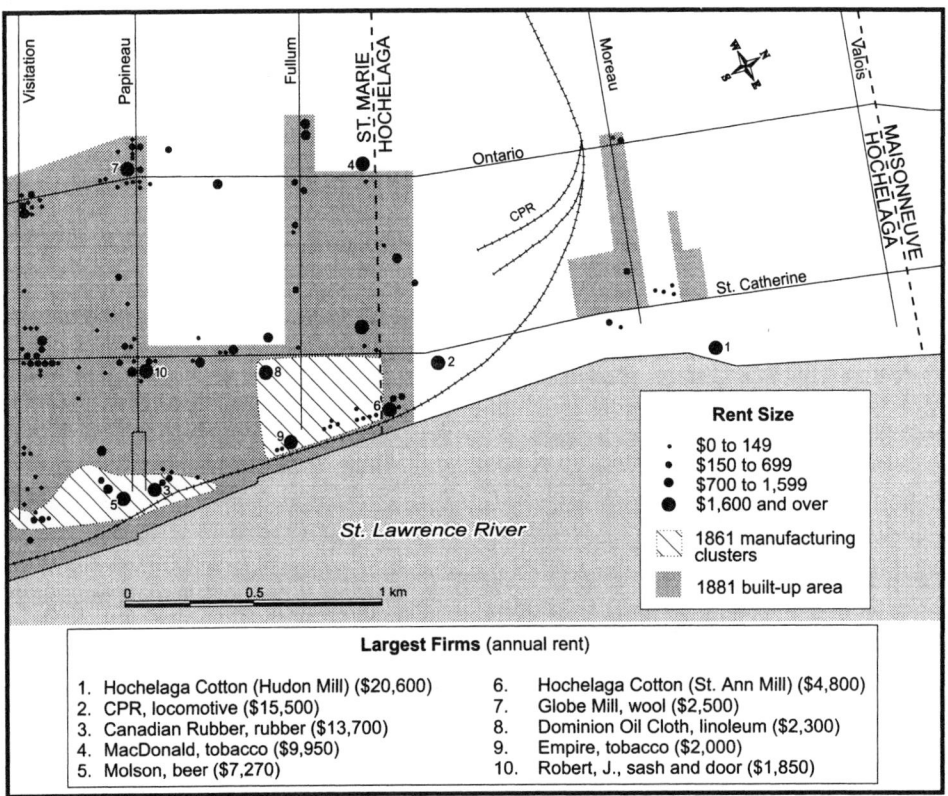

MAP 4.1 SAINTE-MARIE AND HOCHELAGA MANUFACTURING DISTRICTS, 1890. Beginning with a small nucleus around the Molson Brewery, during the second half of the nineteenth century the eastern manufacturing district expanded east into Hochelaga and north to the upper fringes of Sainte-Marie. (Compiled from City of Montreal, Water Tax Rolls, 1861 and 1890.)

gional and national markets. Small firms supplying local residents and businesses included twenty-three bakeries, seventeen blacksmith shops, and three beer bottlers. Many of these firms produced under relatively traditional methods, although some smaller firms did introduce new techniques and machinery and subdivided the labor process.

Persistence of a specialized industrial structure was another distinguishing feature of the East End. Sainte-Marie continued to have a different set of firms than the central districts and retained its traditional specializations, such as rubber and nonmetallic. The establishment of firms from other sectors, however, widened its industrial base as textile, transport equipment, and tobacco firms became important local employers (table 4.1). In the case of beverage and textile manufacture this concentration built upon earlier starts. By the end of the century several firms producing thread, bedding, and wool had formed around a large cotton cloth factory established in the 1870s. An extensive food-processing node developed: bakeries served the local market, and beer bottlers serviced Molson's brewery. In other cases new industries added to the district's concentration. Firms from the tobacco (cigars and pipe tobacco) and chemical (soap, glue, paint, and varnish) sectors added to the original base, which in turn grew because of the advent of oilcloth, glass, metalworking, and cotton thread manufacturers. Finally, a small metalworking cluster of firms including foundries, boiler and engine works, and railway supplies developed around Hochelaga's Canadian Pacific Railway locomotive shops.[4]

Specialization by work methods within firms paralleled Sainte-Marie's sectoral variety. A prerequisite for meeting the imperatives of firm growth was the creation of more intricate organizational methods within the factory. The creation of greenfield sites in Sainte-Marie allowed for the establishment of factory spaces accommodating firms' more segmented work methods and expanding range of new products. Companies producing a range of goods sought out sites where they could build factories with a more sophisticated factory layout arrangement, new and larger machinery, and greater numbers of workers. In 1874, for example, Michel Lefebvre constructed several buildings in which each of his main product lines—vinegar, jam, pickles, and salt—were individually manufactured with modern machinery. The efficiencies that Lefebvre created through horizontal integration by product line in his new plant were supported by the deployment of unskilled workers on a variety of machines in the main manufacturing buildings, with coopers, blacksmiths, and carpenters in shops separate from the main buildings. By 1888 the firm employed three hundred workers and supplied an assortment of products to the national market (fig. 4.1).[5]

The appearance of entirely new firms or those relocating from the central districts fueled fringe growth. Joseph Barsalou, who was involved in a wide range of the city's affairs, exemplifies the intersection of Montreal's bourgeoisie with the growth of an industrial complex in the East End. Beginning as a central-city commission merchant and auctioneer, he branched out into suburban real estate, manufacturing (Canadian Rubber and Dominion Oilcloth), finance (Royal Canadian Insurance Company), and politics (mayor of the industrial suburb of Maison-

FIGURE 4.1 MICHEL LEFEBVRE'S VINEGAR WORKS, 1890. The sprawling suburban works of Michel Lefebvre manufactured a wide assortment of food products for an expanding national market. (*Special Number of the Dominion Illustrated Devoted to Montreal the Commercial Metropolis of Canada* [Montreal, 1891], 165. Reproduced courtesy of Toronto Reference Library.)

neuve). In 1875 Barsalou opened a factory manufacturing toilet and laundry soap, using a new technological process. While the new process had the salutary effect of reducing the odor of soap manufacture, it also reduced the amount of labor needed to produce the same output: what had earlier taken a week was now turned out in an hour and a half.

Ravenhill and Molson's straw works relocated from the central manufacturing districts to Sainte-Marie. Originally located in the heart of the light industry section of Old Montreal, the company specialized in straw hats for local and regional markets. Almost immediately, the need to extend the building and to subdivide the organization of the plant forced Ravenhill and Molson to seek a different spatial strategy than one resting on a central location. In 1877 they built a bleachery in the eastern section of Sainte-Marie and soon after built new premises on Papineau Street, integrating their hat manufacture and bleaching operations at the new site.[6]

Along with new starts, existing firms continued to grow. Canada Rubber, a large and technologically advanced firm, set up new productive practices on the urban periphery. Established in 1853 to produce rubber shoes, its founders included Ashley Hibbard, involved in a local railroad company, and Edwin Chaffee, a manufacturing inventor from the United States. From the outset the firm implemented a high-volume, continuous-process work strategy, building on its managerial expertise in manufacture and transportation (the crude rubber came from Brazil). The company created a dual-labor system, importing experienced rubber workers from the United States and finding a plentiful supply of cheap labor in local women and children.

Canadian Rubber grew rapidly. In 1856 the firm housed 158 workers in five buildings worth more than $55,000; by 1871 the number of workers had risen to 370, and the company produced 650,000 pairs of rubber footwear annually. In 1890 its rent of $13,700 made it the city's fifth largest plant. In the same year it employed more than 1,000 workers in its large four-story factory containing 200,000 square feet of work space (fig. 4.2). An 1889 description gives some idea of its scale and complexity: "aside from this building, they have their engine and boiler houses, wash and drying rooms, varnish and cement house, and heaters, also a three-storey repair shop, 40 x 100 feet, in which a large staff of machinists, carpenters, steam fitters, etc., are constantly employed." Canadian Rubber's strategy of high-volume methods and an intricate series of manufacturing steps distinguished it from most local firms. The manufacture of shoes involved several stages: drying, vulcanizing, rolling, cutting, shoe assembly, varnishing, heating, and packing. As one writer says,

Figure 4.2 Some Views of Canadian Rubber, 1889. One of the largest plants in Montreal, Canadian Rubber is a typical example of a mid-nineteenth-century propulsive firm settling on the urban fringe. Along with a few other large firms Canadian Rubber established the basis for manufacturing growth and working-class residential settlement in the East End after 1850. (*Dominion Illustrated* [7 December 1889]: 364.)

echoing Adam Smith, "a rubber shoe, like a pin, has to pass through a great many hands before finally finding its way into the markets." With few exceptions, such as the heating process, which required experienced workers, operatives functioned as appendages to the firm's machines, minding belts, rollers, calenders, racks, and cars.[7]

Hochelaga: A Nineteenth-Century Industrial Suburb

In the 1870s an industrial suburb emerged as a new manufacturing pole to the east of Sainte-Marie. Like many of Montreal's industrial suburbs, Hochelaga did not begin as a manufacturing district but was initially settled by a small number of Montreal's mercantile elite, who built expensive homes next to the river. Incorporated as a village in 1870, it had no factories, a few artisanal workshops, and just over a thousand inhabitants, most of whom were French Canadian. In the 1870s, however, it was transformed from a sleepy residential rural retreat into a working-class industrial suburb, as local boosters such as the Rolland family and Raymond Préfontaine established the political and financial framework for manufacturing growth. Using leverage gained from property and industrial ownership, they pried profits from land development and built an industrial base by laying down streets, providing public services, and granting bonuses and subsidies to manufacturers and railroads. Building on these changes, large factories and transportation facilities bringing jobs and installing new production techniques were established in the 1870s and 1880s. By 1890 twenty-five firms, some of which were among the largest in the city, had located in Hochelaga. Seeking the same advantages attracting firms to Sainte-Marie, the first companies settling in Hochelaga functioned as magnets for further industrial growth. Established in the 1870s, the Canadian Pacific shops, for example, attracted firms with either direct linkages or with similar manufacturing pathways, including a rolling mill, three foundries, a boiler maker, an engine works, and two railway supply companies. The dynamic interaction of national industrial change and Hochelaga's industrialization can be illustrated through a discussion of three of the suburb's largest firms, Hudon Cotton, St. Ann Spinning, and Macdonald Tobacco.[8]

A principal force behind Hochelaga's growth in the 1870s and a harbinger of the direction taken by the suburb over the following generations was the Hudon cotton mill. Being one of the largest factories in the city when it was built in 1873, Hudon continued to grow over the next half-century. As with many other manufacturers of the day, Victor

Hudon deployed capital accumulated from his wholesaling activities and forged strong connections with Montreal financiers, merchants, and cotton brokers. Longstanding links with the Bank of Montreal enabled him to find the necessary credit for buildings, raw material, wages, and machinery. With extensive knowledge of Montreal and the surrounding municipalities, Hudon and a powerful board of directors found advantages in Hochelaga which were hard to equal elsewhere. With access to cheap and abundant labor, a plentiful supply of affordable land, and harbor facilities and the Canadian Pacific Railway yards for unloading and loading raw materials (coal and cotton) and finished products, most of the locational concerns of a nineteenth-century textile mill were met. In addition to these offerings Hudon prospered with a twenty-year tax exemption, government financed infrastructure, and access to Montreal, the nation's largest market.[9]

Hudon built a typical nineteenth-century industrial mill catering to the demands of a large, high-volume, continuous-process manufacturing operation. He developed an integrated works, like the cotton mills of New England, producing long runs of basic grey cloth. All three major cotton manufacture processes—spinning, weaving, and finishing—were housed under one roof. The five-story mill's relatively open and functional internal space was encased in a stone and brick frame (fig. 4.3). On the first floor was the repair workshop and the storeroom for finished products; the second floor contained three hundred looms; the next two floors had 18,000 spindles for spinning cotton; and the final stages of manufacturing took place on the top floor. A 600-horsepower steam engine powered looms and spindles operated by 600 workers. These employees, many of them women and children, worked in a highly regimented work setting in which continuous-processing machinery dictated the pace of work. Over the last quarter of the century the firm grew rapidly, and by 1901 its 1,275 workers operated 1,580 looms and more than 65,000 spindles.[10]

The establishment of the Hudon mill in the industrial suburb of Hochelaga paved the way for the growth of an extensive textile industry in Montreal. Prior to this date, the only Montreal textile firm of any consequence was the small Woods's cotton cloth factory located alongside the Lachine Canal. Hudon's factory signaled the appearance of a large, corporate-based, and well-developed Canadian textile industry, with Montreal as its financial and manufacturing center. Over the following generation most of the city's new textile firms settled in the fringe industrial districts. One area was in the western industrial suburb of Saint-Henri; another was in Hochelaga. In the early 1880s a second large cot-

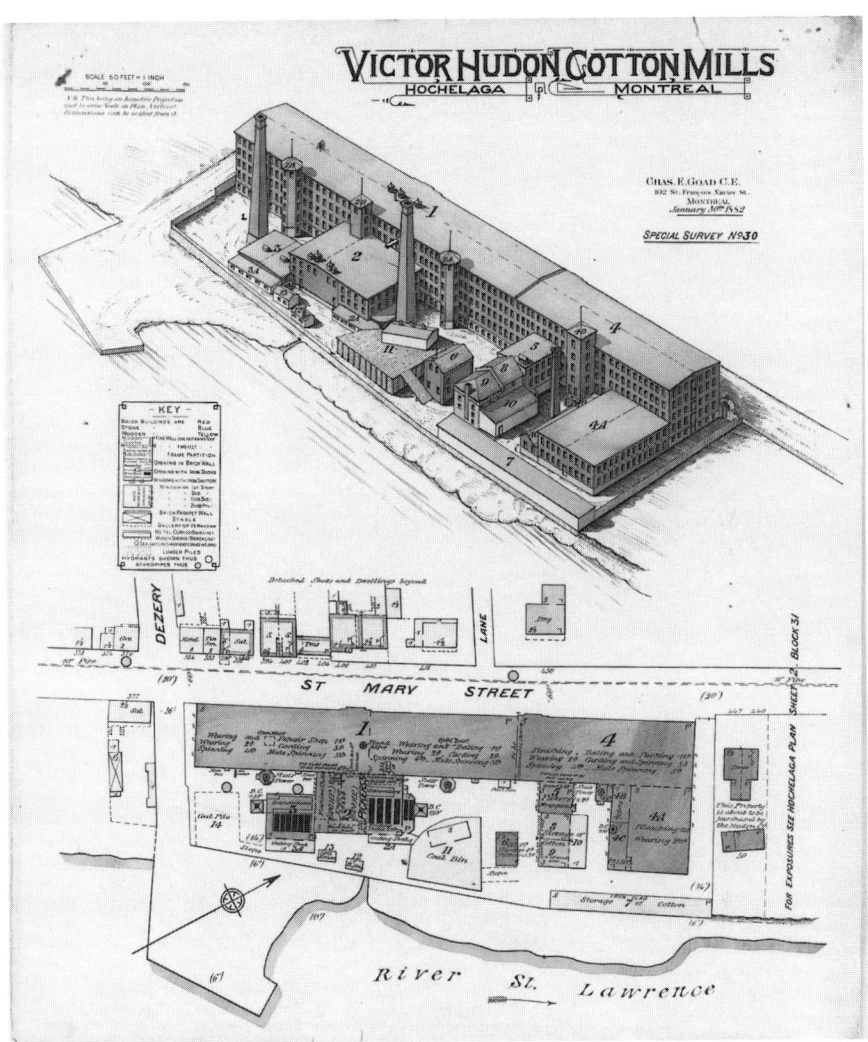

FIGURE 4.3 THE HUDON COTTON MILL, 1882. The fire insurance plan for 1882 shows the large and highly segmented character of textile production. As the city's first large textile firm, Hudon was responsible for the establishment of an extensive textile cluster in the suburban East End. (Charles Goad, *Victor Hudon Cotton Mills Fire Insurance Plan* [Montreal, 1882]. Reproduced courtesy of the National Archives of Canada, negative no. NMC 10602.)

ton mill opened in the East End, fueling the area's growth. Hudon, along with several dry goods merchants and bankers, established St. Ann Spinning in order to reduce dependence upon British supplies of basic cotton lines. Although not as large as Hudon, St. Ann, with its 417 looms and 21,000 spindles operated by 475 workers in 1901, proved to be a significant component of the district's industrial base and followed the manufacturing strategy established by the earlier firm. Once established as a high-volume production cotton manufacturer, St. Ann's growth, just like that of Hudon, accelerated through the application of new machinery, the increased exploitation of labor, and the establishment of the National Policy, which imposed a 30 percent tariff on cotton fabric.[11]

This growth, however, came with a price. By the mid-1880s Hudon and St. Ann's capacity outran domestic demand. In addition, the Hochelaga firms' smaller internal economies of scale, compared to American and British producers, and the restricted character of the Canadian market meant that they had difficulty meeting competition in international and, sometimes, domestic markets. Canadian manufacturers could not compete in fancy lines with American and British firms because of the latter's access to cheaper labor and fuel, colossal chunks of capital, control of world markets, and extensive economies of scale. In an attempt to solve the problems of overcapacity and intense competition, St. Ann and Hudon amalgamated in 1885 to form the Hochelaga Manufacturing Company. Directed by an aggressive Montreal management, the new company initiated mill specialization, widened its product lines, sought out new markets, further stretched out work, and reinvested profits. Although this strategy would not have long-term positive impacts, the merging of the two firms provided a decade of stability and, as a consequence, the continued viability of the Hochelaga industrial base.[12]

Other firms drove Hochelaga's industrial growth. Macdonald Tobacco, another large firm, also employed new continuous-processing productive strategies and settled in the East End. The nineteenth-century Montreal tobacco industry had been concentrated in the central core for several reasons. There tobacco manufacturers relied on the wholesaler for imported raw tobacco leaf from the United States and Cuba; found small workshops able to accommodate a small number of workers, some work tables, and a few molds; and had access to the mass of rail terminals and port facilities bringing in the raw tobacco and shipping finished products to markets. William Macdonald's factory was the exception to the tobacco industry's need for a central location.

In 1858 Macdonald established a plant close to Montreal's central business district, and within the space of ten years he had created the largest tobacco firm in Canada. This was possible, as one writer noted, because "he manipulated an enormous corner in tobacco, reaping most substantial profits and thereby securing the financial standing that enabled him to build up an enormous business" after the Civil War. By the late 1860s the five-story building contained machinery and utensils valued at nearly $40,000, while in 1871 Macdonald had capital to the value of $250,000, employed 550 workers, and produced tobacco goods valued at more than a half-million dollars. In order to continue growing and to remain competitive, Macdonald had to install new machinery and labor processes. By the 1870s the original factory on Water Street no longer served as a viable production space, given the changing conditions of the industry. After searching for a new site, in 1874 Macdonald finally chose one that straddled the Sainte-Marie and Hochelaga boundary. One reason for the choice was Macdonald's need to accommodate expanding demand. Built on an eight-acre lot, the new factory's five floors covered five and a half acres. The move facilitated the installation of more than twice the machinery found in the Water Street factory and the doubling of the workforce. Furthermore, an extensive set of adjacent water and railway facilities were powerful attractions for Macdonald, as they had been for Hudon and St. Ann. The company's dependence upon imported American tobacco and the distribution of its tobacco products to the national market fueled its need for nearby transportation facilities with a national and international reach.[13]

Equally important, Macdonald reorganized the factory's internal production space. Although the Water Street plant had been a success when the company was small and operated under more traditional methods, it could not house the new work processes and machinery required by Macdonald to speed up production, cut costs, and reorganize work methods. Accordingly, the new factory erected on the city's built-up fringe was planned, according to an 1877 description in the *Globe and Mail*, "with a view to economy of labour and perfection in manufacture; with system apparent in every department, and such perfect organization throughout." Even if we take this typical booster description with a grain of salt, the attention to the internal layout of the plant was genuine and representative of many large decentralizing firms. The typical downtown plant formerly occupied by the company was unable to provide the appropriate type of space for competitive production. Macdonald gained an edge by hitching this new factory space with the installation of new machinery operated by large numbers of cheap female and child labor. Alongside the concern with the rationalization of space and production, Macdonald also split manufacture from management.

While tobacco manufacture took place at the new factory, the company's administration was carried out at offices in the heart of the old city.[14]

At a time when most tobacco manufacturers sought a central location, one firm could break with this spatial constraint. Despite the large amount of fixed capital in his Water Street factory, Macdonald pulled up and moved to the Sainte-Marie and Hochelaga boundary on the very edge of the city. He was able to do this by taking advantage of the opportunities engineered by local politicians such as the Rolland family and Préfontaine: tax exemptions, cheap land, transportation facilities, and low-wage female and child labor. These locational assets were especially appealing to tobacco firms employing space-consuming, continuous-process machinery and a large workforce, centered around new work processes, and dependent upon distant markets. The fact that Macdonald specialized in Cavendish tobacco products and used agents for purchasing raw materials added to his ability to construct a new set of productive relations on the urban periphery. From an early date he shifted his plant from the congested core area to a greenfield site, where new forms of plant rationalization could be introduced. Macdonald sought the locational assets that drew cotton and rubber manufacturers to the eastern periphery. In Sainte-Marie and Hochelaga, after 1850, firms established new production practices and created a new spatial milieu to accommodate them. Linked to this was the establishment of a range of firms, scales, and types. From the small firms employing traditional methods to large, sophisticated companies such as Canadian Rubber and Macdonald Tobacco, these districts became the home to manufacturing firms. They, in turn, added to the city's specialized economic geography.

"Plan, Map, and Subdivide": Building a Working-Class District

The large-scale industrial development taking place in Sainte-Marie after the late 1840s and Hochelaga after 1870 was made possible by the transformation of the unsettled parts of the city fringe and the surrounding rural areas into working-class residential neighborhoods. Even though the early settlement of these two areas was tied to the residences of Montreal's merchant elite, after 1850 it was related to the growth in the number of factories. The link between factory and district growth was not lost on contemporaries. In 1894 one commentator noted that Canadian Rubber "affords steady employment at good wages to upwards of nine hundred hands. The value of such an establishment to the

Table 4.2 Population, Place of Origin, and Religion in Sainte-Marie and Hochelaga, 1851 to 1901

	Population		Origin				Catholic (%)	
			French (%)		British (%)			
Year	Sainte-Marie	Hochelaga	Sainte-Marie	Hochelaga	Sainte-Marie	Hochelaga	Sainte-Marie	Hochelaga
1861	10,196	—	69.1	—	29.4	—	85.6	—
1871	13,695	1,061	77.8	62.8	21.1	36.2	89.8	84.5
1881	22,733	4,111	81.3	82.4	17.5	16.2	92.4	89.7
1891	34,746	8,540	—	—	—	—	91.0	93.8
1901	40,631	12,914	80.8	84.9	17.7	13.2	90.0	91.2

Source: Various censuses, 1851–1901.
Note: Place of origin not given in 1891.

community can thus readily be recognized." Likewise, another observer stated that "the growth of manufactories in and around our Hochelaga suburb has given some impetus to real estate in that vicinity, and owners have not been slow to plan, map and subdivide in the most attractive manner for purchasers." The effect was that, from little more than ten thousand in 1861, Sainte-Marie's population more than quadrupled over the the next forty years, while Hochelaga's experienced a twelvefold increase between 1871 and the end of the century. By then the eastern portion of the city and adjacent suburban areas had been transformed into a working-class district (table 4.2).[15]

Along with the changing class composition, another distinguishing feature of the districts' growth was an increasing French-Canadian population. Although more than two million French Canadians left for the textile mills of New England and elsewhere as a result of the changes taking place in rural Quebec, a significant number poured into Montreal. Seeking work in the expanding manufacturing, transportation, and service sectors, many rural Quebecers settled in Sainte-Marie and Hochelaga, where jobs and cheap housing could be found. Accordingly, the French-Canadian (and Catholic) share of Sainte-Marie and Hochelaga's population increased from about two-thirds in the 1850s to more than four-fifths by the end of the century. In contrast, the British share declined from more than 25 percent at midcentury to less than 18 percent in 1901.[16]

The shift from a mixed-class, rural-based area to a working-class district distinguished the East End. Sainte-Marie and Hochelaga started as mixed-class areas geared to rural pursuits as much as urban ones.

With industrialization, however, the two districts became almost entirely working class. Throughout the second half of the nineteenth century workers sought jobs in the districts' factories and homes in the cheap housing constructed by small-scale builders. The exodus of wealthy families such as the Molsons to the salubrious, bourgeois neighborhoods circling Mount Royal reinforced the shifting class composition.

The transition to a distinctly working-class district occurred earlier in Sainte-Marie. By 1861 more than 95 percent of its household heads found work in blue-collar occupations (table 4.3). Rounding out the numbers was a small group consisting of farmers, businesspeople (owners of groceries, saloons, and bakeries), local professionals (doctors), and white-collar workers. The shift was slower in Hochelaga, mainly because extensive industrial growth began twenty years later there and because of its lingering heritage as a refuge for Montreal's elite. As late as 1880, a small number of Montreal's bourgeois still lived in Hochelaga: Henry Morgan, the department store magnate, resided at Milton Lodge; the large-scale builder William Kennedy kept his summer residence on Marlborough Street; and Jean Damien Rolland, a justice of the peace and large landowner, lived at Villa Rolland. The suburb also had a small number of commercial workers, most of whom resided above or close to their shops. The elite and local business elements, however, became insignificant in comparison to the numbers of blue-collar workers. By

Table 4.3   Occupational Structure of Sainte-Marie and Hochelaga, 1861 to 1901

	Households (%)					
	Sainte-Marie	Hochelaga	Sainte-Marie	Hochelaga	City Total	
Occupation Class	1861	1880	1901	1901	1861	1901
Business owner	1.8	13.4	0.6	2.0	10.1	5.4
Professional	1.1	10.6	1.0	1.1	4.4	3.3
Clerical	2.4	6.0	8.4	5.7	11.7	22.4
Blue-collar	39.3	40.0	22.9	21.4	36.2	27.9
Unskilled	55.4	30.0	67.0	69.8	37.6	40.9
Number of households	722	447	4,209	1,133	5,338	22,530

Source: The 1861 and 1901 data are compiled from an approximately 50 percent sample of the City of Montreal, Water Tax Rolls. The 1880 data for Hochelaga is taken from a listing of all households published in Atelier d'Histoire Hochelaga-Maisonneuve, *De fil en Aiguille. Chronique ouvrière d'une filature de coton à Hochelaga en 1880* (Montréal: Atelier d'Histoire Hochelaga-Maisonneuve, 1985), 77–83.

1901 the population composition of both areas was heavily weighed to the lower end of the class spectrum, with more than nine out of every ten households working in blue-collar occupations, the single largest group being unskilled laborers.[17]

The large, mainly French-Canadian, working-class population employed in the eastern manufacturing district sought out nearby housing. A short journey to work was an imperative of working-class life as low wages, long work hours, and expensive local transportation forced workers to live close to their place of employment. As long as jobs were concentrated in the city core, workers sought out adjacent housing. Yet once manufacturing began to settle on the urban periphery, after the 1840s, demand for housing close to the mills and factories grew. In the process the areas once dotted by rural hamlets, the summer resorts of Montreal's bourgeoisie, and a small multiclass population were converted into districts accommodating a large urban working-class population. Land developers in Sainte-Marie and Hochelaga planned, mapped out, and subdivided working-class lots. Following on their heels, small-scale builders constructed row after row of densely packed housing, ranging in quality from poor to reasonable.

The housing conditions of the East End reflected the nature of the population's working opportunities. The combination of low wages, seasonal unemployment, frequent work injuries, and high rents severely constrained the ability of most households to make ends meet. Many firms locating in this district, such as the rubber factories and textile mills, paid extremely low wages. As the evidence given to the Royal Commission on Relations of Labour and Capital makes clear, while adult males' wages barely covered day-to-day subsistence, female and child wages were lucky to reach half that of men. In addition, the seasonal and cyclical nature of work ensured that few families at any one time would be above the poverty line. As William Macdonald told the Royal Commission, even those fortunate enough to have work during the winter had to absorb wage reductions every fall. For other tobacco workers a two- and as much as a six-month layoff was to be expected. The family economy of these fringe working-class areas functioned in the same way as it did in the central districts. To maintain any semblance of an income sufficient to cover the necessities of life, women and children had to join men in the paid workforce outside the home or undertake home-based work outside the formal economy. For most families the normal exigencies of the informal family economy—taking in boarders, washing, and sewing—could not bring them out of poverty. High rents added to the low and irregular wages. Even though they were lower in Sainte-

Marie and Hochelaga than in the inner city, rents for decent accommodation were too high for any but the most skilled worker. The combination of the housing market and precarious work conditions produced high rent, poor-quality housing for most of the local population.[18]

Employment and household income problems were compounded by the development and housing construction policies of the major landowners. The greater part of Sainte-Marie and Hochelaga were owned by a few large proprietors, some of whom, such as Molson, Macdonald, and Hudon, were also the areas' major employers. Over the course of the century land developers subdivided a significant area of these districts, and builders, ranging from small contractors to large manufacturers, constructed housing, most lacking basic sanitary facilities. During the 1867–80 construction cycle, for example, Sainte-Marie, as far east as Papineau Street and north of St. Catherine Street, was built up with duplexes. Builders following the trail of industrial growth on the eastern periphery packed relatively poor quality housing around the brickyards, factories, and mills that were sprouting up in the district by the 1860s. Not only was the housing of poor quality, but rents consumed a large share of the working family's income. Adding to these problems were the seemingly incessant increases in rents; by the 1880s, as one journalist noted, the East End's "rents are going up by leaps and bounds." Local land promoters, industrialists, and speculative home owners profited handsomely from their control over property development, housing construction, and home ownership.[19]

Even though Sainte-Marie and Hochelaga were not company towns in the same sense as Chicago's Pullman or Philadelphia's Tacony, local firms did engage in the construction of worker housing. In Pullman and Tacony two large firms, Pullman Palace Car and Disston Saw, built workers' housing and directly controlled many aspects of local political and social life. While East End manufacturers were not involved in such practices, they provided some workers' housing, especially in the early stages of the area's industrial development. Even though a large number of houses were constructed in the East End in the second half of the nineteenth century, demand outran supply, especially before the 1880s. To compensate for the housing shortage and to serve workers wishing to live close to their workplace, some firms built worker housing close to the factories. In 1850 Molson owned fifty-two company houses in close proximity to the brewery. Michel Lefebvre built homes facing his sprawling manufacturing complex for his superintendent and watchmen. The tobacco pipe manufacturer James Henderson owned eight homes on Colborne Street opposite his factory. A similar building history occurred in Hochelaga. During the rapid industrialization of the 1870s not enough housing was built to match the demand arising from the growing manufacturing workforce. Accordingly, Hudon constructed forty-six homes on St. Germain Street by 1881 which were occupied by newly arrived workers employed as laborers.[20]

Hitched to the growth of a large, working-class population in the East End was the creation of local physical and community infrastructures. Sainte-Marie and Hochelaga's growth as manufacturing districts involved the creation of a selectively located range of infrastructures. In Sainte-Marie the first sewer pipes were laid in 1847 to connect the Molson Brewery with the city system, and by midcentury a few streets had been laid down. Over the next forty years new streets, sanitary facilities, schools, and parishes were opened, but even by the end of the century few houses contained sewerage facilities, and few streets had permanent paving. It was, for the most part, along the retail streets and in middle-class homes and industrial sites that sewer pipes were to be found.[21]

Promoters of Hochelaga, just like their counterparts in Sainte-Marie, realized that their success depended on the ability to link local infrastructures with wider national and international opportunities. Some observers felt that Montreal had the opportunity to become the "Chicago of Canada" and "one of the greatest cattle centres of the world." This would involve assembling transportation infrastructure, abattoirs, stockyards, and processing facilities in Hochelaga. Although the dream of replicating Chicago's stockyard in Hochelaga did not materialize, local developers and politicians, allied with those having a more national reach, constructed infrastructures on which a manufacturing base developed. The beginnings of a modern infrastructure system in Hochelaga dates from the 1870s, when the Canadian Northern Railway placed a track through the suburb. Two years after the construction of his cotton mill, Hudon gave land for the building of the Nativité d'Hochelaga Church. After the first decade of industrial growth and incorporation, Hochelaga possessed a city hall, public market, post office, weighing office, a host of religious buildings (Catholic and Protestant churches, a convent, and a nunnery), and a military barracks.[22]

Suburban authorities, however, were slow to install adequate physical infrastructures for the growing working-class population. As in Sainte-Marie, the villas, mansions, and summer residences of the Hochelaga bourgeoisie received connections to sewerage and water lines, while the vast working-class districts did not. Even those facilities, such as sewers, which were built were frequently defective, often causing as many problems as they purportedly were meant to solve. Behind the lack of proper facilities was the propensity of the city council to siphon local funds to manufacturers and railroad companies in the form of bonuses and tax exemptions. Working-class access to sanitation, sewerage, and water facilities was not a priority for the suburban municipal officers; the interests of business came first. As a result, as more firms moved to Hochelaga and their demands for subsidies and infrastructure escalated, the town was increasingly unable to find enough money to satisfy them. By 1883, because of the disjuncture between the demands on the town coffers and the town's ability to spend, Hochelaga was annexed by Montreal. Under city control expenditures on infrastructures increased, but most monies continued to be pumped into facilities directly benefitting the firm, most notably projects such as the construction of the Sutherland Pier (1891) and the Laurier Pier (1896).[23]

The formation of East End manufacturing and working-class districts outlined here has important implications for the understanding of nineteenth-century suburbanization. According to Robert Fishman, Kenneth Jackson, and Sam Bass Warner, nineteenth-century residential suburbanization was a middle-class phenomenon. In recent years, however, several writers have pointed to another significant element of the nineteenth-century suburban process. The formation of mixed-class suburbs in West Philadelphia, Jamaica Plain, and Cambridge was as much a part of urban fringe development as middle-class settlement. A third type of nineteenth-century suburban development was the working-class peripheral settlement. Generally neglected by historians and geographers, this type, when it is mentioned, is usually associated with the creation of company towns such as Chicago's Pullman or Pittsburgh's mill towns. In nearly all accounts it is considered to be of minor importance. Similarly, industrial suburbanization is commonly held to be a twentieth-century phenomenon, and most manufacturing, regardless of type and scale, is believed to have clustered in the central core of North American cities until after World War I.[24]

Sainte-Marie and Hochelaga provide evidence of working-class and industrial suburbanization from the mid-nineteenth century. In the years following the late 1840s the eastern portion of the city experienced several industrial surges. The large-scale firms established there, first in Sainte-Marie and a generation later in Hochelaga, acted as magnets to a host of smaller firms. These East End districts formed around constellations of industry which had been slowly developing from the late eighteenth century. By midcentury the eastern part of the city at Sainte-Marie had a well-formed industrial base, which became more firmly fixed with the influx of firms in the 1870s. Extensive growth in Hochelaga after 1870 added to the ferment of industrial development, which was initially controlled by a few families. In the second period of industrial growth firms from a variety of industries, attracted by the conditions already created in the early stage of growth, were established and led to the emergence of a diverse industrial district.

Some of the firms making up the new, diverse, yet specialized industrial structure of Sainte-Marie and Hochelaga (rubber, textiles, tobacco, and soap) had several common characteristics: mechanization, a segmented and large labor force, a significant number of female and child workers, a sophisticated marketing system, national and international markets, and large premises. These large firms found in the sparsely settled rural land of Sainte-Marie and Hochelaga a suitable environment for manufacturing. Freed from the confines of the core's physical structures and production relations, they established new manufacturing practices in peripheral greenfield sites. But they were not alone. A significant number of East End firms were smaller in scale, were less innovative in terms of technology and work methods, and catered to more local markets. In other words, the East End was an industrial zone in which a range of modern, high-volume to small-scale, artisanal manufacturing strategies coexisted.

The East End's industrial suburbanization was hitched to working-class suburbanization. Large-scale land developers and small-scale builders built most of the housing, although some company housing appeared before the 1880s. Despite the thousands of dwelling units built, the provision of cheap and adequate housing was not forthcoming for a large number of the districts' families. The laying down of basic infrastructures for industry went full steam ahead, while adequate sanitary facilities for the working-class population were kept to a minimum. Furthermore, employers controlled land development. The Molsons, William Macdonald, and others owned large tracts of land and directed the timing, type, and conditions of housing construction.

CHAPTER FIVE

"The Whirr of Machinery and the Booming Noise of a Thousand Hammers" in the Western Manufacturing Districts

> *The whirr of machinery and the booming noise of a thousand hammers echo on every side, we are within a very hive of human industry. The silent swift flowing waters whose banks we stand have the motive power of eight thousand horses. The Lachine canal, St. Gabriel and Cote St-Paul Locks are golden streams of wealth to Montreal. The factories they keep in operation giving constant employment to 10,000 persons.*
> —Commercial Sketch of Montreal and Its Superiority as a Wholesale Market, 1868

While Sainte-Marie and Hochelaga experienced powerful changes to their economic and social geography after midcentury, a similar transformation took place in the city's western fringe and adjacent suburbs. Building on a history of petty commodity production in the older district of Griffintown, industrialization took place in an area stretching for five miles along the Lachine Canal, where it entered the St. Lawrence River, to the industrial suburb of Saint-Henri. The first full thrust of the new economy was felt on the city's fringe in Saint-Ann ward in the late 1840s and the 1850s. Composed of the Canal and Griffintown districts, this area accounted for 15 percent of the city's manufacturing firms in 1861 and more than a third of its rent. Together these districts formed Montreal's premier example of a mid-nineteenth-century fringe manufacturing landscape. A generation later the canal suburbs of Saint-Henri, Sainte-Cunégonde, and Côte Saint-Paul developed as another hive of industrial activity. Building on Saint-Ann's early concentration of propulsive firms and despite growth in other parts of Montreal, the West End made up 30 percent of the city's rent in 1890. Its large metal, flour, sugar, and locomotive factories differed from the smaller-scale consumer firms clustered in the central manufacturing districts. Although the western complex replicated the East End's structure in some ways, it was distinguished by the scale and technological sophistication of its firms. Tied to Montreal by ever greater bouts of investment in factories, warehouses, utilities, and land, the emerging western industrial districts after 1850 were "golden streams of wealth" for Montreal's manufacturers, merchants, and financiers.

For the area's working class, however, the benefits of the "whirr of machinery and the booming noise of a thousand hammers" were difficult to discern. Low wages, seasonal unemployment, and other travails of nineteenth-century working-class life ensured that the post-1850 industrial landscape was marked by the typical inequalities of industrial capitalism: poor quality housing, high rents, overcrowding, unsanitary conditions, and poor infrastructures. The British, Irish, and French-Canadian workers living in the industrial-residential wedge extending along the canal were geographically segregated from the rest of Montreal society. The residential neighborhoods clustering around the factories of the western industrial belt were built over the orchards and pastures of the Sulpician seminary and the small preindustrial villages dotting the canal edge. With its factories, poor housing, golden streams of wealth, and whirr of machinery, the West End's geography mirrored the unregulated, rapacious industrial capitalism developing in Montreal after 1850.

"A Forcible Impression": The Canal and Griffintown Manufacturing Districts

The West End hive of industry forming after 1850 did not suddenly appear on the urban landscape but emerged from an earlier preindustrial form of manufacture. Before the late 1840s West End manufacturing was characterized by a craft system of production, in which a master artisan employed a journeyman or two and operated under traditional hand methods. At the same time, however, an embryonic industrial base centered on machinery and steam power developed and coexisted with artisanal production. According to an 1831 newspaper account, Griffintown had "more machinery in operation within its limits than any other portion of Montreal." Characteristic of these firms was the Eagle Foundry, where an eight-horsepower engine powered an assortment of turning lathes, grind stones, and trip hammers. Eleven years later the 1842 census cited sixteen manufacturing establishments in Saint-Ann, one-third of the city's total. Small workshops operated alongside these machine-based establishments. By 1848, according to the water tax rolls, the area had thirty-one firms, most of them small and operating under handicraft conditions.[1]

By the mid-1850s a shift had taken place in the character of the area's manufacturing. Factory production, superimposed upon small-scale artisanal production, had become the dominant form of manufacture. The technologically advanced, hydraulically based, energy-intensive system emerging by midcentury was not lost on contemporary observers. Several booster pamphlets and descriptions called attention to the major changes taking place in this fringe section of Montreal. Large capital investments, growing market opportunities, and a multiskilled workforce provided the impetus for industrial development. The burst of industrialization in Saint-Ann between 1847 and 1854 created more than two thousand jobs and rested on the investment of more than two million dollars in the ward's factories. Over the following decades investments made by merchants, artisans, and governments and loans extended by Montreal banks fueled the West End's industrial expansion. State aid, in the form of the reconstruction of the Lachine Canal, supported this growth. By 1901, thanks to aggressive investment by industrialists and state encouragement, capital valued at more than twenty-eight million dollars was invested, and twenty thousand employees worked in Saint-Ann's factories. Expanding markets were important. Entrepreneurs invested in West End factories to capture the new market opportunities of burgeoning industrial capitalism. Growth of regional, national, and, to a lesser extent, international markets provided enterprising merchants and artisans willing to invest in manufacturing with the opportunity to reap profits and growth from their factory interests. The chairs made at William Allen's factory, for example, pushed American imports out of the regional market, while a threshing machine maker's products found export as well as national markets. The influx of an assortment of workers from Britain, the United States, Ireland, and rural Quebec also stimulated factory growth. In Saint-Ann's foundries, sawmills, and rolling mills, work was almost entirely performed by men. Manufacturers in other industries, however, demanded a quite different labor force. In the manufacture of bags and denims at his cotton cloth factory at St. Gabriel locks, Fred Harris employed seventy workers, nearly all women and children. Attracted by the early factories, mills, and workshops, workers, in turn, generated greater manufacturing investment and growth.[2]

Steam- and water-powered factories containing a range of machinery quickly outnumbered Saint-Ann's artisanal workshops. In the wood-processing industry firms such as James Shearer used various machines to produce doors, sashes, blinds, moldings, and architraves for the construction, railroad car, and steamship industries. Employing 790 workers in 1871, the city's largest workplace, the Grand Trunk Railway, constantly updated and experimented with new machines and new forms of locomotive manufacture. The chemical factory of Lyman, Clare, manufactured putty and paints with machinery and equipment of all kinds: rollers, chasers, power presses, power pumps, drug chasers dye woodcutter, circular saw, paint mixer, and a putty mill. According to a mid-1850s report by the Public Works commissioner, firms along the Lachine Canal employed a variety of modern machinery.[3]

These growth pulses laid the basis for the appearance of an extensive range of factories after midcentury. As one observer noted in 1868, "A walk along the banks of the Lachine Canal and the St. Gabriel Locks, will convey to the observer a forcible impression of the extent and importance of the factory interests of the City." One of the most imposing plants of the period, Redpath Sugar Refinery's seven-story stone and brick edifice, produced six thousand barrels of refined sugar monthly. One report noted, "the buildings form a magnificent pile of stone and brick. The main erection measuring one hundred and sixty feet in length, by forty-four, of the same height, exclusive of the boiler house, and charcoal kilns. The whole being by far the largest manufacturing erection in the city." A year after its opening in 1855, Redpath's 100 employees worked with a variety of machines. By 1871 the firm employed 220 workers in the manufacture of sugar worth more than $2.5 million in a plant valued at more than a half-million dollars. So great was the company's growth that ongoing investment in plant and machinery was necessary if the firm were to remain competitive. Its president, George Drummond, acknowledged that, because Redpath is "changing the whole time, it is impossible to conduct a business, such as ours, without an enormous expenditure constantly in changes" to machinery and the refinery process. The company also needed capital for plant expansion (map 5.1). Between 1855 and 1880 the plant grew from a single cluster of buildings bordering the Lachine Canal to a multiblock plant containing three sets of buildings: the refinery and associated buildings on the original site, the raw sugar and bone storage area between Montmorency and Richmond Streets, and the cooperage cluster next to the Inner Basin.[4]

Equally impressive were the sprawling grounds of Augustin Cantin's vertically integrated Montreal Marine Works. Building on his experience in several Montreal and New York shipyards, Cantin started his own yard near the Canal Basin in 1841. Despite failing two years later, he opened a new shipyard further along the canal at St. Gabriel Locks in 1846. Four years later he expanded the yards and, with an investment of $10,000, built a new dry dock. Even though most of Cantin's products went to the domestic market, he did export some steamships. By 1855 his new shipyards covered more than fourteen acres and consisted of two

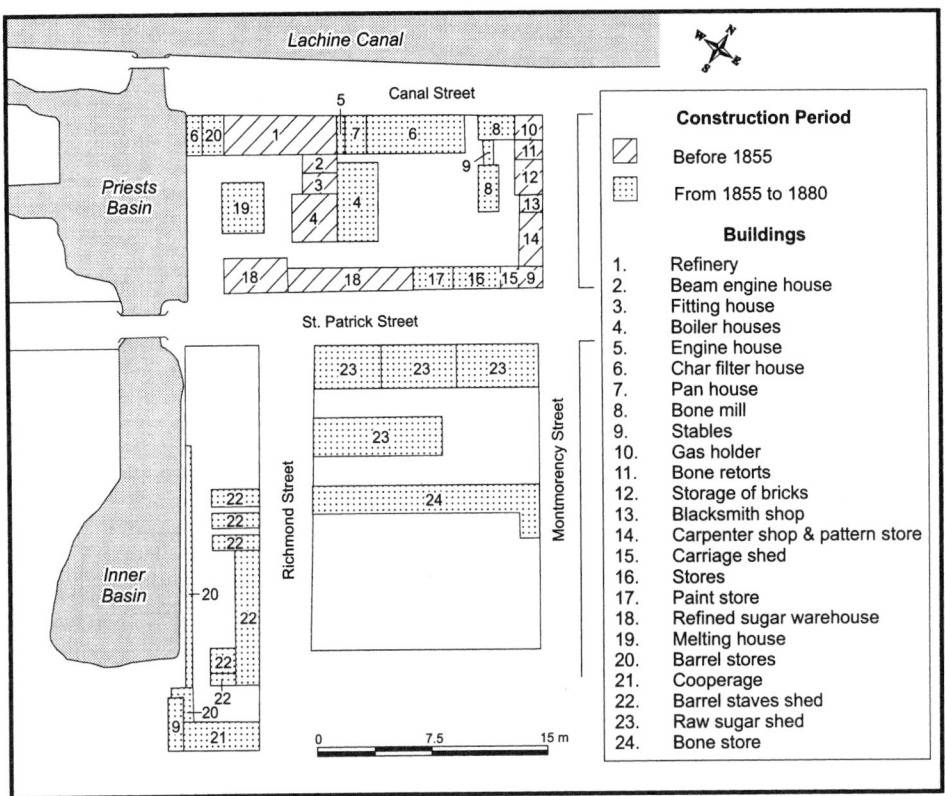

Map 5.1 Expansion of the Redpath Sugar Refinery, 1856 to 1881. From a small set of buildings in 1855, Redpath expanded greatly over the following twenty-five years. Cycles of building construction paralleled Redpath's constant expenditures on more and new machinery. (Redrawn from Richard Feltoe, *Redpath: The History of a Sugar House* [Toronto: Natural Heritage-Natural History, 1992], app. A.)

basins, a sawmill, and an engine foundry in which 250 hands made seven steamships and several smaller vessels. Cantin's shipyard was vertically integrated, capable of performing many of the operations required in ship manufacture, from ship design to the manufacture of individual components. Many of the company's lumber needs were obtained from its own sawmill, with its water-powered upright and circular saws. Workers in other departments operated lathes, planes, and a variety of other machines with the aid of a twenty-horsepower condensing beam engine. Cantin also had drafting and modeling rooms where plans for the ships were drawn up.[5]

Despite the increasing size of firms, the district featured a diversity of scale, industry, and technological organization. The emergence of large firms in leading sectors did not discourage the growth of smaller firms. Capital investment in the West End took a variety of forms. In 1871 almost 40 percent of the ward's firms were small (with rents less than $100), and 13 percent were large (with rents over $800). The interweaving of firms with different degrees of investment in the western landscape was paralleled by the diversity of industries. Fifty-three of the ward's firms (or 84 percent) belonged to the food, metal, chemical, and wood sectors in 1851, and these sectors' share of the area's firms and rent declined very little over the next forty years. But many firms in other industries sprang up. Textile, boot and shoe, broom, clothing, and tobacco factories became more common over the period, but they were generally smaller in size than those in the dominant sectors. Even the four major sectors were characterized by great variety; in the food sector, for example, the flour mills, brewery, and sugar refinery had a mean rent of more than $4,000 in 1871, while the thirteen bakeries averaged only $107.

By 1890 these developments had created a complex of diverse and horizontally disintegrated firms tied together by an elaborate set of linkages in the two Saint-Ann districts, Griffintown and Canal. John McDougall's foundry manufactured railway car wheels for the Grand Trunk Railway shops. Coopers provided barrels and kegs for many firms, and blacksmiths applied their skills to carriages and other trades. Shearer's sash and door factory built wood products for the vicinity's steamships, and the Eagle Foundry manufactured nearly all of the steam engines. In 1876 a government committee found that Redpath Sugar indirectly employed at least seventy-five workers in cooperage and carting besides the three hundred employees in the refinery proper. In constructing his refinery, Redpath ordered cisterns, settling molds, and boilers from local foundries and engine shops. Along with the extensive network in the West End, the districts' firms also had close ties with firms elsewhere in the city. When the Old Montreal company Burland Lithographic was liquidated in 1892, for example, some of its principal creditors included the West End firms P. Dods (inks), G. & J. Esplin (wood products), and Alex McArthur (paper).[6]

The differential growth of Saint-Ann's industrialization was mirrored in the area's geography. In this period two distinct districts are evident. From the early nineteenth-century merchants and small manufacturers in Griffintown, the traditional core of the district, initiated the shift from handicraft production to machine-led, specialized manufacturing. With fifty-three firms and 7 percent of the city's rent in 1861, it had achieved an important position within Montreal's manufacturing division of labor. Although its relative share declined over the century, it remained an important manufacturing node (table 5.1). In 1848 twenty-nine of the thirty-one Saint-Ann's firms were in Griffintown, and as late as the early 1850s most of the ward's firms continued to be concentrated there. For the most part they were small in scale, although some larger establishments were to be found.

From the late 1840s a new manufacturing district, Canal, developed to the west and south of Griffintown. Located along the Lachine Canal and on the rural fringe of both Saint-Ann ward and the city, Canal quickly became the home of the largest and most capital-intensive firms in the city, extending the West End's break with the preindustrial past, which centered on Griffintown. At the time of its redevelopment in the mid-1840s, the Sulpician's land bordering the Lachine Canal was still being used for "the leasing of pasture, the sale of farm produce, and the use of farm workers and horses to haul firewood." According to the Royal Commission of 1887 looking into the leasing of water power at the Lachine Canal, "at the date [1851] of the lease of the power at . . . [Saint-Gabriel] lock, it was at the outskirts of the city." But during the 1850s this agricultural land on the urban periphery was transformed into the single most important manufacturing area of the city. Being an agricultural area at midcentury, in 1861 Canal's firms accounted for more than a quarter of the city's rent (table 5.1).[7]

The establishment of mills and factories in Canal, first on Mill Street and later at St. Gabriel Locks, created a modern manufacturing district featuring some of the city's largest and most sophisticated firms. This territorial expansion occurred very rapidly. Between 1851 and 1856 Canal underwent an infusion of capital investment and jobs: from five firms with an aggregate rent of $3,000 to twenty-nine firms with an aggregate rent of $21,960. Its share of the ward's total rent rose from 28 to 64 percent, and its share of the city rent rose to 26 percent by 1861. Firms such as Montreal Rubber, Ostell's sawmill, Redmond's foundry, and Redpath Sugar appeared, while those established in the late 1840s, such as Gilbert and Bartley's foundry, Gould Flour, and Peck Nail, expanded. As a result, the median rent for Canal's establishments rose to five hundred dollars in 1856, more than four times the city median. Canal, however, was not homogeneous: large flour mills, foundries, and nail factories dominated Mill Street, while firms at St. Gabriel Locks were smaller and from a more varied set of industries. The scale of this burst of industrial growth after

Table 5.1 Manufacturing Structure of Griffintown and Canal, 1861 and 1890

	Griffintown				Canal				Griffintown and Canal	
	1861		1890		1861		1890		1861	1890
Industry	No. of Firms	Rent Share (%)	No. of Firms	Rent Share (%)	No. of Firms	Rent Share (%)	No. of Firms	Rent Share (%)	City Rent Share (%)	
Metalworking	11	41.6	20	40.1	15	21.2	15	20.8	67.4	29.0
Wood	11	22.3	11	9.5	4	5.9	6	7.2	80.9	34.6
Leather	5	11.0	1	2.6	1	0.7	4	0.4	8.4	2.5
Food	5	5.6	7	16.0	7	34.5	9	24.1	69.2	32.4
Clothing	1	3.6	0	0	0	0	5	0.3	1.4	0.3
Transport	4	3.3	5	1.0	2	25.1	4	26.6	76.1	41.7
Chemical	4	1.7	3	4.3	4	1.8	6	5.9	20.9	31.5
Tobacco	1	0.9	4	9.0	0	0	0	0	10.5	13.6
Furniture	0	0	3	3.9	1	1.0	1	0.1	6.4	44.4
Textile	0	0	3	1.4	3	2.2	4	9.3	46.7	14.2
Rubber	0	0	0	0	1	1.9	0	0	40.0	0
Other	11	10.0	10	12.2	2	5.7	14	5.3	2.3	8.5
Total	53	100	67	100	40	100	68	100	33.5	17.3

Sources: Water Tax Rolls for Montreal and surrounding municipalities, 1861 and 1890.

midcentury was not to be repeated. Despite a steady increase in firm numbers and volume of rent over the following forty years, the district never equaled its earlier growth. Yet Canal remained the nineteenth-century locational core of the city's large and technologically advanced firms, whose scale of operations remained large relative to the rest of the city (map 5.2).[8]

The rolling mill of Pillow and Hersey illustrates some aspects of West End manufacturing district development and the logic of firms' spatial strategy before 1900. The company's origins go back to a small nail workshop opened by John Bigelow in St. Lawrence ward in the 1790s. Using primitive American nail machinery, Bigelow manufactured cut shingle nails by combining hand and machine labor and was one of very few producers in Canada, most nails being imported from Britain (wrought nails) and the United States (large cut nails). In the early 1850s he moved his operations to the recently constructed Canal Basin on the Lachine Canal. He found several attractions there, including hydraulic power and the advantages of being part of an emerging industrial nucleus. By 1856 the new Canal basin factory had thirty-two water-powered machines, employed sixty men, and produced twenty-five thousand tons of nails and spikes annually.[9]

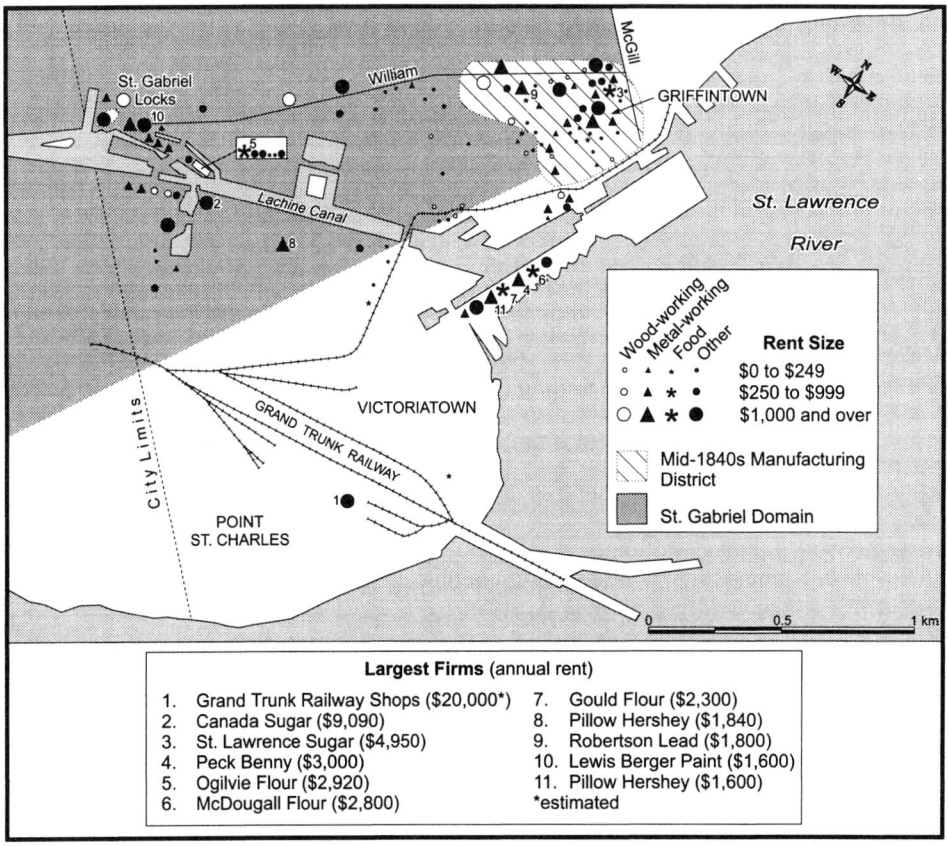

MAP 5.2 GRIFFINTOWN AND CANAL MANUFACTURING DISTRICTS, 1881. Large-scale industrial growth along the Lachine Canal led to the development of Montreal's largest manufacturing complex in Griffintown and Canal. (Compiled from City of Montreal, Water Tax Rolls, 1861 and 1881.)

Bigelow's move to the Canal district coincided with the construction of rolling mills by several neighboring nail makers. The introduction of rolling mill technology, a critical step in the substitution of machinery for labor power, accelerated Canal's industrial power. The new system of supplying puddled iron lowered shipping costs, cut waiting time for shipments, and gave better control over the quality of iron bars. The greater range of iron also allowed for diversification among the finishing trades. While Bigelow did not build his own rolling mill, he acquired the Mansfield Holland mill in 1863 and took on John Pillow and

Randolph Hersey as partners. At his death in 1868 the partners took over the running of the firm and renamed it Pillow and Hersey. By 1871 the company employed 157 workers and had capital of $150,000 to produce goods valued at $225,000. An expanding market and increased competition forced the firm, a year later, to build a second rolling mill on St. Patrick Street, a few blocks along the canal from its first plant. Its specialization in several lines of hardware was rationalized by plant: the Mill Street premises concentrated on the manufacture of nuts, bolts, and tacks, while St. Patrick Street specialized in horseshoes, nails, and spikes. Further specialization and a sign of Canal's integration into citywide networks took place when the firm opened an office and warehouse on St. Peter Street in the heart of Old Montreal's financial district.[10]

By taking advantage of the locational assets of the districts, some of Canal and Griffintown's other metal manufacturers who had started out small in time built themselves a substantial fortune. One was James Robertson, who came to Montreal to establish a warehouse and retail branch of a Scottish lead pipe firm. Quickly striking out on his own, he went into partnership with William Brown in 1858. Success came quickly. After five years he bought Brown out with his savings of four thousand pounds. At the back of his Queen Street warehouse he built a small foundry for the manufacture of lead pipe and later extended his product line by adding the manufacture of saws. In 1871 fifty workers produced lead piping and saws valued at $156,000. By the 1870s competition from local and foreign firms forced Robertson to expand his Montreal factory, add new lines such as lead traps, and open other factories and warehouses in Toronto and Baltimore (fig. 5.1). Over time he brought in his four sons, who retained control after his death and turned the firm into a joint-stock company. Robertson's success, like that of Pillow and Hersey, resulted not only from business acumen and the ability to capture expanding markets by extending control of a specialized product line but also by embracing the benefits of the western districts.[11]

An assortment of other metalworking firms found the locational assets of Canal and Griffintown extremely beneficial. The establishment of the Grand Trunk Railway shops in Point Saint-Charles in 1857 provided a strong impetus to the development of the districts' metal industry, while the growth of the railroads gave entrepreneurs an opportunity to build a reasonable competency centered on railroad trade specialization. Vulcan Iron, for example, manufactured frogs, switches, and diamond crossings for the Canadian railroads. Dominion Brass's speciality was steamship work, but, to reduce its dependency on one line, it also manufactured brass products, copper work, and fittings for the railroads

FIGURE 5.1 THOMAS ROBERTSON'S LEAD PIPE WORKS, 1890. The Robertson Lead Pipe Works illustrates the importance of the growth of the metalworking sector for Montreal's industrial expansion and the type of factory common in the western part of the city after midcentury. (*Special Number of the Dominion Illustrated Devoted to Montreal the Commercial Metropolis of Canada* [Montreal, 1891], 149. Reproduced courtesy of Toronto Reference Library.)

and plumbers. Similarly, other metalworking firms took advantage of the agglomeration economies that were being created in Saint-Ann.[12]

These economies were especially important for firms seeking to base their manufacturing upon an array of products. Diversity was a sound strategy for the districts' firms, given the small size of the market, and, as was the case for the founders of Ives and Company, it "obliges them to keep a large variety of manufactures ... instead of confining themselves to a few articles." Ives started off in 1859 producing small hardware but ran into difficulties in the 1860s. In response, the company undertook product diversification: to provide greater balance to the company's output they added large-scale production of stoves to their traditional lines. In response to the growing opportunities of western Canadian expansion, the firm added barbed wire manufacture in 1880.[13]

Ives's manufacturing strategy consisted of maintaining an assortment of products to meet the fluctuating demands of numerous small markets. This did not mean that there was no rationalization of production: as one of the largest metal firms in the city, Ives was able to split production into two workshops. At Griffintown the firm located a general foundry, along with pattern, machine, and finishing shops to make bedsteads, railings, and crestings (fig. 5.2). In the foundry three heavy cranes and an elevator facilitated the work of the two hundred workers. Ives also opened a hardware and stove plant on Montreal's South Shore at Longueuil in 1881. Large orders, such as the thousand tons of heavy castings made by Ives for the new St. Lawrence Sugar refinery in Montreal's East End, helped the city's foundries introduce some rationalization into their plants. Despite this improvement, metalworking firms of all scales and markets remained tied to fluctuating and uncertain demand, yet the locational assets of Saint-Ann minimized the difficulties associated with the problems of nineteenth-century manufacture.

Building a Working-Class District below the Hill

The large-scale industrial development taking place in the West End between the 1840s and 1890 was hitched to the formation of a working-class area running in a wedge extending west from the city's built-up limits along the canal. Between 1851 and 1891 Saint-Ann's population more than tripled, from 7,455 people to more than 23,000, while that of Saint-Gabriel, a working-class residential suburb to the west, grew rapidly from the 1870s (table 5.2). More than 90 percent of all household heads in both wards were manual workers, half of them unskilled (table 5.3). The reformer and shoe manufacturer Herbert Ames, in his survey of the

FIGURE 5.2 FOUNDRIES AND WAREROOMS OF IVES AND COMPANY, 1890. Built gradually over thirty years, the Griffintown factory of Ives and Company consisted of a sprawling complex of foundries, workshops, and showrooms. Using diversification as a manufacturing strategy, Ives produced a large variety of metal products for the national market. (*Special Number of the Dominion Illustrated Devoted to Montreal the Commercial Metropolis of Canada* [Montreal, 1891], 195. Reproduced courtesy of Toronto Reference Library.)

"city below the hill," described it as "the home of the craftsman, of the manual wage-earner, of the mechanic and the clerk, and three-quarters of its population belong to this, the real industrial class. This area is not without its poor, and, as in other cities, a submerged tenth is present with its claims upon neighborly sympathy." Saint-Ann and Saint-Gabriel had concentrations of workers in skilled occupations such as machinists, molders, and coopers, most of whom were from Britain, Ireland, and the United States. With homes clustered around Canal and Griffintown's factories, a close journey to work distance was a necessity to those facing long hours, expensive transit, and insecure employment. Although the majority of the population was of British origins, close to half in Saint-Gabriel were French Canadians. Concentrated in sectors such as textiles and woodworking, French Canadians tended to be concentrated in the lower end of the occupational scale. Together, this ethnically diverse, working-class population provided a multiskilled labor force for the area's factories. Along with the extensive residential areas developing in the western areas, manufacturers also drew upon workers from the working-class areas of Saint-Antoine ward just to the north of Saint-Ann.[14]

Land subdivision and housing construction paved the way for the West End's population growth. The earliest residential subdivisions can be traced back to the beginning of the nineteenth century, when John McCord obtained the Nazareth fief for ninety-nine years from the Sulpician Seminary. The area of the fief directly west of Old Montreal and north of the canal in Griffintown was the earliest section to be built up,

Table 5.2 Population, Place of Origin, and Religion in Saint-Ann and Saint-Gabriel, 1851 to 1901

	Population		Origin				Catholic (%)	
			French (%)		British (%)			
Year	Saint-Ann	Saint-Gabriel	Saint-Ann	Saint-Gabriel	Saint-Ann	Saint-Gabriel	Saint-Ann	Saint-Gabriel
1851	7,455	—	23.0	—	66.0	—	—	—
1861	16,200	—	20.5	—	76.5	—	70.2	—
1871	18,639	—	26.5	—	72.2	—	70.1	—
1881	20,443	4,506	28.6	40.2	69.7	58.2	71.5	73.1
1891	23,003	9,986	—	—	—	—	71.4	63.2
1901	21,835	15,959	29.7	49.1	66.4	48.8	71.5	65.7

Sources: Various censuses, 1851–1901; and Alan Conter, "The Origins of a Working-Class District: A Portrait of Saint-Ann's Ward in the 1850s" (Undergraduate paper, McGill University, 1976), 6.
Note: Place of origin is not given in 1891.

Table 5.3 Occupational Structure of Saint-Ann and Saint-Gabriel, 1861 and 1901

	Households (%)				
	Saint-Ann	Saint-Ann	Saint-Gabriel	City	
Occupational Class	1861	1901	1901	1861	1901
Business owners	1.9	1.7	1.3	10.1	5.4
Professionals	0.9	0.9	1.4	4.4	3.3
Clerical	6.2	7.5	11.5	11.7	22.4
Blue collar	29.5	30.7	26.7	36.7	27.9
Unskilled	61.6	59.2	59.0	37.6	40.9
Number of households	1,153	2,228	1,473	5,338	22,530

Source: Compiled from an approximately 50 percent sample of households taken from the City of Montreal, Water Tax Rolls, 1861 and 1901.

and with the construction of the canal, in-migration, and the beginnings of factory production the area's population increased significantly from the 1830s. Land subdivision and house building proceeded rapidly during the 1840s, as both the area's petty commodity and budding industrial manufacturing base developed. By 1853 Griffintown had been settled and was the residential core of the West End, many of whom were Irish immigrants. Most of the population lived in poor quality and densely built housing, much of it wooden shacks lying close to Griffintown and Canal's foundries, rolling mills, and sawmills. At the same time, some residential growth took place just to the north, in Petite Bourgogne, after the establishment of Cantin's dry docks and further industrial expansion in the 1840s. This residential development before the middle of the century provided the nucleus of a multiskilled labor force that found work in the area's workshops, factories, mills, and warehouses.[15]

Industrial development occurring along the Lachine Canal at midcentury did not trigger immediate full-scale residential development elsewhere in the West End. In 1861 most of Saint-Ann's population continued to live in Griffintown, and only 13 percent (341 households) lived south of the canal. Over the next few decades, however, the building opportunities afforded by the opening up of the St. Gabriel domain by the Sulpicians in the early 1850s led to housing construction south of the canal. Twenty-five new streets were added in the ward between 1851 and 1871 to accommodate this growth. At midcentury new Irish immigrant arrivals leapfrogged over the earlier Griffintown Irish settlement and moved into the new areas along the Lachine Canal. By the 1870s settlements formed in the Canal areas of Saint-Gabriel, Point Saint-Charles, and Victoriatown: more than 1,400 households south of the canal lived in residential developments in 1881 whose land, a decade earlier, had been pasture and orchards. Saint-Ann and Saint-Gabriel remained culturally diverse: Irish Catholics and Protestants, French Canadians, English, and Scottish were residentially intermingled.[16]

Equally important for large-scale industrial growth area after mid-century was the creation of a built environment geared to increasing capital flows. The government, by leasing hydraulic canal sites, and the Sulpicians, by selling the St. Gabriel domain, played a central role. Industrial expansion depended upon the availability of manufacturing land with the appropriate infrastructure and services, and both the colonial government and the seminary ensured that they would be available. The commissioning of a plan by the government in 1844 to lay out hydraulic lots at the Canal Basin (Mill Street), close to where the canal emptied into the St. Lawrence River, marked the beginning of Saint-Ann's landscape of large-scale industrial firms. In 1846 lots along Mill Street were auctioned, and by the mid-1850s each of the lots had been leased as manufacturing sites, mainly to metal and milling concerns. The seminary's initial strategy also involved converting its rural land along the canal into manufacturing subdivisions. This plan was only partially successful, and the seminary eventually turned to residential subdivisions for its profits. Nevertheless, when the seminary land at St. Gabriel Locks was sold to the Young clique in the early 1850s, it began to be used for manufacturing. In 1851 John Young, a grain and wholesale merchant, and Ira Gould, a miller, leased five lots at St. Gabriel Locks. Joined by John Ostell and Jacob DeWitt, both of whom were active in manufacturing along the canal, the group subdivided the original lots into twenty and subleased them to other manufacturers. In 1853 Peter Redpath obtained the land for his refinery along the canal bank from the Seminary and from Ostell, who also happened to be the plant's architect. Not only were Young and his partners interested in controlling the land around the canal; they also financed other manufacturers. In 1853, for example, they extended a thousand-pound cash loan to Fred Harris for the construction of a three-story cotton mill at St. Gabriel Lock. By the end of the first major flush of manufacturing in 1856 most of the new, large-scale firms of the previous ten years were built on this newly created manufacturing land at Mill Street and St. Gabriel Lock.[17]

The success of manufacturers on these industrial lots depended upon government's ability to maintain the quality of the canal's structure and to expand the flow of water as demand grew. As the commissioner of Public Works wrote in 1859, "such is the increase in machinery, and the use of water, by the mills established on the line of the canal, that it is very much feared, when the river falls again to its ordinary level, there will not be a sufficient supply of water to keep them all going." During 1860 the demand for water increased by 25 percent, and the canal superintendent, John Sippell, complained of having difficulty keeping water levels up. To meet the recurring problem of the demand by manufacturers for canal facilities, the government undertook extensive redevelopment. Between 1840 and 1881 it poured $7.5 million into new construction. During the 1840s it spent more than $1.4 million enlarging the canal. Over the next quarter of a century (1849–73) there was a lull in construction. In the last half of the 1870s and the early 1880s, however, a massive work's program was undertaken worth almost $6 million, for an average of more than half-million dollars a year. By 1883, at the end of the new round of redevelopment, the canal had been reconstructed in a variety of ways: new locks were built, the old basins were enlarged and deepened, and Montreal's harbor had a new entrance.[18]

Even though Griffintown was the original core of the western manufacturing complex, other districts in the West End were being incorporated into the industrial and residential fabric of the expanding metropolis. In the early 1850s the Canal district, with its whirr of machinery and booming factories, was the center of Montreal's large-scale industry. Over the course of the second half of the nineteenth century a working-class manufacturing complex developed below the hill, first in Saint-Ann and later on in the suburb of Saint-Gabriel. At the same time, farther west, rural villages and farmland stretching out along the Lachine Canal were transformed into new suburban manufacturing districts. As early as the 1870s, Saint-Henri, Sainte-Cunégonde, and Côte Saint-Paul were autonomous suburban municipalities in name but functioned as working-class residential and industrial districts of Montreal.

Côte Saint-Paul: Becoming an Industrial Village

The beginnings of Côte Saint-Paul lie in the early 1850s, when William Parkyn, the former owner of the St. Mary foundry in the city's East End, bought 110 acres of land alongside the Lachine Canal. Although he intended to develop an industrial and working-class residential area, neither people nor industries were immediately forthcoming, and Côte Saint-Paul remained geared to small-scale agricultural production. Miles from Montreal and isolated from workers, supplies, and markets, Côte Saint-Paul was terra incognita to most of the city's businesses—yet from this inauspicious start a small industrial village developed. By 1859 a few mills had been established, and the village had four hundred residents, most of whom depended on local factory employment. According to one newspaper report: "St. Gabriel Lock was formerly considered far out of town. It is now an important part of the city, though the space intervening between it and the old part of the town has not yet been filled up. But far beyond St. Gabriel Lock, at Cote St. Paul, on the highest lock of the Lachine Canal, a new set of factories is springing up, which promises to be as important as any of those nearer the city." Côte Saint-Paul had several attractions for manufacturers, from cheap land to a canal location. One contemporary booster argued that the area's location was highly beneficial to new firms; for existing firms a site alongside the Lachine Canal "enables them to bring coals from Jersey City in the same boat they are embarked on at that place, without any transshipment or breaking of bulk. Grindstones can be brought from Ohio in the same way, and iron and steel have no very great distance to pass through the canal from the ship to the factory."[19]

In spite of the advantage afforded by the canal, the industrial development of Côte Saint-Paul was inhibited. Partly this was because of the fact that it was miles from the city and also, despite claims to the contrary, it was isolated in terms of its factors of production. Manufacturers did not see the same advantages as the village promoters, and both industrial and population growth continued to be slow. By 1881 Côte Saint-Paul had only 949 people, two-thirds French, one-third British (table 5.4).[20]

Côte Saint-Paul's industrial potential was also hindered by a group of Montreal merchants and manufacturers who, through their control over the suburb's land and industrial development, created conditions inimical to large-scale growth. Two Montreal hardware merchants oversaw the development of Côte Saint-Paul, its industrial base, and its harnessing of the Lachine Canal's water power. Established as a partnership in 1836, the hardware firm of John Frothingham and William Workman became the largest wholesale hardware company in Canada. Typical of the Montreal bourgeoisie, they had a range of interests. Both were promoters and shareholders in railroad, shipping, and financial companies; they were involved in the Montreal Board of Trade and the Association for the Promotion of Canadian Industry; and they had real estate interests throughout the city. Adding to this economic power was political muscle; Workman was mayor of Montreal between 1868 and 1870. Tak-

Table 5.4 Population, Place of Origin, and Religion in Selected Western Suburbs, 1871 to 1901

Year	Population			Place of Origin								
				French (%)			British (%)			Catholic (%)		
	Saint-Paul	Saint-Henri	Sainte-Cunégonde	Saint-Paul	Saint-Henri	Sainte-Cunégonde	Saint-Paul	Saint-Henri	Sainte-Cunégonde	Saint-Paul	Saint-Henri	Sainte-Cunégonde
1871	—	2,467	3,656	—	—	—	—	—	—	—	—	—
1881	949	6,415	4,849	67.8	86.0	86.9	29.8	13.2	12.3	81.7	92.3	93.5
1891	842	13,413	9,291	—	—	—	—	—	—	78.5	93.5	86.6
1901	1,496	21,192	10,912	59.4	82.4	77.8	36.1	16.7	20.5	72.9	88.6	85.4

Source: Various censuses, 1871–1901.
Note: Place of origin not given in 1891.

ing over the Côte Saint-Paul concern in the early 1850s the two entrepreneurs added manufacturing and land ownership to their range of business interests and oversaw the development of the suburb's factories. With direct interest in the local edge tool factories, Frothingham and Workman's activities created a cluster of linked manufactories regulated by their control over land, production, and distribution. By inhibiting other forms of manufacturing development and keeping land off of the market, the hardware merchants stunted the village's growth for a generation. The result was that Côte Saint-Paul remained a small industrial village with a few small plants producing edge tools, flour, and nails and employing a local population that depended on these industries. For example, of the 170 residents who had an occupation listed in the 1871 directory, 61 were in the metal trades (shovel, auger, axe, and scythe makers), while another 23 were laborers, many of whom did the dirty, physical labor required in the plants.[21]

Under the merchant-industrialists' control a new suburban manufacturing node appeared in Côte Saint-Paul, albeit one restricted in size and importance. In 1855 the frontage of 4,000 feet along the canal was home to factory sites that remained the nucleus of the village's industrial development over the next fifty years (fig. 5.3). By 1866 there were ten factories at Côte Saint-Paul. Flour, edge tool and nail industries dominated the village's growth. The two flour mills, while not large by the standards of Montreal mills, ground 460 barrels of flour daily and stored 105,000 bushels of grain and 6,000 barrels of flour. The Mount Royal flour mill was so up-to-date that one writer exclaimed, "to visit the process of manufacture in one of the old-fashioned mills and the process as now carried on in these mills, the contrast is so striking." One striking feature was that "from the time the wheat is being taken out of the vessels until the flour is shipped, no process in the manufacture is even handled by manual labour, every thing being done by machinery," such as separators and scouring, brush, milling, and bolting machines. Likewise, Dunns's nail factory employed large amounts of machinery and had a rolling mill, while in his axe factory J. Higgins used a variety of modern methods. For example, the bitumen furnaces in Higgins's edge tool factory had been replaced by ten furnaces, all of which "vomited a white bright flame produced by anthracite coal," and bellows had been replaced by fans. Serving these factories was Paxton's barrel factory, where a combination of machine and hand work produced nail casks for local firms. The district grew little after this initial settlement, and firms remained relatively small in scale. As late as 1888, Henry Warren's Vulcan Works employed only forty workers in a small (80 by 120 ft.) factory.[22]

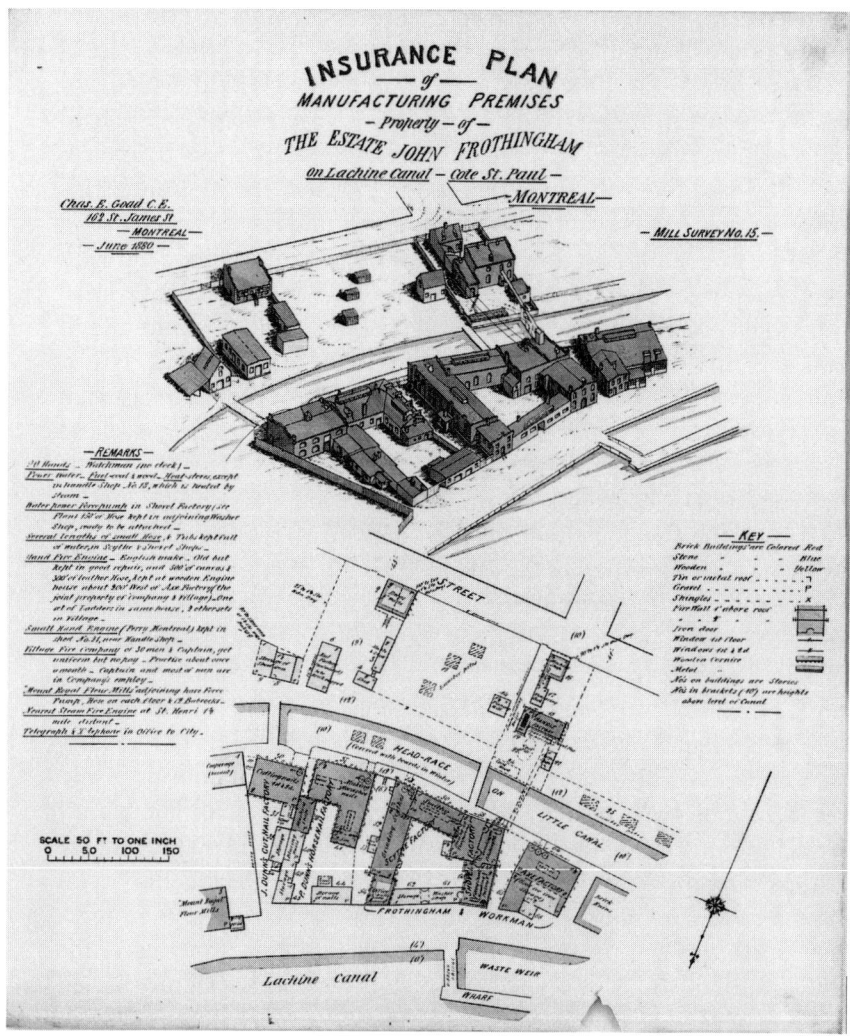

Figure 5.3 Frothingham and Workman's Côte Saint-Paul's Factory Complex, 1880. As the 1880 fire insurance plan shows, the medium-scale Côte Saint-Paul factories were controlled by the merchant-manufacturers, Frothingham and Workman. (Charles Goad, *Insurance Plan of Manufacturing Premises, Property of the Estate of John Frothingham* [Montreal, 1880]. Reproduced courtesy of the National Archives of Canada, negative no. NMC 10549.)

Even though manufacturers could obtain advantages from locating closer to the city, the pull of the suburban fringe was compelling. After the early 1850s a small industrial enclave developed along the banks of the Lachine Canal in Côte Saint-Paul, molded by Frothingham and Workman, who through control of markets, credit, and land development created a company village geared to manufacturing augers, axes, nails, and screws for their hardware business. Over the next forty years there would be little or no growth: firms changed hands, but no new factories appeared. In part this was because canal facilities were not being used adequately. The canal commissioner reported in 1864 that there was a depth of only six feet in the canal at Côte Saint-Paul and that "it should be deepened to correspond with the enlarged canal, and suitable wharfage accommodation provided for these manufacturers." Manufacturers never realized the full potential of the water power at the canal. With more than a thousand horsepower available, they never used much more than half. Finally, contemporary industrial geographic dynamics worked against Côte Saint-Paul. Most of the advantages accruing to the suburb's manufacturers could be obtained elsewhere, closer to the city's large labor force, lower transport costs, and large financial and commercial institutions. As long as Côte Saint-Paul remained under the tutelage of Frothingham and Workman, it was to experience slow growth; a small nucleus developed, but the area did not realize its full potential. By the turn of the twentieth century Côte Saint-Paul, its traditional industries in decline, had become a working-class dormitory, while manufacturers seeking sites for their factories looked to other parts of the city.[23]

Saint-Henri and Sainte-Cunégonde: Two Industrial Suburbs

Two districts that manufacturers increasingly sought out were the adjacent villages of Saint-Henri and Sainte-Cunégonde. Their development, as with Côte Saint-Paul, took place under the active guidance of a small group of Montreal and local business and political elites. In contrast to Côte Saint-Paul, however, the development of the two villages was not inhibited by these entrepreneurs. Before the early 1850s Saint-Henri and Sainte-Cunégonde were distinct villages separated from Montreal by the open countryside of the Sulpician lands. Saint-Henri, a small leather working village, had a population of 466 in 1825, climbing slowly to 600 in 1852. Prior to the 1850s it was a tanning and shoemaking village centered on artisanal production. With the transformation taking place in the regional shoemaking industry during the 1840s and 1850s, Saint-Henri's industry declined, and the Montreal-controlled putting out system replaced independent artisans. Concurrently, new industries developed which would determine Saint-Henri and Sainte-Cunégonde's future paths. After 1850—with the development of residential neighborhoods, infrastructures, and manufacturing plants from a specific set of industries—the two villages became suburban satellites deeply integrated into the orbit of the Montreal industrial complex.[24]

Extensive land development dates from midcentury. In the late 1840s a large block of land in lower Saint-Henri (or Saint-Augustin, as it was then known) was ceded to three associates who, after a complex series of property transactions, sold it to Arthur Webster, who in turn started selling off lots in 1855. The remaining lots were sold to another group in 1860. Over the next ten years the group sold 161 lots. At the same time, to the northwest of Saint-Augustin the Turcot family disposed of lots in the old village of Saint-Henri des Tanneries. From these midcentury land deals the shift from small preindustrial villages to a working-class industrial suburb is apparent. Across the municipal border, in Sainte-Cunégonde, the fur merchant and politician Frédéric Auguste Quesnel started selling off land in the late 1840s. From the beginning Alexandre Delisle, William Workman, and Charles-Joseph Coursol, who between them had powerful financial, commercial, manufacturing, and political connections in Montreal, were instrumental in helping Quesnel sell and subdivide land, lay out manufacturing sites, build roads, and put in sewers. As was common in the nineteenth century, the developers left the business of house construction to small builders. As in the East End, the small builder, restricted by the small incomes of working-class families, constructed low-quality working-class housing. Land development, housing construction, and street openings led to population growth. From just six hundred in 1852, the two suburbs had a combined population of more than thirty-two thousand in 1901 (table 5.3). In contrast to the largely British and Irish composition of Saint-Ann and Saint-Gabriel, these two industrial suburbs were populated mainly by French Canadians.[25]

With the gradual laying down of the western districts' residential and infrastructure components, there was increasing clamor for political autonomy. The incorporation of the separate villages into the town of Saint-Henri in 1875 and Sainte-Cunégonde a year later laid the basis for the political regulation of urban growth on Montreal's western periphery. The composition of both local governments consisted of local property owners and manufacturers who shared a similar view of industrial urbanization, which entailed a program of industrial tax exemptions harnessed to minimal infrastructure provision. In the first years after incorporation the two towns established a few essential infrastructures: fire and police services, street lighting, and streetcar services (after giving City Passenger Railway a tax exemption). Even though Sainte-Cunégonde provided tax exemptions to metalworking, cotton, and pottery firms, it was more liberal with the funneling of municipal funds to infrastructures for the local population. It established its own aqueduct in 1879, and by 1890 nearly all of the town's streets had water pipes. But conditions for the working-class population remained poor. The policy of industrial promotion pursued by both governments ensured that the basic infrastructures of urban development intended to provide decent living conditions were not forthcoming. This joining of industrial subsidies and infrastructure neglect remained the basis on which the two towns developed until 1905, when large debts and inadequate services led to their annexation by Montreal.[26]

After midcentury Sainte-Cunégonde and Saint-Henri experienced a wave of industrialization that integrated them into the Montreal economy (map 5.3). Before 1890 Saint-Henri became the home of firms such as Merchants Manufacturing (cotton) and the Union Abattoirs, while Sainte-Cunégonde received the Singer Manufacturing Company, McCaskill's varnish works, William Rutherford's door and sash factory, and Robert Mitchell's brass foundry. Several firms illustrate the dynamics fueling the early growth of the new suburbs. The sewing machine firm of Williams Manufacturing Company, established in Montreal in 1861, moved to Saint-Henri in the late 1870s as part of a company policy of expansion. Saint-Henri's large pool of labor, industrial bonuses, large sites at cheap prices, and excellent transportation facilities gave Williams a set of locational assets that differed from those of the central manufacturing districts. In the 1880s the company employed 150 workers in its four-story building. Attracted by the suburbs' locational assets, other firms followed. The move of Dominion Wadding from Sorel to Sainte-Cunégonde in 1885 gave the firm access to a large metropolitan market as well as modern machinery and production methods in a large factory close to shipping facilities. Likewise, to facilitate changes in operating methods and factory layout resulting from a reorganization of the industry, Joseph Luttrell moved his biscuit interests to larger premises in Sainte-Cunégonde. Even though the company's works were "situated beyond the city limits, they are accessible by the telephone or the city cars which run to Vinet Street on Notre Dame Street; and from this situation they claim that they can produce at less cost than city manufacturers."[27]

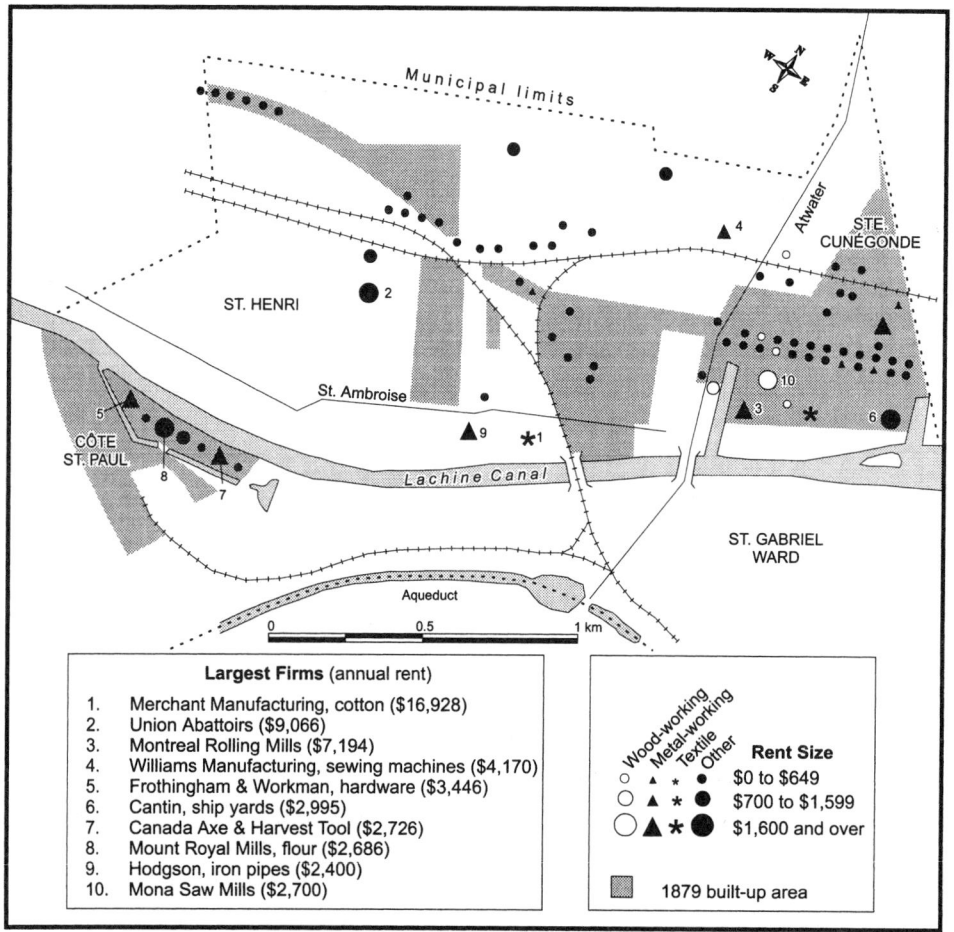

MAP 5.3 THE WESTERN SUBURBAN MANUFACTURING DISTRICTS, 1890. By 1890 an extensive manufacturing base dominated by metal, textile, and food industries had developed in the western suburbs. (Compiled from Ville de Saint-Henri and Ville de Sainte-Cunégonde, Water Tax Rolls, 1890; Town of Côte Saint-Paul, Assessment Rolls, 1895.)

Saint-Henri and Sainte-Cunégonde's industrial landscape in 1890 was built upon bursts of investment over the preceding two decades. Most firms were owner-operated enterprises, manufacturing an assortment of products—nails, engines, stoves, leather, and crackers—for a range of markets. These firms coexisted with others such as the Montreal Rolling Mills and St. Lawrence Glass, bankrolled by Montreal financial, mercantile, and manufacturing capital. Leather making, Saint-Henri's traditional trade, was transformed with the elimination of craft work and transmuted into the putting-out system or incorporated into in situ factory production. By 1886 there were ten shoe firms, two tanneries, and one saddle factory. The tanneries and saddle factory were medium-size firms employing steam-driven machinery and were linked to Montreal through the offices they kept there. The boot and shoe firms were small in scale; all ten firms employed only seventy-four workers. The small scale of these firms does not signify the persistence of artisanal production but, rather, the integration of the trade into the wider Montreal economy through subcontracting. By the 1870s the Saint-Henri firms were firmly linked to Montreal firms that passed out work to the small suburban shops. Other firms were also linked to local and national markets. The Canada Engine Works, for example, manufactured engines and boilers, heavy forgings and castings, iron ships and bridge work, and various types of machinery for a national market. Moreover, these early enterprises, especially the larger corporate ones, created a powerful set of agglomeration economies, which attracted other firms to the suburbs. Some of these early firms were magnets because of the direct physical linkages they offered to other firms, while others drew new firms by introducing new uses of space and creating comparative industrial advantages in a new suburban milieu.[28]

The Montreal Rolling Mills reflects one strand of the western suburbs' transition to industrial capitalism. In the 1850s and early 1860s the Montreal hardware merchants of Thomas Morland and Charles Watson faced several problems. Not only did they have difficulties over price and quality with imported hardware lines; their share of the local hardware markets was threatened by other Montreal manufacturers building rolling mills. In 1868, in order to counteract these threats, Morland and Watson, in association with some powerful merchants, financiers, and industrialists, sank $150,000 into a joint-stock rolling mill on the banks of the Lachine Canal in Sainte-Cunégonde. With the influx of capital, the participation of a section of Montreal's bourgeoisie, large credits from the Bank of Montreal, and tariff walls, the Montreal Rolling Mill became a leading Montreal factory. Over the years the firm grew rapidly and became more technologically sophisticated and more specialized. By the late 1860s it employed 350 in the manufacture of iron bars, sheets, and nails, and in 1871 its fixed capital stood at $200,000 (fig. 5.4). With three steam engines operating an array of machines that made wrought, shingle, and horseshoe nails, the mill was divided up into a rolling mill and lead, nail, and horseshoe nail departments. As Thomas Raphael observed in 1867, "standing in the centre of the yard, and looking around, one begins to appreciate the magnitude of the business and the enormous amount of capital requisite to successfully carry it on." Covering two and a half acres, the rolling mill was a large suburban factory producing a variety of goods and controlled by some of the leading industrialists of the day.[29]

From these early beginnings the mill continued to grow. The company built Canada's first iron pipe mill in 1881. Six years later, in response to increasing demand for wrought iron pipe, a new pipe mill with the most modern gas producers and furnaces was built, doubling capacity. By 1890 the mill had a mostly male workforce of 625, $620,000 in fixed capital, and an annual rent of more than $7,000, making it the tenth largest plant in the region. Then, in response to the 1890s depression, the company widened its product line and introduced new technology. To complement wrought nail and nail plate production, it installed wire mill and wire nail machines, and it introduced steel manufacture alongside that of iron. In the Montreal Rolling Mill the introduction of new technologies by Montreal capitalists in the form of a large, multi-unit rolling mill and finishing works on the suburban fringe of the city represented an important adaptation to the requirements of early industrial capitalism. It was also the western suburbs' first large-scale factory. As such, it provided a node around which other manufacturing firms developed, both by establishing direct linkages to foundries and machine shops and by bolstering the more general suburban economies that anchored Saint-Henri and Sainte-Cunégonde to the rising fortunes of a swelling metropolitan economy. The mill's growing number of jobs also contributed to the growth of the property market and the formation of a working-class residential area.[30]

Another firm seeking the western suburbs' advantages was the St. Lawrence Glass Works. In contrast to the Montreal Rolling Mills, St. Lawrence did not have such a long-term and deep influence on the western suburbs; nevertheless, it illustrates another element of their development. While its existence in Sainte-Cunégonde was short—established in 1867 it closed down in the early 1870s—it put into place a major element of Montreal's industrial growth in this period, the replacing of imports with domestic-made products. Glass manufacture was organized into shops, with teams of skilled glassblowers, finishers, and their helpers working as a unit, being paid by the piece, and being supervised by a superintendent. In 1871 the 117 employees of the company worked in four buildings with a total fixed capital of $50,000. There imported

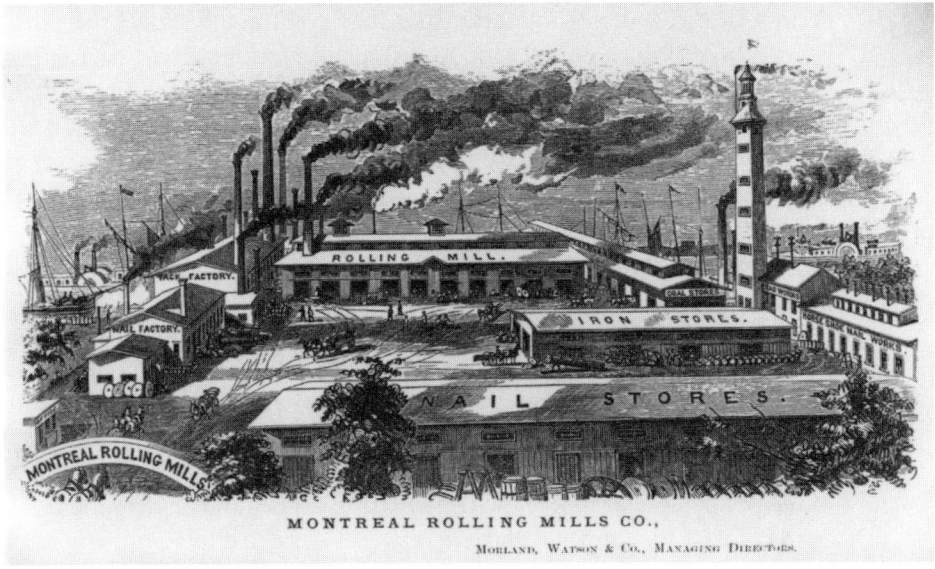

FIGURE 5.4 THE MONTREAL ROLLING MILLS, 1868. The Montreal Rolling Mills, a pioneer suburbanizing firm, became the nucleus of a large manufacturing cluster in the industrial suburbs of Saint-Henri and Sainte-Cunégonde. (*Commercial Sketch of Montreal and Its Superiority as a Wholesale Market* [Montreal, 1868], 8. Reproduced courtesy of Toronto Reference Library.)

white sand from the Berkshire Mountains in Massachusetts was dried, mixed with other compounds in pots in a furnace, and then manipulated into lamp chimneys, vials, and goblets. As with the Montreal Rolling Mills, St. Lawrence Glass exemplifies the penetration of corporate capital into Montreal's industrial suburbs. Its board of directors included some of the major players in Montreal industrial and political life: William Workman, Alexandre Delisle, and Peter Redpath pumped $20,000 into the plant before it was closed down.[31]

Industrial growth along the Lachine Canal during the boom years following the depression of the 1840s was a major break with the past. Large-scale, energy-intensive firms incorporated new labor processes, organizational structures, and modern machinery. They also located in a new part of town, representing industry's move from the core area. The state, land developers, builders, merchants and manufacturers were instrumental in industrializing the canal; their decisions functioned alongside processes at work at the international and national levels. Industrial capitalism, technological change, larger capital investments, and Montreal merchants' strategies to overcome the crisis of the dismantling of the colonial economy all played an important part in the creation of industry along the Lachine Canal, both in Saint-Ann and the industrial suburbs of Côte Saint-Paul, Saint-Henri, and Sainte-Cunégonde. Not only were the West End districts distinct in the sheer scale of the enterprises locating there; they also contained a specialized array of firms. With their multitude of rolling mills, metalworking shops, food processing mills, chemical factories, and sawmills, these districts stood in sharp contrast to the central manufacturing and East End districts.

Over the course of the century the city's western area remained the most important manufacturing zone of Montreal. There a built environment was constructed that provided the basis for the development of capitalist industry. Local alliances parceled out land for the construction of working-class homes, built the basic infrastructures, and established lax regulations for the construction of the built environment. As in the East End, new productive relations were tied to a built environment providing locational assets for firms seeking manufacturing sites. Their histories, separate yet similar in many respects, reflect an attempt by Montreal's bourgeoisie, in alliance with local entrepreneurs, to create new manufacturing relations within the context of new spaces. The different array of social relations and the timing of industrial development underlying the growth of the western suburbs ensured that the new districts emerging outside the city were not clones of Canal and Griffintown; they differed in their industrial structure, scale, and technology and the manner in which their manufacturing space was created. Between the late 1840s and the end of the century a distinct and specialized industrial milieu had been created on the western side of the city.

Part II

Montreal's Manufacturing Districts, 1890 to 1929

CHAPTER SIX

"One of the Most Magnificent Cities"

Montreal's Economy, Planning, and Housing, 1890 to 1929

> Historically, commercially and financially, Montreal is to-day the metropolis of Canada and one of the most magnificent cities on the American continent.
> —Montreal Board of Trade,
> *The Board of Trade Illustrated Edition of Montreal*, 1909

By the last decade of the nineteenth century the factory and generating plant of the Royal Electric Company symbolized Montreal's position as the industrial metropolis of Canada. Heralding the new age of electricity, it grew rapidly from its small beginnings in 1884, then with twelve workers and capital of $50,000. By the end of the century its expansive, well-equipped plant built a range of large electrical machines for an expanding national market. During the Federated Canadian Mining Institution's convention in 1897 hundreds of visitors crowded into the company's works to view the array of large and gleaming turbines, generators, and other equipment (fig. 6.1). Electricity and the Royal Electric Company, symbols of the new industrialism, reflected the broad front of growth taking place in Montreal as well as the city's national importance. As William Browne, the company's general manager, told the visitors, "the successful progress of the Royal Electric Company is a component of the increased commercial prosperity of the City of Montreal, and through the use of its manufactures the Dominion will obtain similar advantages."[1]

Aligned with this sense of long-range opportunities, the development of vast new markets, and the changing character of manufacturing was the formation of a more elaborate industrial geography. The metropolitan area's nineteenth-century pattern of locational sites after 1890 was extended and deepened, as a larger selection of districts emerged and existing ones underwent changes (table 6.1 and map 6.1). Old Montreal and the Outer Core experienced a dramatic reduction in their share of the city's rent but remained the locus of a wide range of sectors deploying an assortment of manufacturing strategies. Similarly,

FIGURE 6.1 THE ROYAL ELECTRIC COMPANY, 1897. Inside the Royal Electric factory the visitors to the 1897 convention of the Federated Canadian Mining Institution crowded around the array of large-batch product equipment manufactured by the company. (*Canadian Engineer* 4 [March 1897]: 338.)

the mid-nineteenth century suburban districts—Canal, Griffintown, and Sainte-Marie—declined in importance, though they continued to be significant manufacturing clusters dominated by industries established in the first burst of growth in the nineteenth century. Nonetheless, the advantage of being greenfields in the 1840s had largely disappeared, and the legacy of increasingly obsolete and congested older industrial space undermined their ability to meet the needs of firms seeking new types of spaces. At the same time, new manufacturing nodes emerged as firms sought out new greenfield sites on the metropolitan fringe. In the west satellite towns such as Lachine developed around the establishment of large locomotive, glass, and steel plants owned by non-local corporations (some Canadian, some foreign) employing the latest technologies and work methods. In the east the earlier manufacturing complex expanded with the growth of Maisonneuve (metal, food, and shoes) and Mercier and Montréal Est (cement and oil refining). Finally, after 1900 a manufacturing district characterized by an assortment of paper, food, clothing, and chemical firms emerged along the railroad lines in the metropolitan area's North End.

Many historians and historical geographers have asserted that the geography of industry in the first few decades of the twentieth century was characterized by a simple polarization of scale and industrial type between core and suburb and that only large, capital-intensive firms employing the latest production methods moved to the suburbs to take advantage of cheap land and technological advances in transportation. This was probably not the case anywhere, and certainly not in Montreal. While it is true that large propulsive firms employing the latest manu-

Table 6.1 Manufacturing Districts in Montreal, 1890 and 1929

Manufacturing Districts	Date of Emergence	Number of Firms		Share of Rent (%)		Median Rent ($)	
		1890	1929	1890	1929	1890	1929
Old Montreal	pre-1840	378	303	23.6	8.3	300	900
OUTER CORE		600	1,042	31.3	25.6	200	800
Saint-Antoine	1840s	355	451	24.6	15.8	280	1,500
Saint-Lawrence	1850s	123	446	4.2	7.6	150	650
Saint-Jacques	1850s	119	129	2.4	1.7	100	450
WEST END		239	379	30.0	33.7	600	1,800
Griffintown	pre-1840	67	73	5.3	2.4	300	1,500
Canal	1840s	68	59	12.0	7.3	310	2,400
Saint-Henri	1870s	88	140	10.5	12.3	150	1,400
Lachine	1880s	13	43	2.2	9.8	60	1,800
EAST END		142	309	15.1	23.0	120	900
Sainte-Marie	pre-1840	123	131	7.2	6.3	120	700
Hochelaga	1870s	25	75	7.8	3.2	145	600
Maisonneuve	1870s	—	70	—	4.5	—	950
Montréal Est	1900s	—	18	—	9.0	—	11,800
NORTH END		—	411	—	9.2	—	700
North End	1900s	—	176	—	6.1	—	700
Plateau	1890s	—	116	—	2.1	—	600
City Total		1,365	2,444	100.0	100.0	200	800

Source: Water Tax Rolls for Montreal and surrounding municipalities, 1890 and 1929.

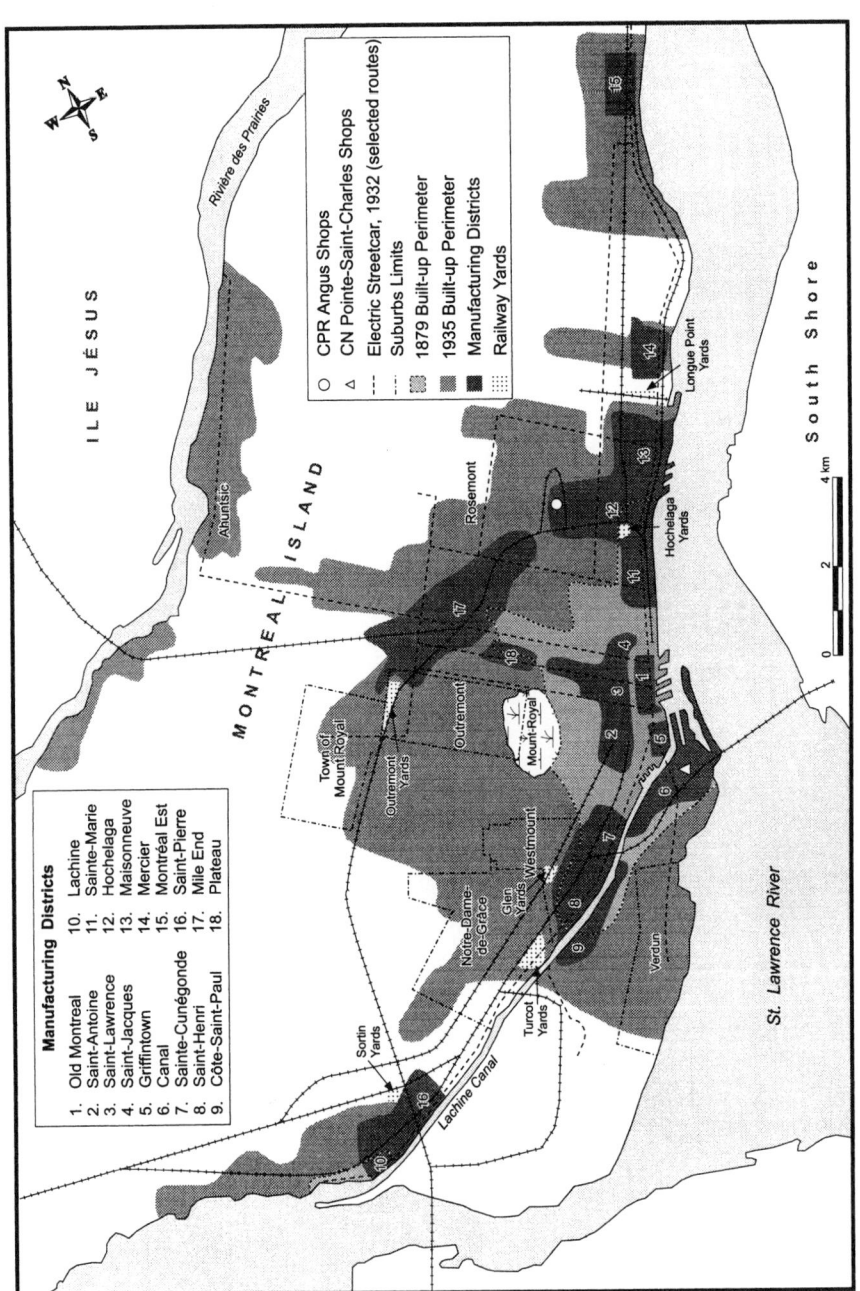

MAP 6.1 MONTREAL AND ITS MANUFACTURING DISTRICTS, 1929. In the forty years after 1890 metropolitan Montreal's built-up area and its manufacturing districts had extended greatly.

facturing methods frequently formed one element of industrial suburban development, the evidence for Montreal indicates that firms deploying quite different production methods and investment capabilities also moved to suburban locations. The new manufacturing districts forming on the urban periphery featured a range of firms and scales, from small independent workshops to large corporate branch plants. Adding to the complexity of Montreal's industrial geography between 1890 and 1929 was that the mid-nineteenth-century fringe manufacturing districts (Canal, Sainte-Marie, and Hochelaga) were enveloped by the expanding city. In the process an elaborate metropolitan geography was created.

The Leading Center: Economic Change and Manufacturing Pathways

The broad front of economic change sweeping through North America after the depression of the 1890s brought a new industrial order in its wake. While this did not involve a structural transformation of Canadian capitalism, it did mean a change in the way that capital was used. One key element was the increasing scale of industry. By the turn of the century corporations, many of them American, operating new and intricate forms of machinery, new production processes, new forms of labor, and new energy sources, became an important avenue of the city's manufacturing growth. Between 1890 and 1929 firm scale increased as industries restructured to meet the changing demands of national and international markets. Growth through increased demand was linked to a larger battery of machines and new and faster forms of continuous-flow work methods. While Montreal's mean firm employment size was 20 in 1891, it had more than tripled, to 62, in 1929. Over the same period mean manufacturing capital (in constant dollars) increased more than sixfold, from $38,320 to $244,788. Growing scale was not a new phenomenon, but after the 1890s it took on a new importance. In 1919, for example, the largest mean employment size was in locomotives (990), cotton (634), and steel mills (476), while other large-scale ones were electrical apparatus (254), chocolate (201), meatpacking (184), breweries (159), and shoes (129).[2]

Contributing to increasing scale was corporate consolidation. As early as 1888, a Federal Select Committee discovered combinations in the sugar, biscuit, and rope, and twine industries. Many of the 505 consolidations in Canada between 1900 and 1929, involving 1,068 enterprises, took place to Montreal firms. Dominion Textile, for example, grew out of several Canadian plants, and its flagship plants—Hudon and Merchant Manufacturing—were located in Montreal. Canada Rubber Company underwent the same process, expanding into a trans-Canadian corporation, with other plants spread throughout Quebec and Ontario. Its 1850s factory, greatly expanded over the years, remained the company's flagship plant, while the head office was located in Old Montreal. Rationalization through mergers occurred for many reasons. The chief objective for the Montreal-based Canada Cement Company was to devise a new organizational structure to regulate distribution. For Dominion Cotton it was to reduce costs through the rationalization of production facilities. Regardless of these differences, the drive to the large-scale corporate firm established the basis for important cost savings to be gained from sites outside the older manufacturing districts.[3]

Direct foreign investment fueled Montreal's industrial rationalization. During the second half of the nineteenth century most capital flowing into Canada was British portfolio capital channeled into the construction of a national railroad network. In the last decades of the century, however, there was a shift in the origins and type of foreign investment in Canada. By the turn of the century American direct investment had become more important in the manufacturing, forestry, and mining sectors. Cities such as Montreal, Toronto, and Hamilton welcomed a growing number of American firms opening manufacturing operations to jump tariff barriers, gain access to imperial markets, and capture lower manufacturing costs. As early as 1897, American companies had invested thirty-five million U.S. dollars in Canadian manufacturing. Many, including Singer, American Tobacco, and General Electric, located their plants in Montreal. By World War I investment from the United States into Canada's manufacturing sector had reached one hundred million U.S. dollars, and new firms moving to metropolitan Montreal included United Shoe Machinery, Allis-Chalmers, American Locomotive, and American Can. Frequently, these new branch plants sought out greenfield sites away from the older congested manufacturing districts established in the earlier wave of expansion. In many cases the American and, to a lesser extent, British firms became the dominant manufacturer of their lines in Canada. The American silk manufacturer, Belding Paul, for example, established a plant on the canal banks (fig. 6.2). It grew rapidly and came to control silk production in Canada. Not only did it expand its Montreal plant, but it built two others and established offices and warehouses in Toronto, Winnipeg, and Vancouver.[4]

New forms of technological change initiated work restructuring and established new impulses for spatial relocation. Implementation of

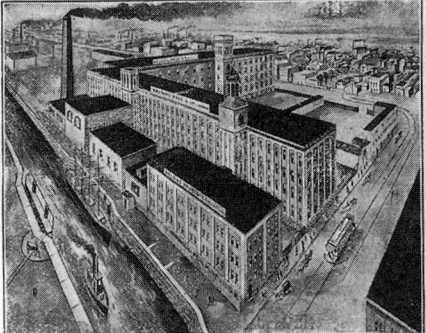
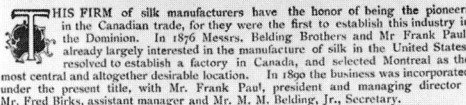

FIGURE 6.2 BELDING, PAUL, 1905. The American silk manufacturer Belding, Paul was typical of the branch plants established in Montreal by the beginning of the twentieth century. (Ernest Chambers, *The Book of Canada* [Montreal: Book of Canada Co., 1905], 310.)

faster and more sophisticated machines contributed to sectoral change and the emergence of new cost structures and product lines. So intense were the changes to the technological basis of industry that not keeping up with the changes frequently meant failure. In the electric industry, as William Browne told Royal Electric's shareholders at their annual meeting in 1899, "inventive genius was so prolific that the plant or apparatus for manufacturing electricity soon became so obsolete. Hence, during the existence of this company, $1,800,000 of equipment had become useless, or superseded by improved machinery and methods." In other sectors, such as metal, glass, tobacco, and chemical, machinery also changed the organization of production, boosted productivity, and provided manufacturers with new ways of controlling labor. In steel and locomotive manufacture, for example, deploying more intricate machines and introducing scientific managerial methods undermined traditional industrial work. Sophisticated labor management strategies were hitched to new technologies. Firms increasingly substituted managerial control for traditional informal methods, machinery directed the labor process to a much greater degree than in the past, and the workplace became more bureaucratic. Growing labor specialization, allied with increasing technical sophistication, freed manufacturers from dependence on skilled labor and created new labor demands. At the same time, the expansion of semiskilled work made labor a much more replaceable commodity, which in turn created more flexible labor markets. Firms faced greater locational possibilities, as managers could further differentiate activities over space. The telephone reduced the need for face-to-face contact within and between firms, permitting the head office to be separate from the factory, and the opening of suburban working-class districts created new labor markets on the urban periphery.[5]

The introduction of electric power into work operations supported these changes. The first major industrial user of electricity for powering machinery dates from the late 1890s. Discarding its antiquated steam power system, Dominion Cotton installed 27 electric motors driven by 3,412 horsepower. Other factories followed suit. The two new East End locomotive factories—the Locomotive and Machine Company and Canadian Pacific Railway—both installed their own electric power generators and purchased power from the local utility company. By the early twentieth century greater numbers of firms adopted electric motors because they offered better speed regulation, higher power supply savings, and greater floor space efficiency. The application of electricity to manufacturing was not an easy matter, but the introduction of centralized power generation solved many problems. As the cost of electric power fell, demand from industrial users grew. In 1901 the 572 electric motors employed by Quebec's manufacturers generated less than 10 percent of the province's motive power; a decade later it had become the major power source. By the end of the 1920s electricity generated more than two-thirds of manufacturing power.[6]

The large firm employing modern production methods played a decisive part in the industrial and geographic structure of the city after 1890. Just as important was the assortment of firms by scale and manufacturing trajectories. As one writer noted in 1909:

> Today Montreal is the centre of the cotton manufacturing of Canada; the great proportion of sugar consumed in Canada is refined here; the largest paint mills are here located. Its mills produce iron and steel bridges, and structural work which finds a market throughout the Dominion. It is the leading centre of the boot and shoe industry. The country's entire white lead industry is to be found here. It is the chief producer of rubber goods, and the production of nails still remains a large and important industry. It has wallpaper, oil cloth, sewing machines, typewriter, telephone, electrical apparatus, glass, jewellery, ready-made clothing, tinware, trunk and many other of the largest industries of Canada.

This was reflected in Montreal's industrial structure, which experienced both stability and change between 1890 and 1929 (table 6.2). Labor-intensive consumer sectors remained important, although their share of the manufacturing rent declined: clothing, food, beverage, leather, tobacco, textile, and printing accounted for 50 percent of the city's rent in 1929, down six points from 1890. The nineteenth-century heavy industries of metalworking, chemicals, and transportation equipment grew slightly over the period. While Montreal did not have the same range and size of metal firms as Pittsburgh or Chicago, it had an impressive set of rolling mills, foundries, nail factories, and locomotive and machinery makers. Montreal became Canada's major chemical producer, and other new propulsive sectors were established, including paper, electrical, and nonmetallic. Together, these sectors' rent share grew from 4 percent in 1890 to more than 17 percent in 1929.[7]

Along with these sectoral changes were changes to the city's rent distribution. Most noticeably, a larger share of the total rent was concentrated in a smaller number of the largest firms. As in the earlier period, the mean, median, and rent distribution of individual sectors varied greatly. At one end lay blacksmithing, clothing, printing, furniture,

Table 6.2 Rent Characteristics of Selected Montreal Manufacturing Sectors, 1890 and 1929

	1890						1929					
	Share of	Rent ($)		Percentage of Rent		No. of	Share of	Rent ($)		Percentage of Rent		No. of
Industry	City Rent (%)	Mean	Median	<$300	>$1,000	Firms	City Rent (%)	Mean	Median	<$1,000	>$4,999	Firms
Metalworking	15.5	815	400	6.9	69.7	147	14.5	5,115	1,000	6.5	83.5	199
Clothing	14.2	393	200	18.7	41.5	281	14.8	1,664	900	26.1	25.1	623
Food	11.4	551	160	14.1	65.8	160	11.2	2,914	700	11.4	70.3	268
Textile	8.3	2,222	450	1.3	92.4	29	7.4	4,294	1,400	8.3	74.5	121
Transport equipment	7.8	801	150	11.1	78.3	75	8.6	10,044	950	3.3	91.3	60
Leather	7.3	511	200	15.2	63.0	110	4.0	2,010	1,000	15.2	46.6	139
Printing	6.1	487	200	15.7	56.5	97	5.6	1,420	570	28.1	45.5	278
Furniture	4.7	483	200	14.9	58.0	76	2.0	2,005	650	19.4	48.7	70
Beverage	3.9	852	200	10.1	79.0	35	4.9	7,349	1,000	4.3	88.0	47
Wood	3.9	564	300	8.0	56.1	53	1.8	1,709	600	19.1	48.3	72
Tobacco	3.3	799	280	8.3	71.0	32	2.5	3,770	500	10.2	79.2	46
Chemical	3.1	675	380	11.4	61.8	35	4.8	2,488	1,200	11.6	52.1	36
Paper	1.7	Too few firms				13	2.1	4,293	1,800	8.5	69.6	35
Nonmetallic	1.3	492	315	15.5	40.3	20	7.2	5,725	600	5.3	85.7	88
Blacksmithing	0.9	71	60	100.0	0	102	0.3	252	200	100.0	0	94
Electrical	0.8	Too few firms				6	2.8	8,130	2,300	1.6	83.3	24
All Firms	100.0	567	200	12.9	63.6	1,365	100.0	2,866	800	12.7	65.6	2,444

Source: Water Tax Rolls for Montreal and surrounding municipalities, 1890 and 1929.

wood, and leather, all having low mean rents and less than 50 percent of their rent in firms with $5,000 or more of rent. At the other end, transportation equipment, electrical, beverage, nonmetallic, and metalworking firms had high median rents and more than 80 percent of their total rent in firms with rents of $5,000 or more.

The widening manufacturing rent scale reflected the increasing diversity of organizational structure. Despite the widespread effects of economic expansion, sector growth did not follow any one single trajectory. From the small-scale blacksmith working under traditional conditions to the giant textile or steel mills in which masses of semiskilled workers labored under scientific managerial methods, there were numerous ways to organize production. Rather than having a dualistic structure of large and small, corporate and proprietary firms, there was a wide variety of productive strategies. While some firms and industries were able to maintain control over markets and their profits through oligarchic techniques and government support, others faced obstacles associated with the implementation of new technologies, changes to the labor process, and the expansion of markets. Where production required the application of heat and involved chemical rather than mechanical methods, there was greater potential for high-volume production techniques. The emergence of a modern chemical industry—characterized by high-volume and continuous processing of complex chemicals, larger plants, more sophisticated machinery, standardized production, hierarchical managerial structures, and laboratories for research and development—underpinned one element of change. A similar trajectory was found in others industries, such as cigarettes, grain, soap, and canning. A high-volume production strategy was not an option for other industries, even with new machinery and new sources of energy. In the clothing, cigar, shoe, and furniture industries the possibility of accelerating the velocity of production was constrained by the nature of the product and the way that it was transformed. In other words, despite firms' increasing scale, the pathways open to firms continued to vary within and between industries.[8]

The tobacco industry, for example, had an extremely complex structure, ranging from the small hand-rolling proprietorial cigar shop to the large, mechanized, multiplant cigarette corporation. The range of tobacco manufacturing pathways arose out of differences in the organization of production, the gender- and skill-based division of labor, the degree of mechanization, and the nature of distribution networks. The cigar industry was defined by a highly developed division of labor and little mechanization. Neither of the two major steps of cigar manufacture, stemming and rolling, were mechanized, and the preparation of the tobacco leaf—stripping and stemming—required little skill, so workers were easily replaced. Rolling, on the other hand, was highly skilled, but some of its skill content was reduced as the mold replaced hand bunching. Different models of the mold and suction wrapping "machine" introduced since the 1860s were in almost all cases hand powered and rarely used and then only in large factories. It was not until the 1920s that machinery became a staple of cigar making. These changes were associated with a reorganization of the gender structure of the labor force. Working with the mold, women made the cheaper lines, while skilled male workers continued to hand-roll the expensive ones. Cigar manufacture was also characterized by a degree of product differentiation; cigars were defined by the type of tobacco used, method of production, nature of the workforce, and scale of retail prices. Finally, because selling directly from the factory was unprofitable, cigar manufacturers relied on jobbers, grocery wholesalers, and wholesale tobacconists to distribute their products.[9]

Cigarette manufacturing stands in contrast to cigar manufacture. Although it too was a skilled trade that relied on skilled workers to roll the leaf, cigarette manufacture was transformed with the introduction of the Bonsack machine in the 1880s, a continuous-process automatic cigarette machine able to make 120,000 cigarettes a day. American Tobacco quickly captured the large share of the American market by investing heavily in new plants and sales offices, creating an extensive advertising campaign and distribution network, and lowing prices. Introduced into Montreal in the late 1880s, this model of production and distribution set up a pathway that was quite different from that of cigar manufacture, characterized as it was by semiskilled machine tenders, standardized products, large capital investments, patent control, and national distribution and advertising networks. Cigarette manufacture, however, was not homogeneous. Several smaller firms specializing in handmade cigarettes coexisted with the large companies. At least two distinct pathways emerged from the division between the various segments of the cigarette industry, based on a series of organizational differences involving distribution methods, the character of the labor force, the extent of mechanization, and access to capital. While high-volume firms catered to growing mass demand, smaller firms made hand-rolled, refined cigarettes at a slightly higher price.

The differences characterizing the tobacco industry were common in most sectors. Many textile firms were small and operated under highly competitive conditions in fragmented markets. In contrast, a few multi-unit corporations such as Dominion Textile and Belding Paul employed large numbers of semiskilled labor and a high volume of output of basic cotton and silk lines by carefully coordinating every element of production between and within individual mills. Metalworking had an even greater assortment of pathways than either tobacco or textile. Producing a multitude of products from rolled iron and huge bridges to small screws and complicated machinery, Montreal's metal sector was greatly differentiated. Large corporations such as Canada Car and the Steel Company had massive markets, large capital inputs, an array of machinery, a managerial hierarchy answerable to stockholders, and thousands of workers operating under a variety of labor conditions. Metalworking was also populated by an assortment of small and medium-sized firms: those operating on the fringes of product markets dominated by large-scale firms, those serving as a "satellite" to a large corporation, and those occupying a niche and competing through product differentiation. Many of these firms employed modern methods but functioned in a different set of markets or operated in a separate realm from the metal corporations. These types of pathways were replicated in nearly all manufacturing sectors, and between 1890 and 1929 they shaped Montreal's manufacturing geography.[10]

"One Harmonic Whole": Social Geography and Manufacturing Districts

Over forty years, from 1890 to 1931, the city and suburb's population almost quadrupled, topping the million mark in 1931 and allowing Montreal to remain among the ten largest metropolitan areas in North America (table 6.3). Annexation of rapidly growing suburban areas contributed greatly to the city's growth, leading to a 1909 description: "At your feet spreads out the city proper—with its wilderness of spires, domes and towering structures, its miles of wide tree-adorned streets, its wonderful harbour, and its numerous outlying towns or suburbs, that form one harmonic whole." Of the city's more than 600,000 new inhabitants, 90 percent (540,000) lived in the suburban areas annexed to the city after 1883, and only 10 percent (60,000) lived within the pre-annexation limits. In 1931 another 180,000 people resided in the adjacent suburbs. The city at its 1881 limits accounted for more than two-thirds of the metropolitan area's population ten years later, while this number had fallen to just over a quarter by 1931. Within the city different areas grew at different rates. The old industrial ward of Saint-Ann experienced continuous decline over the period, losing almost 8,000, or 33 per-

Table 6.3 Population of Montreal and Suburbs, 1891 to 1931

Place	Year of Annexation	1891	1901	1911	1921	1931
Island of Montreal	—	277,525	360,838	544,761	724,205	1,003,868
Montreal and suburbs	—	271,285	352,557	549,190	721,184	1,000,661
Montreal at the census date	—	216,650	267,730	470,480	618,506	818,577
Montreal at its 1881 limits	—	182,695	203,078	225,132	230,978	242,984
Suburbs outside 1881 limits	—	88,590	149,479	324,058	490,206	757,677
Longueuil	—	2,757	2,835	3,972	4,682	5,407
Saint-Lambert	—	906	1,362	3,344	3,890	6,075
Verdun	—	296	1,898	11,629	25,001	60,745
Lachine	—	3,761	5,561	10,699	15,404	18,630
Ville Saint-Pierre	—	—	505	2,201	3,535	4,185
Saint-Gabriel	1887	9,986*	15,959*	18,961*	—	—
Saint-Henri	1905	13,413	21,192	30,335*	—	—
Sainte-Cunégonde	1905	9,291	10,912	11,174*	—	—
Nôtre Dame de Grace	1910	2,305	2,225	5,217*	—	—
Emard	1910	842	1,496	6,179*	—	—
Outremont	—	408	1,148	4,820	13,249	28,641
Westmount	—	3,076	8,856	14,579	17,593	24,235
Saint-Laurent	—	1,184	1,390	1,860	3,232	5,348
Saint-Jean Baptiste	1886	15,423*	26,754*	34,561*	—	—
Côte Saint-Louis	1893	2,972	9,025*	45,670*	—	—
Saint-Louis de Mile End	1910	3,537	10,933	37,000*	—	—
Hochelaga	1883	8,540*	12,914*	28,597*	—	—
Longue Pointe	1910	2,445	2,519	5,531*	—	—
Maisonneuve	1918	1,226	3,958	18,684	—	—

Sources: Canada, *Census of Canada,* various years; Paul-André Linteau, *Montréal* (Montréal: Boréal, 1992), 40, 160, 314.
*Population after the suburb was annexed to Montreal. The census up to and including 1911 provided the population of annexed suburban municipalities, but this was discontinued after 1911. The suburbs included here are those with a population of more than five thousand at any of the census dates.

cent, of its residents. The eastern portion of Sainte-Marie, on the other hand, grew steadily, from 34,000 in 1891 to more than 76,000 forty years later. Population increased in all of the other pre-annexation districts over this period, finally reaching a plateau by the early twentieth century. The greatest absolute and relative growth in the city at any one time took place in those areas closest to the built-up fringe. Before World War I Saint-Henri, Saint-Denis, and Laurier accounted for more than 60 percent of metropolitan population growth, but in the interwar period the population frontier moved farther out. Newly annexed districts such as Ahuntsic, Rosemount, and Notre Dame de Grâce had significant growth in the 1920s. Similarly, the greatest population change occurred in a gradient leading out from the city boundary, with suburbs such as Verdun,

Outremont, and Westmount experiencing tremendous expansion over the period.[11]

A significant number of Montreal's working class clustered in the central districts and the older fringe manufacturing districts. Low-paid service employment in the financial, retail, and commercial sectors and a large number of jobs in manufacturing attracted workers. For many working-class families dependent upon more than one income, a central location was often the most rational choice given their employment needs. For the large proportion living in poverty, with low wages or little job security, the price of the trolley was out of reach, and low rents were a primary concern. Most low-rent housing was found in areas adjacent to the central business district, although this became less common as the central housing market underwent changes. The most important dynamic leading to housing shortages and rent squeezes in the central districts was the invasion of the older residential districts by commercial, financial, retail, industrial, and transportation functions. The expansion of the central business and factory districts resulted in housing demolition, the deterioration of dwelling conditions, and pressure on rent levels.[12]

Several districts experienced these changes. In his survey of "the city below the hill" in Saint-Ann Ward, Harold Ames found that it was characterized by poor sanitation, rear tenements, high population densities, and few open spaces. A similar set of conditions developed in the once middle-class district of Dufferin, where single-family homes were converted to multi-occupancy residences or else torn down to make way for the construction of new office and manufacturing loft buildings. Ironically, these changes led to a situation in which rents per square foot may have been greater in the central districts than in the suburbs but, because of the high densities, those for individual dwellings were lower. The central districts were typically home to native-born and immigrant single men and seasonal workers, many of whom dwelt in the large number of boarding and lodging houses encircling the business district. Even though some of these men married and moved closer to employment sites on the urban periphery, new population streams from rural areas, other Canadian cities, and Europe found the cheap rents and close proximity to work a unique attraction.[13]

As the central district changed, many wealthier residents left the city for the more salubrious suburbs. Local suburban government and property developer alliances converted farmland on the city's periphery into solid middle-class areas. Pushing out from the existing built-up area of the city, more prosperous suburban districts swung northwest through

Westmount, Notre Dame de Grâce, and Montreal West and north through Outremont and the Town of Mount Royal. Housing and environmental conditions in these new suburbs contrasted sharply with large swathes of the central district. In Notre Dame de Grâce, for example, the trolley, developers, and the local government were responsible for building a white-collar, Anglophone suburb after 1900. The suburb's government constructed infrastructures and parks, installed essential services, and passed bylaws restricting the types and costs of housing. More systematically, and following the principles of the City Beautiful and the Garden Suburb, the Canadian Northern Railway built a model corporate suburb north of the city in the Town of Mount Royal. Complete with wide diagonal boulevards, curving side streets, an extensive system of parks, and large lots, this model suburb came to be a home of Montreal's English-speaking professionals. To ensure that it remained so, the town council, as in Notre Dame de Grâce, implemented strict site and design regulations, established modern sanitary services, and created community functions.[14]

The contrast between middle-class suburbs such as Town of Mont-Royal and Notre Dame de Grâce and the large working-class districts of the central city is a dominant image associated with North American urban social geography before World War II. But the middle classes were not the only ones moving to the suburbs. Although the scale of working-class deconcentration after 1900 has not been documented, evidence from various North American cities demonstrates that a significant number of suburban residents were working class. In some cases workers moved to the suburbs to be close to their place of work, while in others factories settled where a working-class labor force was already located. Without detailed and systematic case studies it is difficult to determine the exact causal relationship between manufacturing and working-class suburbanization in Montreal. Evidence from cities as different as Toronto, Knoxville, and Chicago, however, indicates that fringe manufacturing employment attracted workers' settlement. In Chicago, for example, from the formation of Packingtown around the suburban Union Stockyards after 1865, the building of the Calumet steel mills and Pullman in the 1880s, to the growth of the industrial suburbs of Clearing and Cicero in the early-twentieth century, workers followed firms out to the urban fringe.[15]

The importance of existing suburban labor pools for manufacturing, however, was not lost on contemporaries. As Hamilton's (Ontario) industrial commissioner, C. Kirkpatrick, pointed out in 1919, "the first concern of practically all large manufacturers is labor. The availability of a labor supply concerns him more than anyone other thing." For manufacturers an existing suburban labor pool created the possibility of a loyal and captive labor force and ensured that wages did not have to include transportation costs. This, as Kirkpatrick noted, was true for workers as well as for manufacturers. A worker, he opined, would prefer to live close to industry but not where factories were "staring him in the face every time he opened his front door." Workers wanted cheap and decent suburban housing, especially given the worsening conditions and rent levels of the central districts, and, frequently, it was to be found in industrial suburbs, although in some cases working-class residential suburbs developed in districts with little or no manufacturing. As builders of worker housing followed the exodus of factories to the metropolitan fringe, they also established working-class suburban communities that remained almost entirely residential. In these cases highly paid workers used the trolley to commute between their suburban residences and a central workplace, while others walked to their suburban workplace.[16]

All of these different types of working-class suburbs were to be found in Montreal before 1929. Even though the degree and type of separation of industrial and working-class residential areas in Montreal did not fulfil the requirements laid down by Kirkpatrick, a labor force coalesced around industrial nodes, as the city's housing market produced a highly differentiated and specialized set of suburban housing markets. In Montreal the older industrial-working-class suburbs such as Saint-Henri, Sainte-Cunégonde, and Hochelaga continued to grow (table 6.3), and between 1883 and 1918 many became part of the city through annexation. Completely new industrial working-class suburbs also appeared. Pulling workers from the central city and attracting migrants, new industrial suburbs such as Ville Saint-Pierre, Maisonneuve, and Montréal Est grew after 1900. In other cases older villages such as Lachine, with an existing labor force, experienced a transition from a rural-based economy to a manufacturing one. As these manufacturing districts formed on the urban fringe, workers sought out housing close to suburban factories.

A suburban mosaic characterized metropolitan Montreal, as an assortment of large development companies, small builders, street railway companies, and suburban municipalities transformed vast tracts of rural land into working-class housing. In some cases workers moved into housing built by the formal construction industry under its typical anarchic rules. Commonly, these suburbs were highly unregulated, although some, such as Saint-Henri, Sainte-Cunégonde, and Lachine, implemented municipal policies favoring business interests over those of their residents. In other cases land developers and civic leaders laid down a systematic policy of land development, infrastructure, and services. In contrast to planned middle-class suburbs, planned working-class districts such as Maisonneuve developed around an industrial policy based on widespread advertising and financial inducements. Large development companies were often the active force behind the opening of new peripheral housing developments in areas such as Saint-Jean-Baptiste, Saint-Louis de Mile End, and Maisonneuve, which provided extensive tracts of cheap working-class housing. After 1908, for example, the Park Realty Company sold more than three thousand lots in its Park Avenue Extension, in the city's north end, to an ethnically mixed working-class clientele, while the Rosemont Land Improvement Company developed the mainly British working-class suburb of Rosemont.[17]

Even though there was extensive suburban working-class house building, workers frequently had to resort to building their own homes. In Toronto self-building of a small shack on the urban periphery was the first step to home ownership and better housing conditions for many workers. Cheap land, few building restrictions, low building costs, and low or no taxes allowed workers to grasp a modicum of independence from the high rents and poor conditions of the central city. Although there has been no study of working-class self-building in Montreal, there is anecdotal evidence indicating that it was a part of suburbanization. Writing in 1911, Dr. E. Pelletier, the province's Health Council secretary, stated that Montreal had peripheral "areas of shack and sheds constructed of planks, covered in tar paper and held together with iron." Turn-of-the-century Kensington, at the western extremity of Notre Dame de Grâce, was a shack town. Italians in northern Mile End after 1900 built their homes out of discarded material and added on as means allowed. Further north, alongside the lines of the Montreal Park and Island Railway Company, workers were "buying bush lots, clearing the sites and building houses for themselves. Forty or Fifty humble dwellings, erected close together in one spot, resemble more a clearing in the backwoods than the embryonic suburb of a great city."[18]

Although suburban housing tended to be newer and in better condition than that found in the core, this was not always the case. Slums were found in the city fringe and the older suburbs as well as in the city's central districts. In the industrial sections of Notre Dame de Grâce and Rosemount the same type of slum-producing processes were at work as in the central districts. The economics of suburban land development, the poor quality of working-class housing construction, low wages, and job insecurity fed the process of poor housing and environmental conditions. Adding to these problems was that building space had been utilized without much thought to open space. Both in the center city and farther out, according to Mary Davidson's 1933 account, "homes are built close to the streets, with little space in the rear, just enough for the clothes lines. Thus the houses, built back to back, not only provide little air space, but absolutely no play space." Things were no better in 1924, as builders continued to "use almost the whole surface for their building and to cramp the open spaces surrounding them." Verdun, the fastest growing suburb in the first decades of the twentieth century, featured shoddy-quality, poorly designed housing and long and dull streets.[19]

Working-class suburbs were differentiated by ethnicity. The influx of a large number of immigrants after 1890, coupled with their position in the labor force, laid the basis for segmentation within the urban milieu. Contemporaries argued that Montreal's inner-city slums, among the worst in North America, were caused by the overcrowding of immigrants. There is no doubt that concentrations of Chinese, Italians, Syrians, as well as other ethnic groups clustered in the inner city. Significant numbers of immigrants, however, also lived in the older nineteenth-century suburbs and the new residential and manufacturing areas built after 1900. In the late nineteenth century, for example, Italian settlements developed around five nuclei, two of them on the western (Goose Village) and eastern (Tetreaultville area of Hochelaga) fringes of the city. By World War I an extensive Italian settlement in the north end along the Canadian Pacific tracks developed to take advantage of land for growing vegetables on the city outskirts and to be close to the Angus car shops, the city's largest employer of Italian labor. Likewise, Ukrainians, most of whom worked in the metal sector, had settlements close to the Grand Trunk shops in Point Saint-Charles and the Angus shops in the West and East Ends, respectively. Meanwhile, the large Jewish population moved north along the St. Lawrence Boulevard corridor, out of the central districts into the manufacturing district of Mile End and the adjacent residential district of Park Extension.[20]

The timing of new working-class residential construction was linked to the electrification of the city and its surrounding districts. Poor commuter links between the various parts of the city had impeded suburban expansion during the nineteenth century; as one commentator pointed out at the end of the nineteenth century, "there is plenty of land in the vicinity of Montreal available for choice residence lots, and it is plain that the question of better transit facilities must soon be taken hold of as one of the important elements of the land market." Ridership increased through the first decades of the twentieth century, allowing greater residential choices for some working-class families. Intra-urban electrification, however, was not problem free. Problems of municipal control over the trolley companies, a small number of lines, and unequal access to the trolley reinforced and accentuated existing patterns of residential segregation. In particular, low wages kept a large segment of the working class from using public transit. Even after the introduction of special workingman tickets the trolley remained inaccessible to many workers. For some, especially those who lived close to their place of work, this was not a problem. For a large number, however, the extending metropolitan boundary, coupled with industrial suburbanization, meant that their options were restricted.[21]

The housing and land markets' ability to satisfy working-class demand must not be overstated. Cheap and decent housing was largely unavailable, both in the inner city and the new suburban developments. Uninterested in providing low-rent housing, the local building industry sought the higher profits of the upper end of the housing market. The gap between demand and supply was obvious in the northern reaches of the city. There the construction of the gigantic Canadian Pacific locomotive shops produced several thousand jobs but did not immediately generate extensive working-class house building. Even when building was undertaken, it consisted "very largely of manufactories, places of business, public establishments and dwelling houses of a greater cost and size than is suitable to the needs of a workingman's family." Nonetheless, working-class suburbanization was a principal feature of Montreal's social geography in the first decades of the twentieth century, continuing a trend established in the previous century. While the central districts contained a large working force, a significant share of Montreal's working-class population lived on the city periphery and in new suburban municipalities. The territorial clustering of suburban factories, working-class residential settlement, and occupational and ethnic specialization after 1890 characterized the expanding urban fringe. Montreal's property market tied together the industrial and residential functions of the peripheral districts. In the process these clusters of workers provided an essential labor pool for manufacturers throughout the metropolitan area.[22]

"A Fringe of Painless Ugliness": Manufacturing Districts and Planning

For contemporary Montrealers the fact that the city was one of the most densely populated in North America was a matter of grave concern. As G. Ferguson brought out in his 1924 study of Canadian cities: "The endless conglomeration of annexed suburbs, where building had proceeded in hap-hazard fashion, is an expensive and most unscientific way to build up a city. It not only develops a fringe of painless ugliness near the city's rim that must later on be re-shaped to the general configuration of the urban area, but also adds seriously to the city's problems; especially those of transportation and public utilities." Many reformers, planners, and politicians looked to employment and population suburbanization as the solution to the overcrowded and deteriorated conditions of the city, as described by Ferguson. Even though Montreal experienced both population and manufacturing deconcentration, the filling up of the fringe belt districts after 1890 did little to alleviate overcrowding, inadequate sanitation, and poor housing. As one editorial writer put it in 1892, "it is unfortunate that the era of suburban development now promised from the modernizing of the street car service, does not also promise all that is desirable in the way of change." Similar sentiments were echoed in 1913 by Montreal's mayor, Arsene Lavallé, who announced the need to adopt a "definite and harmonious plan of extensions." In 1929 George Stephens, a leading local politician, repeated the litany of problems which had been sounded for decades: "We are letting [Montreal] grow by itself, without direction. Nearly all the major problems are ripening at the same time, City planning, zoning, railway terminals, underground means of transportation, street widening, building, reconstruction, traffic problems, all these have sooner or later to be faced."[23]

Montreal's metropolitan expansion, as in other North American cities, was not orderly and scientific, nor had suburbanization since the 1890s provided a satisfactory solution to the inherent problems of a fast-growing industrial metropolis. With few serious impediments to unregulated speculation and development, early-twentieth-century suburban growth continued to follow the haphazard patterns of the nineteenth century. According to one critic in the interwar period, for example, the new South Shore middle-class suburbs were "just like most of Montreal's suburbs . . . a jumble of sub-divisions, promoted by various real estate operators, whose only desire has been to profit by their operations." In his 1924 overview of Canadian industrial suburbanization G. Ferguson pointed to the lack of coordination of suburban development. Metropolitan growth proceeded hurly-burly as population and economic activities spilled haphazardly over the edge of the existing city boundaries and set up home, office, and factory in new suburban municipalities.[24]

Large-scale and haphazard growth was not new in Montreal; it had been taking place since the mid-nineteenth century. What was new was the extraordinary scale at which this was happening and the refusal of the city and the suburbs to regulate the emerging metropolitan district, either by developing coordinated planning methods or by imposing strict subdivision and building codes. Even though the province had granted Montreal the right to enact legislation to regulate and shape public health, transportation, residential construction, and town planning, it had done little to implement any formal regulation. According to the city's Junior Board of Trade, "Montreal . . . already has considerable powers over its own growth; it remains to correlate these powers and endeavour to bring out a planning scheme suitable to the needs of the City." Percy Nobbs, a leading Montreal architect and town planner, echoed this point, arguing that there was sufficient national and provincial legislation to ensure better urban conditions in the city. Montreal's problems, he believed, lay in municipal political and business elites' unwillingness and inability to enact the necessary changes. The absence of any systematic planning controls on the part of Montreal and its industrial suburbs before World War II ensured that unregulated and uncontrolled development remained a hallmark of the city's growth.[25]

Despite general agreement that the city needed to be remodeled in order to accommodate economic growth, there was little consensus among private and public elites about just how this was to be achieved. The emergence of a reform movement at the end of the century had little impact on the implementation of any systematic planning. Composed of business elites, professionals, technocrats, and middle-class women, the city's reform movement before World War II could not bring about rational planning. From its early stirrings in the 1890s, with the establishment of the Volunteer Electoral League, to the winning of local elections in 1900 and 1910, it was unable to address effectively the major problems facing the city. While reformers implemented some changes, including improving the aqueduct, reorganizing local administration, and regulating utilities, Montreal's development was only minimally controlled. Given legal loopholes, the small building department staff, and an absence of political will, building regulations were poorly implemented.

The fundamental problem was that reformers were deeply implicated in the contradictions of the city as a growth machine. They believed that to be competitive the city had to function as an efficient economic machine, and local government's main task was to keep taxes and expenditures down. But they also acknowledged the problems inherent in industrial expansion and the social inequalities of urban-industrial growth. Typically, the tensions between business interests and social equity were resolved in favor of business. Being members of, or working closely with, dominant business organizations, including the Board of Trade, the Real Estate Owners Association, Montreal Businessmen's League, and the Chamber of Commerce, reformers were locked into serving the needs of financial, manufacturing, and land interests. Their attempts to negotiate between the opposing views of utilities, industry, developers, and citizens often met with hostility from all quarters. The question of providing services, for example, was finessed by relegating the public good to private interests for the public good, almost always with public funds and state backing. The divisive nature of the reform movement, coupled with the predominance of private enterprise, often grouped together in growth alliances, prohibited any advancement in either curtailing haphazard urban growth or proposing measures to ensure its orderly development in the long term.[26]

Throughout the nineteenth century the only formal land-use controls introduced by Montreal and surrounding municipalities dealt with fire and nuisance industries. In some cases nuisance bylaws created to ensure the segregation of the most noxious manufacturing concerns led to manufacturing decentralization. Bylaw 129, for example, abolished slaughter yards within city boundaries, leading to the establishment of abattoirs, meatpacking, and ancillary industries in eastern and western suburban districts (Hochelaga and Saint-Henri). Other than these bylaws there was little direct and formal planning for industry in the nineteenth or early decades of the twentieth centuries. Indeed, the city's planning department was only created in 1941, while the first zoning bylaw was implemented in 1929. As late as 1944, only 19 of the city's thirty-five wards were subject to general zoning bylaws, many of which had been adopted as short-term measures. Consequently, the state's role in directing urban development was minimal, and private interests undertook development without planning controls. The considerable movement of factories from central to suburban areas took place without any coordinated planning at the metropolitan level, and the fact that Montreal and its suburbs neglected any form of organized or coherent planning had important repercussions for Montreal's manufacturing districts. As Montreal town planner James Ewing saw it, the lack of planning controls led to "factories springing up promiscuously all over the place." Factories were "spreading in their neighbourhood ugly, malodorous and grimy conditions, forcing up the cost of land but depreciating residential values with the inevitable consequence that buildings fall into disuse, decay and delapidation [sic], and the foul reeking cankerous slum is created."[27]

The lack of a plan or zoning was not for want of trying on the part of a group of Montreal's middle class who had been pushing for the adoption of a comprehensive plan and strict zoning codes since the early twentieth century. In a 1921 special issue of the *Journal of the Town Planning Institute of Canada* E. Deville, a Montreal doctor and chair of the institute, stressed the need for protection by separating residential districts from factory and commercial establishments. James Ewing agreed. A comprehensive plan, to his mind, should segregate "industrial, business and residential property to the localities where each would be mutually advantageous and least detrimental to the others." For these writers zoning was an imperative, not only because too many problems had already been created by the impact of factory districts on residential districts but also because the future of the rapidly growing metropolitan area was at risk. Similar to New York, Chicago, and Pittsburgh, Montreal's large-scale industrial growth reinforced a polarized society and intensified spatial inequalities within the city. In Montreal, as Ewing stated in his typically hyperbolic and distasteful manner, "the factories downtown have driven out all of the respectable residents leaving a lot of tumble-down shacks that soon become infested with foreigners and the improvident, dens of misery, squalor and vice, open cankerous sores, that eat into the body politic." Moreover, industrial suburbanization was compounding the city's problems, "for the factories soon find that the sources from which they have been drawing their labour supply is disappearing, and they start after it, invading the upper town districts and leaving in their trail the slum areas they have created, and the large derelict tracts that we see of unoccupied and unproductive land." For Ewing and Deville it was obvious why the city needed a comprehensive plan and zoning codes. Before 1929, despite reformers' efforts, Montreal, more than other North American cities, remained an unplanned metropolis.[28]

Montreal also had the reputation of having the worst conditions of North American cities. In the field of public health, as several contemporaries noted, Montreal was "a hygienic disgrace to civilization," and its high death rate was "due to the indifference of the public to all matters appertaining to hygienic engineering, and absolute weakness of the Legislature in refusing to frame laws of any hygienic value to the community." Behind the lack of decent and affordable housing were inadequate building bylaws. As Nobbs bemoaned in 1937, the bylaws needed a "radical overhauling. . . . The existing building bylaws of Montreal are saturated with obsolescence and crystallize at every turn the practice of about 1880." The same was true of the city's roads. In 1907 only 33 of the city's 200 miles of city streets were paved with permanent materials, and the state of the sewers was abysmal. Although this situation would get better over the next twenty-five years, Montreal continued to lag behind most other cities. As late as 1929, less than two-thirds of the city's 668 miles of streets were paved. The dismal nature of the city's environmental conditions was outlined by Julia Schoenfeld, the field secretary of the Parks and Playground Association of America. In 1913 she was given the task by the city's local association of making a survey of the city's recreational facilities. Her report was far from flattering. She found that Montreal's per capita park space was very low compared to large American cities. Whether it was in the area of public health, transportation systems, or park space, Montreal continued to lag behind most other North American cities.[29]

While there was little formal planning taking place within Montreal metropolitan area's ongoing urban-economic growth, the building of infrastructures continued to shape the location and nature of the city's industrial districts. One infrastructure contributing to the changing parameters of manufacturing location was the electric trolley. Paralleling trends in other cities, the Montreal Street Railway Company's system grew from just 13 miles of electric track at the end of 1892 to 80 in 1895, reaching 262 in 1918 and 305 in 1926. Similarly, the number of passengers increased from 46 million in 1901 to almost 240 million in 1928. With a much greater passenger capacity and a doubling of its speed, the trolley quadrupled the area reached by the intra-urban transportation network. The rapid extension of lines radiating out from the city core to fringe areas enabled the reorganization of urban settlement. Electrification and the extension of the trolley lines permitted expansion of the urban frontier, integrated adjacent suburban areas more tightly into the central city, and increased urban-wide construction activity and land values. Moreover, the trolley enabled many manufacturers to move from the city core to take advantage of the extended urban perimeter. The cheaper land prices of the periphery, coupled with the larger expanses of available land, made suburban locations an extremely attractive option to those manufacturers able to free themselves from the ties of the more central districts.[30]

The channeling of large chunks of capital into new infrastructure provided the basis for the development of new nodes of economic and residential activity. The harbor, extending from Saint-Ann in the west to Montréal Est in the east, underwent periodic remodeling as federal and provincial monies were secured by the harbor commissioners. Granting long-term contracts to private utility companies gave the Montreal Street Railway Company and the Montreal Light, Heat, and Power Company de facto monopolies in the Montreal regional market and paved the way for the extension of the urban built-up area. Developers and builders of middle-class and, increasingly, working-class areas laid down new sewers and water pipes as they opened up new developments. The creation of new infrastructures and the remodeling of existing ones established the direction of urban growth and opened up new options for industrial location, as they had in the nineteenth century. In the long period of expansion from the late 1890s through the 1920s, all types of firms moved farther out from the city core. In many cases large, propulsive firms led the way, while in others manufacturing clusters developed around existing small villages.

Manufacturing decentralization occurred despite the unorganized and ad hoc manner of metropolitan expansion. Even though new municipal leaders after 1890 "took pride in Montreal's growth," they "never entertained any arcadian ideas that growth should be tempered." To this end they advocated suburban residential expansion, the remodeling of infrastructures, limited regulation of local utilities, and further annexation. The ascendancy of French Canadians to municipal power and their search for individual profits through city institutions and urban growth resulted in the creation of new local institutions and alliances, battles organized along the lines of reformism and bossism, and more active intervention in the administration of metropolitan growth.[31] The machinations and profit seeking of utility capitalists, land developers, politicians, and corporate executives underwrote the creation of a new urban landscape between the 1890s and 1929. Although limited in their achievements, the local elites' support of relatively unbridled urban liberalism and expansion provided the political context for further industrial and working-class suburbanization. As a result, Montreal continued to develop as a large manufacturing complex composed of specialized and locally managed manufacturing districts.

CHAPTER SEVEN

"Pierced by Another Giant Skyscraper"

The Changing Fortunes of the Central Manufacturing Districts

> Montreal's ever-changing skyline is to be pierced by another giant skyscraper.
> —Montreal Gazette, 9 July 1929

Even though Montreal remained a formally unplanned city throughout the first decades of the twentieth century, the city's central manufacturing districts continued to be remade. One celebrated case that epitomized the upward thrust of the city's business district's buildings was the two-million-dollar Aldred Building. With a steel-frame exterior, it soared 23 stories and 316 feet into the sky—and it was not alone. In 1929 just before the big crash undermined the hopes of many office developers, several other large structures were constructed in central Montreal. Built only eight years earlier, the two-story Marcil Trust Building on St. James Street was demolished in 1929 to make way for a new nineteen-story building. Developers in Montreal, like those carried away by New York and Chicago's 1920s speculative office tower binge, sought the best returns of increasingly expensive central land.[1] While skyscraper construction in the 1920s symbolized Montreal's increasing financial and commercial importance, it also signaled the changing fortunes of the central manufacturing districts. The city's spatial division of manufacturing deepened between the last decade of the nineteenth century and the Great Depression. Even though many of the old trends continued, the economic geography of the central manufacturing districts changed.

"Moving Uptown"

In his address to the city's Civic Improvement League in 1925, James Ewing, vice president of the Canadian Town Planning Institute, pointed to the shifting nature of the central manufacturing districts' land use:

"The residential, commercial and industrial life of Montreal is moving uptown, and I ask you to note that it is not simply extending uptown, it is shifting, leaving behind the blight district, the noxious and costly slum areas and derelict abandoned lot." Ewing, who had long been concerned about the deteriorating position of central Montreal, wrote extensively after World War I on what he believed to be the hollowing out of the core and the "cancerous infection" forcing residential, commercial, and industrial functions to move uptown and to other parts of the metropolitan area. Although he exaggerated the situation, Ewing had, nonetheless, identified a trend that was well under way by the mid-1920s: Old Montreal had lost a significant amount of its economic variety and had increasingly become an office-based employment node. Even with the agglomeration economies spun off by the area's many small producers, the core mainly attracted office employment, reducing its appeal as a manufacturing center.[2]

Montreal's central district, like those elsewhere, continued to undergo a highly competitive process of land-use specialization, in which manufacturing was the loser. The finely textured set of nineteenth-century financial, commercial, retailing, and manufacturing functions clustered in the small central area leading out from the waterfront was unraveling by the turn of the century. Increasingly, an array of producer services associated with the national dominance of the city's financial sector took over space in the central district. So great was their demand for central space that vertical growth became increasingly important in Old Montreal. Beginning in the late 1880s, the form of central Montreal's skyline took a new turn with the development of the skyscraper. From the New York Insurance Company, Montreal's first skyscraper, constructed in 1887, to the Aldred Building, many low-rise business and residential buildings were torn down and replaced by high-rise ones. Not only could office functions outbid manufacturing for these locations, but the new multistoried and specialized buildings were unsuitable for most manufacturing activities.[3]

Together, these processes had the effect of increasing office employment and reducing manufacturing jobs in the core. Even though Old Montreal continued to be an important component of Montreal's industrial base in 1929, its superiority as a manufacturing center had waned appreciably. Between 1890 and 1929 the number of firms in Old Montreal dropped from 378 to 303, and the district's share of the city's rent fell from almost a quarter to just over 8 percent. These changes were not new to the twentieth century. Old Montreal's declining position as Montreal's premier manufacturing center over the first decades of the

Table 7.1 Manufacturing Structure of Old Montreal and the Outer Core, 1890 and 1929

Industry	Old Montreal 1890 No. of Firms	Old Montreal 1890 Rent Share (%)	Old Montreal 1929 No. of Firms	Old Montreal 1929 Rent Share (%)	Outer Core 1890 No. of Firms	Outer Core 1890 Rent Share (%)	Outer Core 1929 No. of Firms	Outer Core 1929 Rent Share (%)	Central Manufacturing District 1890 City Rent Share (%)	Central Manufacturing District 1929 City Rent Share (%)
Clothing	89	25.2	58	15.5	160	24.8	486	42.5	96.4	82.3
Printing	64	19.0	88	30.7	29	5.0	121	8.6	99.0	84.3
Metalworking	33	9.9	22	4.4	48	8.9	47	3.4	34.7	8.6
Food	29	8.7	29	18.4	73	15.6	41	14.4	45.5	32.4
Leather	23	8.5	16	6.9	59	14.1	53	5.6	88.4	50.5
Furniture	19	5.5	3	0.9	44	9.9	29	2.6	92.4	37.3
Jewelry	21	5.2	12	1.1	12	2.8	19	1.2	100.0	80.6
Tobacco	15	3.4	10	3.3	9	1.7	16	1.1	42.0	22.7
Paper	5	3.2	5	2.5	4	1.0	18	2.3	61.9	36.5
Wood	8	3.1	5	1.0	15	2.3	12	1.3	37.2	23.9
Chemical	5	1.9	18	6.1	11	3.3	32	4.0	48.4	31.9
Carriages	10	1.6	2	0.4	40	4.7	8	0.6	81.3	33.9
Textile	7	1.3	11	3.7	5	1.3	55	4.1	5.2	18.5
Blacksmithing	20	0.9	3	0.2	39	1.1	24	0.4	58.3	34.4
Electrical	—	—	3	1.0	3	0.3	8	1.6	12.9	17.5
All Firms	378	100.0	303	100.0	600	100.0	1,042	100.0	54.7	33.9

Source: Water Tax Rolls for Montreal and surrounding municipalities, 1890 and 1929.

twentieth century continued a process that was well under way by the last decades of the nineteenth century. In some cases, such as clothing, furniture, and jewelry, the drop was precipitous (table 7.1). Old Montreal's role as the city's major garment manufacturing center, for example, already evident by 1890, was gone by 1929; the fifty-eight firms accounted for less than 10 percent of the sector's total rent. Similarly, its share of the city's jewelry firms' rent had fallen from more than half in 1890 to 18 percent in 1929. On the eve of the Great Depression, Old Montreal continued to be home to a diverse set of industries, but, given the currents of metropolitan change, it contained fewer. While the top three industries (clothing, printing, and metal) in 1890 accounted for just over half of the district's rent, in 1929 the top three (printing, clothing, and food) accounted for more than three-quarters of the rent.

Despite its declining importance and Ewing's claim about the exodus of manufacturing firms to other parts of the metropolis, many companies remained in Old Montreal. This was especially true of industries such as light chemicals reliant on the chains of production which had made the area such an impressive manufacturing district in the first place. The pharmaceutical and druggist supply firm of Evan and Sons, for example, had begun with an assortment of functions. Established in 1860 as a retailing and wholesaling venture, it discontinued retailing and began manufacturing in 1884. Growing demand for drug preparations and photo supplies over the years led the company to enlarge its mill, laboratory, and shipping plant close to the harbor section, adjoining the financial and retailing districts. Even though Evan failed, in 1929 the plant, with its established manufacturing facilities, was operated by James Tellier, a dye and color manufacturer. Tellier was only one of the several chemical manufacturers that clustered on the eastern edge of Old Montreal; medicinal and perfume manufacturers such as Frosst and Kerry, Watson, for example, remained there to be close to the warehousing, harbor, and railroad facilities.[4]

Similarly, printing, clothing, food, metal, and leather companies in search of the same locational assets crowded along the edges of Old Montreal's central business area. Agglomeration forces continued to have a magnetic appeal, as in the case of Daoust, Lalonde's shoe plant and its move to Montreal from the small town of Acton Vale in the early twentieth century. Despite the advantages of Acton Vale's low wages, tax exemptions, and cheap land, management felt isolated from Montreal's bustle, information, and labor force. Deciding that Montreal's attractions outweighed the town's advantages, the company moved in order "to be in touch with other manufacturers and to be in a better position to secure labor." A few years later it moved into the old Ames-Holden factory on Victoria Square on the edge of Old Montreal. Daoust, Lalonde found the Victoria Square plant to its liking and in 1929 continued to produce shoes in a large factory, with a rent of $10,000.[5]

One of the areas to gain from the relative decline of Old Montreal was, as Ewing pointed out, the uptown districts of the Outer Core.[6] Building on their ability to attract a large number of firms before 1890, the three districts making up the Outer Core remained a major manufacturing magnet over the next forty years (table 7.1). Its 1,042 firms in 1929, an increase of 442 over 1890, accounted for 41 percent of the total increase in the metropolitan area's firms in this period. Despite this growth, the Outer Core's share of rent declined from almost a third in 1890 to a quarter in 1929, as suburban manufacturing districts received a substantial share of the city's large manufacturing firms. Growth among the three Outer Core districts was uneven. Both Saint-Antoine and Saint-Jacques experienced an increase in the number of firms and a decline in their relative share of the city's rents. In contrast, the number of firms in Saint-Lawrence increased by 262 percent (123 to 446), and the district's share of the city's rent almost doubled (4.2 percent to 7.6 percent). As with Old Montreal, the structure of these districts was not static. Industries shifted in importance over the period. In some cases—notably in clothing, jewelry, and printing—the Outer Core's share of industries increased (map 7.1). In other cases it experienced significant declines, as in furniture, carriages, and baked goods.

The move by the city's major retailing node to the Outer Core added to the importance of its manufacturing base. Even though Old Montreal retained a share of the city's retailers, land-use dynamics forced the development of retailing outlets to other parts of the city. Most notably, by the end of the century a new retail district was created farther north in Saint-Antoine, close to the wealthy neighborhoods encircling Mont Royal (map 7.2). A range of specialized retailers, including jewelry, furniture, shoe, and clothing, clustering around the large department stores settled on St. Catherine Street. By World War I the strip running along St. Catherine west of Guy Street was the major retailing section of the city, while a minor one had developed further east near Visitation Street. Following on the heels of the retailing dispersal to the Outer Core, several consumer industries gathered in the vicinity of the department stores. The new retail core attracted manufacturers because of cheaper rents, access to a wealthy and fashionable clientele, a citywide labor force, and proximity to the retail market while remaining within walking distance of Old Montreal's financial and wholesaling activities. Cochenthaler, for example, began operations in the 1870s as a manufacturer and dealer in watches, jewelry, plated ware, clocks, and diamonds in the heart of the financial district. In 1906 he moved from Old Montreal to the new retail district. Specializing in diamonds, cut glass, and solid silverware, Cochenthaler decorated the new premises "with the latest style of fixtures, and much after the pattern of the jewelry stores on Fifth Avenue, New York." Like McGarvey before him, Cochenthaler combined retail with manufacture: in the store's basement skilled workers combined handiwork and machinery in a modern factory.[7]

As in Old Montreal, the growing importance of office employment in multistory structures formed part of the changing geography of the manufacturing in the Outer Core. As the city's national importance grew, central business district functions spread to the Outer Core, especially the Saint-Antoine district. In their search for a central location and adequate work space for their white- and pink-collar workers, an increasing number of commercial, producer service, and financial firms

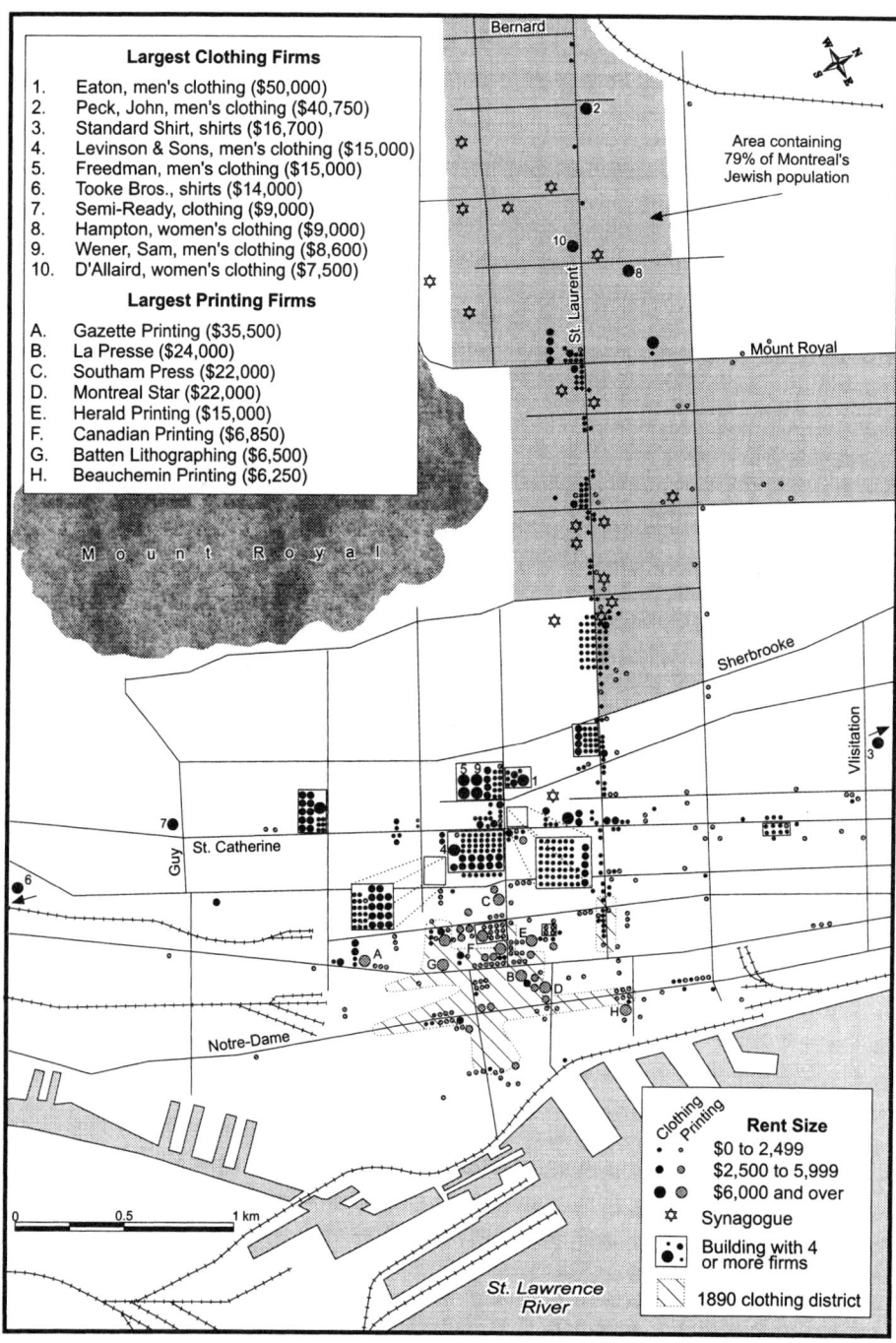

sought locations outside Old Montreal's high land values. As one commentator noted at the height of the prewar boom, "during the past two years Montreal has had a building boom. This has been accompanied by extensive speculation in real estate, enormous subdivision of outlying land, and a very pronounced advance in the price of city property." Rapid growth continued through the 1920s building cycle and further intensifed the pressures on the central city. Financial, legal, and commercial growth spurred the price of land and, in the process, squeezed out many manufacturers, both large and small.[8]

Bell Telephone, established in 1880 in two small rooms of the British Empire Building on Notre Dame Street, was a small, but rapidly growing, company, moving several times in its first fifteen years. In 1895 Bell built its first major office structure: a five-story building in the heart of Old Montreal. Thirty years later the growth of the telephone industry and the company's monopoly of the Canadian market forced it to seek new premises. Facing a combination of large space requirements and the need to escape Old Montreal's high land values, it built new headquarters on Beaver Hall Hill in Saint-Antoine district. With twenty stories and two basements, for the first time all of the company's executive and engineering functions were brought under one roof. Retaining the old building for public service functions, Bell made the new one the symbol of the modern company. Many of the company's sixteen thousand workers operated out of the new Beaver Hall building or out of the old Notre Dame one, both of which contributed to Montreal's ever-changing skyline.[9]

The growing number and increasing specialization of high-order functions forced a change in the geography of the central manufacturing districts' production space. Associated with the land squeeze were the revised space demands of manufacturers. By the end of the century the changing character of industry and the need to facilitate the faster circulation of capital through the firm led to a demand for new forms of plant layout. The attraction of the nineteenth-century manufacturing-warehouse district, with it cramped lofts, declined as firms looked for

---

MAP 7.1 MONTREAL'S CLOTHING AND PRINTING DISTRICTS, 1929. The geography of the clothing industry greatly changed between 1890 and 1929. Moving out of Old Montreal, it formed a major cluster around Bleury and Ste. Catherine Streets and stretching along the St. Lawrence Boulevard corridor. The printing industry, on the other hand, remained rooted to its nineteenth-century location. (Compiled from City of Montreal, Water Tax Rolls, 1890 and 1929.)

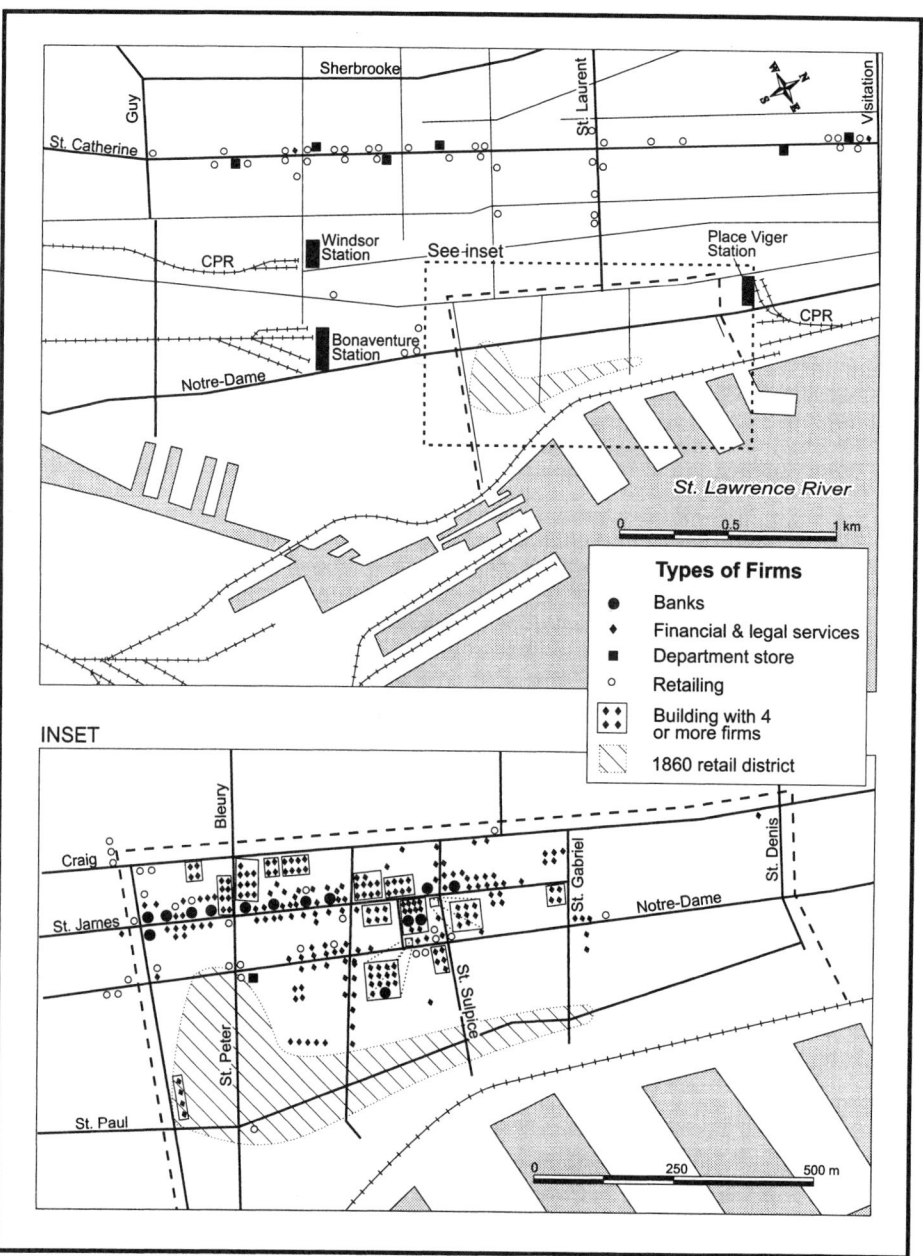

Map 7.2 Central Montreal's Retail, Financial, and Legal Functions, 1909. By the early years of the twentieth century the central business district had two nodes: retailing to the north along Ste. Catherine Street and financial and legal services in Old Montreal. (Lovell, *Montreal City Directory, 1909–1910* [Montreal: Lovell, 1910].)

larger loft space or a horizontal layout. Finally, a growing number of Montreal firms divorced office activities—accounting, buying, selling, and executive functions—from production and, in the process, fostered higher employment in the service sectors in Old Montreal, at the expense of blue-collar jobs. The offices of manufacturers with large factories in noncentral locations, such as Ogilvie Flour, Canada Sugar, and Canadian Rubber, clustered in or close to the financial district. William Clendenneng, for example, produced stoves, furnaces, and ornamental architectural products at his three-acre Griffintown foundry but located his salesrooms on the edge of the financial district. The two large Lachine firms, Dominion Wire and Dominion Wire Rope, had their offices close to Clendenneng's salesroom. Foreign companies also established their offices in the core. Babcock and Wilcox, the large American boilermaker, opened a branch office at Place d'Armes Square without opening a factory. The silk manufacturer Belding Paul undertook suburban manufacturing, while its office and salesroom were set up in the financial district.[10]

Slums and Boulevards: Making Way for Manufacturing

"The average civic authority must be made to understand that slums are far more expensive luxuries than boulevards, and that careful segregation of residential, business and industrial establishments will leave them more money to spend than they ever had before." In a pointed rebuff to the advocates of the City Beautiful movement, James Ewing's comments on the segregation of Montreal's economic and residential functions introduced the major themes of the City Efficient movement. Rather than taking the aesthetic and socially irresponsible approach in the Burnham style of broad boulevards, cities such as Montreal should be planned in a rational, comprehensive, and cost-conscious manner. Comprehensive zoning, to keep economic and residential functions separate, was central to this idea; it was the only way to provide the long-term savings and efficiency necessary to bring order to the burgeoning twentieth-century city. What Ewing and other planners of his ilk were responding to were the powerful thrust of soaring land values, extensive but uncoordinated construction, ever-shifting and congested land uses, and deteriorating residential districts. Continually subject to the property industry's boom and bust cycle, the city had difficult periods of growth and stasis. Montreal planners worried about the problems set in motion by the pre–World War I boom, one example being that the dynamics of the property market produced worse congestion and housing conditions. Another problem was overbuilding; according to a 1914 editorial in *Contract Record and Engineering Review*, "the number of large offices is out of proportion to the demand, with the result that both in the up-town districts and down-town districts there is an abundance of this class of accommodation."[11]

But offices were not the only form of work space in great supply in this period. The great boom in multistory building construction in the central manufacturing districts before the war and in the 1920s produced a significant amount of manufacturing loft space. As the shift of firms from Old Montreal to the Outer Core after 1900 reorganized the central districts' geography, the demand for light manufacturing work space in the districts adjoining Old Montreal increased dramatically. So great was the demand for manufacturing space in the Outer Core that older structures, usually dwellings and one- to two-story business buildings, were demolished to make way for buildings containing manufacturing firms. The multistory Jacobs, Blumenthal, Kellert, and Wilder Buildings were built on St. Catherine and Bleury Streets before World War I. This kind of construction, according to one writer, marked "the progress of the ... transition of the uptown section from the two and three story era into the eight and ten story era." Similarly, a burst of building in the 1920s resulted in the extensive construction of loft manufacturing space and furthered the transition of the uptown section.[12]

While the development of multistory buildings in the Outer Core had been taking place in the nineteenth century, the extent of verticality and the speed at which conversion took place in the twentieth century accelerated greatly. In Old Montreal multistory buildings to provide offices for the expanding corporate economy were constructed, spurring the development of manufacturing space in the Outer Core. In particular, small commercial enterprises and working-class dwellings were demolished to make way for the march of manufacturing firms out of the central business district. In some cases speculative developers built multistory buildings containing a large number of wholesaling and manufacturing spaces and rented them out to individual firms. In other cases large corporate manufacturers built a single-use factory suited to the specific needs of the industry. The Benson and Hedges firm, for example, moved into a specially built five-story building taken up entirely by the manufacture of cigars, cigarettes, and tobacco in 1929. Unlike the tobacco company, some firms frequently occupied a part of the building and rented out the rest of the work space to other firms, usually in the same or allied industries. Large-scale building demolition followed by manufacturing construction resulted in a patchwork of loft manufacturing space forming a crescent around Old Montreal. Although they did not reach the heights and densities of Manhattan or the manufacturing districts adjacent to Chicago's Loop, commercial and manufacturing buildings in the Outer Core went higher and occupied more of the lot after 1900 than ever before. In the process parts of Saint-Antoine, Saint-Lawrence, and Saint-Jacques stretching around Old Montreal were transformed, and, as manufacturing extended uptown, the residential character of the zone declined.[13]

Buildings containing a mix of dwellings, retail establishments, and small manufacturers along a stretch of St. Maurice Street were torn down to make way for the building of the Salvador plant of National Breweries. Having a specific role in National Breweries' interplant task division, the new Salvador plant deployed new machinery and older brewing equipment transferred from its Dawes subsidiary to manufacture specific product lines. Over the years the company extended the building and increased its land coverage from 35,000 square feet in 1907 to more than 65,000 in 1929. The Wire and Cable Company constructed a plant in 1899 with a layout "more efficient, certain, rapid and more easily manipulated than those of old-fashioned manufacture" on land vacated by nineteenth-century residential and commercial buildings. To the east, in Saint-Lawrence district, houses on St. Alexandre Street were pulled down to make way for two large buildings. One was the Gillette Safety Razor Building, where the American company kept its Canadian works and offices. The other, the publishing giant Southam Press, produced a range of publications and a newspaper on the top floors of the nine-story Herald Building (fig. 7.1). Located right in the heart of the major printing cluster in the city, Southam rented space to other firms, especially those in the printing, ink, and publishing industries. Finally, although it lay due north of the central business district, Vitre Street throughout the nineteenth century had remained impervious to business intrusions: in 1901 it contained forty working-class dwellings, a few small retailers, and no manufacturing firms. By 1907, however, the houses had been razed, and J. Black, a subsidiary of the large clothing corporation Canadian Converters, built a five-story loft building housing its women's clothing factory as well as other printing, clothing, cigar, paper box, and textiles firms.[14]

Extensive land-use conversion in the Outer Core and, to a lesser extent, Old Montreal provided working space for a large number of manufacturers. Although individual firms had their own particular reasons for locating centrally, the importance of the central manufacturing dis-

THE HERALD BUILDING, LARGEST NEWSPAPER PLANT IN CANADA.

FIGURE 7.1 HERALD BUILDING, 1915. One example of the multistorey loft building boom spreading through the central districts after 1900 was the Herald Building, constructed in 1915. Along with Canada's largest newspaper plant the building housed an assortment of small, light industry firms. (Lorenzo Prince, *Montreal Old and New* [Montreal: N.p., 1915], 374.)

tricts continued to rest on the reconstruction and expansion of critical infrastructures. In keeping with Montreal's business and political elites' general disdain for any systematic planning of the city, there was little long-term control over the central district's form. The building of the city core through public and private capital investment was undertaken in a piecemeal manner. Some basic services, such as the laying of modern water and sewer systems, were provided at a citywide level after 1900. The redevelopment of streets, however, was less systematic. On the city fringe streets were laid down by developers in the typical grid fashion, yet in the central manufacturing districts the old street pattern, conditions, and widths remained an obstacle to creating a modern built environment. The new multistory buildings and increasing congestion forced the construction of a new system of streets. Unlike in other areas of the older city, local government sunk capital on a piece-by-piece basis into the redevelopment of central streets—a process that involved destroying many older streets and neighborhoods and opening up those central areas "cramped in between Dorchester street and the river, and between McGill street and Dalhousie square." Thus, unlike other areas of the older city, local government, encouraged by local businesses, demolished many of the area's "narrow lanes, with their dull mediaeval aspect" and replaced them with "broad streets." Individual bylaws, for example, were made for widening St. Lawrence and Notre Dame Streets, the costs covered equally by the city and the proprietors. Despite these changes, many of the inner-city's areas continued to be extremely "compact" due to the lack of concerted efforts, the high costs of redevelopment, and the general environment of inaction. But, as long as land costs and space demands from a variety of users continued to exert pressure, this problem would continue.[15]

The harbor remained an important locational asset of the central districts. Even though expansion and remodeling had been a frequent feature of the port since the 1840s, changes taking place after 1900 added another layer of modern harbor facilities, attracting an assortment of manufacturing interests. In 1898 Israel Tarte, the Public Works minister, initiated a major redevelopment of the harbor. Completed by World War I, the federal government funneled more than $32 million into harbor improvements. The aim, as in earlier bouts of redevelopment, was to maintain Montreal's position as one of the leading North American ports. With the support of most segments of Montreal's commercial, financial, and manufacturing bourgeoisie, the harbor commissioners acquired the necessary capital funds and funneled a significant amount into the area bordering Old Montreal. Incessant competition from other cities led to the construction of a seawall to keep out floods and to building more quays. Fourteen permanent steel and concrete storage hangars were built between 1904 and 1908 and another seven between 1911 and 1914, which allowed the port to store grains and other commodities coming from the West, via the Saint-Lawrence River and bound for markets in Europe. Addressing the growing need for new and more effective installations to transfer the shipped raw materials, antiquated floating grain elevators were replaced by two fixed ones. And, by taking over the management of the railway tracks crisscrossing the wharves, and extending them, the commission was able to provide a more efficient circulation of goods and people throughout the harbor complex. Finally, the commission extended or updated the port's other facilities, including building a large dry dock, a coal tower, a sawmill, and a floating crane and buying new tugboats. The harbor changes implemented by World War I allowed Montreal to remain competitive as a major North American shipping center for another generation. More important for the story here, harbor rationalization reinforced the central districts' manufacturing locational assets.[16]

Adding to these assets was the linking of central railroad facilities to suburban railroad sites. As the railroad center of Canada, Montreal had an extensive network of freight yards, manufacturing works, and repair shops. Many of the newer yards and shops were located out on the urban fringe, their locations dictated by the requirements of large amounts of land, new layouts, and faster access to nonlocal markets. The central core, however, benefited from the decisions made by the two giant railroad companies, Canadian Pacific and Canadian National. As in other North American cities, the firms kept their major passenger terminals and services on the edge of the central core. While the advantages of centrality underpinned this, railroad facilities located there also provided advantages for firms remaining or wishing to locate in the city core. On the western and eastern boundaries of Old Montreal Canadian Pacific Railway's two passenger terminals (the Windsor and Place Viger Stations) were linked to the Sortin and Hochelaga freight yards and the harbor commissioner tracks strung along the harbor front.[17]

Private utility companies distributed the blessings of new electrical technologies throughout downtown and uptown, connecting the central manufacturing districts with the wider metropolitan, regional, and national scenes. The telephone, for example, reinforced the core's control over the area by immediately linking it with different parts of the city. The use of the telephone grew slowly over this period; by 1915 there were only 51,201 stations. Even though the number of residential lines increased, their share of the market remained very small, at about 4 percent of households in 1915. Most of the lines linked bourgeois neighborhoods with one another and with downtown offices. Bell's main purpose until the 1920s was to facilitate business control by linking businesses to one another. In the process these investments in wiring the city consolidated Old Montreal and the commercial-manufacturing districts of the Outer Core as the metropolitan area's communication center. As early as the twentieth century, businesspeople could easily communicate with one another and between their offices in the multistory buildings lining St. James Street, Place d'Armes Square, and Beaver Hall Hill and their warehouses and factories throughout the metropolitan area.[18]

Similarly, the electrification of the streetcar added to the consolidating tendencies of the central manufacturing districts. In particular, the growing reach of the trolley after 1892 made possible longer commutes by white-collar workers employed in the central office towers; the electric street railway facilities converged at Place d'Armes Square in the heart of the banking district. For the mass of blue-collar workers, however, the impact of the streetcar was minimal. To cope with increasing rents, overcrowding, and dismal sanitation conditions, there was growing clamor for the Montreal Street Railway to extend its lines as far as possible into the outlying districts, with the object of attracting those from the crowded central parts of the city and to build up the suburbs. As one commentator stated in 1892, "a complete suburban system of railways to all the surrounding parishes, north, south, east, and west, is rapidly becoming a necessity, in view of the high cost of living in the city." Twenty-two years later the calls for "a means of rapid transit throughout the city of Montreal and connecting up with points outside the city" were still being made, especially the need for bridges and tunnels to connect the expanding suburbs of the South Shore with the city core and for the construction of a rapid approach between the inner city and districts to the north and the west. Nevertheless, poor transit links between the inner city and the outlying districts continued to hinder commuting and to force the poorer segments of Montreal's population, dependent upon centrally located jobs, to remain in the inner districts.[19]

Before the Great Depression Montreal's central districts remained very densely populated, causing residential congestion, overcrowding, and high rents. "The majority of tenement tenants" in Montreal, as one commentator noted, were "paying far more for their two floors than people in Toronto pay for a 'self-contained' home and bit of garden." Moreover, as the 1935 *Report on Housing and Slum Clearance* clearly showed, the largest mass of slum housing encircled Old Montreal. Interspersed with the central districts' harbor, railroad terminals, and manufacturing areas were vast stretches of working-class neighborhoods. Composed of workers with an array of skills and from a variety of nationalities, these neighborhoods offered some of the cheapest (and poorest-quality) housing in the city and easy access to the office and manufacturing jobs of the core. Even though the central districts' population grew very little between 1901 and 1931 (remaining stable, at just over 140,000), it provided central firms with a large and accessible manufacturing labor force. Despite the static nature of the population, different areas experienced change; while some grew, others experienced a sharp decline. The population of Saint-Lawrence ward, for example, fell 18 percent between 1911 and 1931, as dwelling units were razed to make room for office and manufacturing use. The assorted ethnic character of the central districts continued to be a defining feature. In 1931 more than 71 percent of the population were of French and British origin, the rest mainly from southern and eastern Europe. Although there were no ghettos, inner-city neighborhoods, as in the late nineteenth century, differed in their ratio of French, English, Italian, Jewish, and Eastern European residents. While Saint-Jacques was more than 90 percent French, Saint-André and Saint-George were more than two-thirds English, and Saint-Louis was more than half Jewish. Significant clusters of Eastern Europeans were found in Cremazie, Saint-Laurent, and Saint-Louis. Together with the redevelopment of the built form and an array of transportation facilities, a large labor force underlay the agglomeration economies critical to the central manufacturing districts.[20]

Moving the Tobacco and Clothing Industries Uptown

Tobacco manufacture illustrates many of the dynamics behind the changing geography of Old Montreal and the Outer Core. Beginning in the nineteenth century, large tobacco manufacturing firms moved to suburban sites to acquire certain locational assets. In the 1870s Macdonald Tobacco moved out to an East End greenfield site, where it installed a high-volume, mass-market productive pathway on cheap land. In the early years of the twentieth century Imperial Tobacco built Canada's largest cigarette factory in Saint-Henri. Yet very few firms followed Macdonald and Imperial out to the suburbs. For most the advantages of the central districts remained paramount. In contrast to Montreal's two large decentralizing tobacco firms, the cigarette firm of Landau and Cormack and the Samuel Davis and Louis Grothé cigar factories illustrate the blending of the central manufacturing districts' productive and spatial strategies.[21]

Registered in 1907 by two experienced tobacco men, Charles Landau and James Cormack, and a tobacconist from Halifax, who provided some of the working capital, Landau and Cormack Company worked exclusively on the manufacture of high-grade cigarettes. To the owners, given their experience in the industry, it would have been obvious that the firm could not compete with high-volume producers deploying a massive distribution network, licenses to modern machines, and a large number of standard product lines. But producing high-grade goods could give them a niche in the cigarette market. This decision had several important implications for the organizational structure of Landau and Cormack. In the first place it required that Landau and Cormack's high-grade brands—Celando and Kalifa—be handmade, although this did not rule out the use of machines. As a writer describing the company in 1907 put it, "all the work is done by hand with the exception of the cutting and sifting, which is done by the latest improved machinery." In the cutting room two machines separated and cut the Turkish tobacco leaf. But it was the actual making of the cigarette, not the separating and cutting, which was the critical stage: hand bunching and rolling constituted the part of the production process differentiating high-grade from the standard lines of the high-volume producers. Landau and Cormack's principal concern was "to supply the dealer with a cigarette equal to the best imported." In other words, they attempted to reap profits by concentrating on a particular niche of the market through batch production of high-grade cigarettes and the lower cost of Canadian cigarette manufacture.[22]

This required that skilled workers form the core of the labor force. The majority of Macdonald and Imperial's labor force consisted of semiskilled women and men working as adjuncts to the high-volume Bonsack machines. Landau and Cormack, in contrast, built their reputation and cigarettes with skilled immigrant workers. Recruited from Egyptian factories, the company's expert cigarette makers rolled as many as 2,000 cigarettes a day. Although it was a small number by the standards of Macdonald or Imperial, where a team of three operating one machine produced 160,000 cigarettes a day by the first decade of the twentieth century, the high retail cost of Landau and Cormack's cigarettes more than made up for the small output. Moreover, not only were the Egyptian skilled rollers extremely proficient at making a high-grade cigarette; they were also experienced working with Turkish tobacco, the primary raw material for the company's cigarettes. A third component of their productive strategy consisted of specialization in Turkish and Egyptian blends, a market segment not covered by the big companies. Landau and Cormack blended their own brands, hiring experienced blenders to supervise this stage of production. The firm's specialization in turn led to its securing the agency of Mirama and Company in 1908. The Cairo manufacturer was unable to compete in the Canadian market because of the high import duty. To circumvent this barrier, it sent its tobacco to Landau and Cormack, which manufactured cigarettes under the supervision of one of Mirama's representatives. Finally, unlike high-volume producers, Landau and Cormack specialized in a market niche and concentrated their efforts upon a small range of products. As late as 1916, the company produced only ten brands of cigarettes and four brands of cut tobacco.[23]

Landau and Cormack's production pathway shaped its spatial strategy. Unlike the large firms setting up their plants in the industrial suburbs, the company located in the heart of the city's old cigar-making and warehouse district. The Lemoine Street plant, located in a typical four-story warehouse building, had 6,000 square feet of floor space (fig. 7.2). On the first floor were the shipping and office facilities. The second floor was made up of the cigarette making, inspecting, and packing departments. On the third floor were the rooms for separating, grading, and blending tobacco leaf, while raw leaf was stored on the top floor.[24] The building met the needs of this traditional and layout; leaf entered the firm at the top floor and made its way down in well-defined sequences to the shipping department on the ground floor. Furthermore, the agglomeration economies found in the core were important considerations in Landau and Cormack's locational strategy. Despite increasing land values, office construction, growing congestion, and the move of manufacturing firms out of Old Montreal, the company relied heavily on the core's linkages to wholesalers and financial institutions, access to a large labor market, and transportation facilities. In contrast to large suburban manufacturers, it found Old Montreal's advantages difficult to pass up.

The same pressures compelled cigar firms to remain rooted to a central location. As late as 1890, most cigar plants congregated in Old Montreal. In the following years, however, many firms moved out to those parts of the Outer Core that were shifting from residential to manufacturing uses, and by World War I these areas formed the major cigar center of the city. Two of the city's largest cigar firms, Louis Grothé and Samuel Davis, experienced the pressures giving rise to this new geography and moved uptown because they felt compelled to expand their

FIGURE 7.2 LANDAU AND CORMACK CIGARETTE FACTORY, 1907. Landau and Cormack manufactured a specialized assortment of high-grade cigarettes in a recycled building in the heart of Old Montreal's warehouse district. (*Canadian Cigar and Tobacco Journal* 13 [July 1907]: 62.)

works and to restructure the internal layout of their factories in the face of growing competition.

Louis Grothé learned the trade by apprenticing with a local cigar company. Starting off on his own in 1877 in a small shop in Old Montreal, he moved in 1888 to allow for expansion. Very soon, however, the limitations of a central location drove him to seek factory space elsewhere. In 1901, after a series of meetings, the northern suburb of St. Louis de Mile End granted him a $16,000 bonus, but, for reasons that remain unclear, Grothé did not accept the offer. Yet he was still itching to move out of his old plant. Five years later Grothé bought 22,000 square feet of land in the Saint-Lawrence district, where he built "one of the largest and finest cigar factories in Canada" (fig. 7.3). Five floors high, it housed six hundred workers and more than 100,000 square feet of floor space; the first floor he used for offices and storage, while the rest of the building was devoted to cigar manufacture. The new factory was built to accommodate expanding production and to double the workforce of the old plant, but, just as important, it was meant to create a more efficient flow of material and use of labor. One writer in 1913 exclaimed that the plant "is celebrated for the economic arrangement and splendid equipment.... A visit to their plant constitutes a liberal education in modern cigar manufacturing methods." An important strategy, given the industry's absence of extensive mechanization, involved stepping up production and exploiting labor more intensively through increased specialization of tasks and greater speed of the transfer time between tasks. Grothé's new factory was built to maximize this process. So successful was this strategy that the company not only extended its workforce and output but expanded by opening a branch plant in the nearby industrial town of St. Hyacinthe in 1911 and a cigar box factory just around the corner from the main factory in 1917.[25]

The largest cigar manufacturing firm of the day was Samuel Davis and Sons. Davis moved uptown at the turn of the century for the same reasons as Grothé but in 1884 a fire forced Davis to find new premises. He purchased an old church property in Saint-Lawrence and built a seven-story factory complete with elevators, telephones, and new equipment. In 1895 American Tobacco took over the factory, and Davis moved to a new location in the West End. Two years later, unhappy with the conditions there, he moved to the Saint-Antoine district. The five-story premises presented the firm and its eight hundred workers with "facilities which were impossible in the old quarters," including more space, the ability to implement greater task segmentation, and the opportunity to be close to the company's box factory in Saint-Lawrence. Another fire

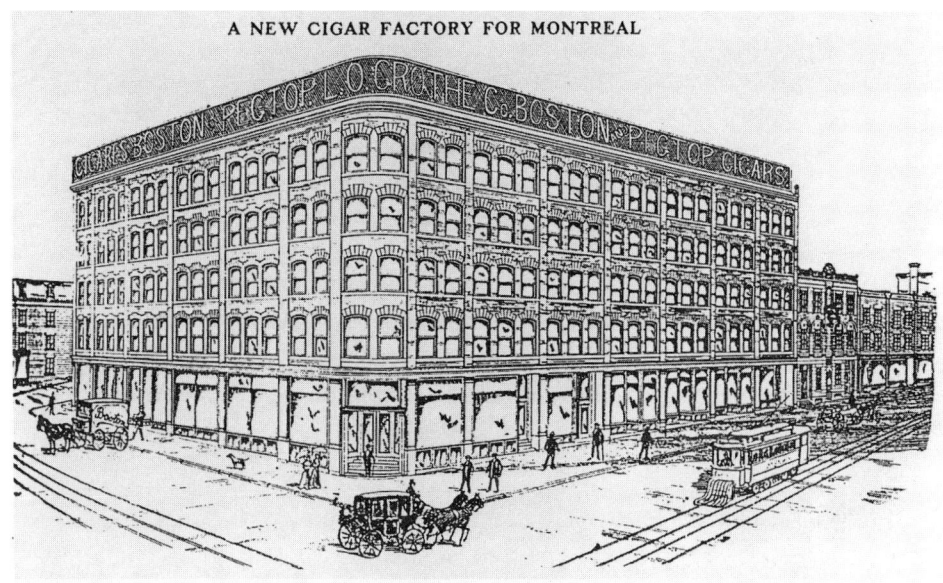

FIGURE 7.3 L. O. GROTHÉ CIGAR FACTORY, 1906. One of the leaders in the move uptown, Grothé built a large and modern factory where he could manufacture cigars under new work conditions. (*Canadian Cigar and Tobacco Journal* 22 [June 1906]: 19.)

in 1905 allowed the company to seek larger and more suitable premises. The new Saint-Antoine factory had the typical requirements of a large-scale cigar factory of the time: a goods elevator running from top to bottom and job specialization by floor (fig. 7.4). What makes the Davis company even more interesting is that the second floor of the building was "entirely shut off from all communications with the other factory [the rest of the plant], practically making two factories under one roof." The second floor housed the Cuban department, where skilled, immigrant hand-rollers made high-priced "Havanas." Having more clout than other workers there, they were more likely to go on strike and thus were kept separate from the other eight hundred workers. Accordingly, "the Cuban factory is reached through an entirely separate entrance, a separate staircase being provided for the employees of this branch."[26]

Garment manufacturers faced a similar set of structural pressures as cigar manufacturers. Most important, the organizational dynamics of the industry changed, and by the end of the century the clothing industry underwent the transition from rural outwork to the contract system.

FIGURE 7.4 SAMUEL DAVIS CIGAR FACTORY, 1898. After several moves Davis settled in an newly built, modern, five-story mill building in Saint-Antoine District. (*Canadian Cigar and Tobacco Journal* 4 [June 1898]: 151.)

This involved a much more detailed division of labor and the decline of a wide assortment of products manufactured by individual firms. It was still common in the 1880s to find firms, such as Montreal Whitewear, manufacturing "innumerable designs and varieties of novelties"; their selection of aprons, for example, "ranges from the cheapest cotton to the finest lace and satin, embracing about 100 varieties." Similarly, hundreds of women employed on sewing machines at the Small factory made "every style of goods, from the lowest classes up to high class goods which heretofore had been only procurable of custom make, and were never attempted in ready-made lines." The volume of in-house production by these firms was low, however, compared to the amount done through rural outwork. Aside from those in his Montreal factory, Small also employed "several hundred in the surrounding districts who do their work at home." According to Ernest Chambers, outwork was still common practice in 1903: "the Montreal wholesale clothing houses have availed themselves of . . . extremely industrious and very adept needlewomen . . . and, as a result, thousands of farm houses throughout the province of Quebec may now be described as branch workshops of the great Montreal distributing houses."[27]

By the end of the century, despite Chambers's claim, the vast assortment of products made by individual firms and the dominance of rural outwork were in eclipse. The dynamics of the industry were changing, and rural outwork waned as the contract system became the predominant mode of production. The transition from outwork to contracting occurring at the end of the century was driven by four factors. First, contractors produced a better-quality product than home workers. Second, not only was the distribution of raw materials and the collection of finished goods cumbersome; costs were also high. Third, contractors took over responsibility from manufacturers for organizing, training, and supervising the workforce. Finally, subcontracting permitted greater specialization of work by allowing better use of machinery through internal economies of scale and a more refined technical division of labor. As long as manufacturers continued to face the problems of seasonal demand, rapidly changing styles, and great competition, the contracting system proved to be a viable organizational form. To establish oneself as a contractor was easy, as starting up only required a small amount of capital: with trained labor, tight task specialization, and control over raw material and distribution costs, the contractor could go into business. By the end of the century many people, especially Jewish immigrants, did just that, and very quickly there developed in Montreal numerous small firms that manufactured for the larger factories and wholesalers.

At the same time that a shift took place from outwork to contracting, the large clothing factory (inside shop) began to play a more important part in the city's industry. Specialization in specific lines (especially shirts), the ability to contract work out, tapping the expanding mass market for ready-made clothing, and creating an extensive set of internal economies drove the development of large factories, dating back to 1870s and 1880s, when they began to compete with firms practicing outwork. By the end of the century the large factory functioned both as a competitor with and as a supplier to the contracting system. According to one observer in the 1880s, the business of the machine button-holer was "slowly departing" because the large firms "now run their own machines and no longer send their work out to be done."[28]

The changing industrial dynamics and the resulting different organizational forms after 1900 forced firms to deploy a spatial strategy different from that of the nineteenth century. Old Montreal was the locational center of the industry when the small custom shop, outwork, and the wholesale-manufacturer dominated the industry. In 1890, with a third of the city's firms and 42 percent of its rent, Old Montreal was an important locus of the city's clothing and ancillary industries. Beginning in the last decades of the nineteenth and accelerating after 1900, however, the industry moved out of Old Montreal. The major thrust of the industry's shifting geography was to move uptown. By 1929 the industry clustered in a central node at Bleury and Ste. Catherine Streets and from there followed a corridor running along St. Lawrence Boulevard north to the North End district of Plateau (map 7.1). This area in 1929 accounted for 80 percent of the city's clothing firms and 78 percent of the city's clothing firms' rent. Old Montreal, with its 9 percent of firms and rent, had declined appreciably.

Many clothing firms were forced out of Old Montreal by the expansion of the central business district. Dependent upon central agglomeration economies, many clothing firms found a home in the new six- to ten-story loft structures of the Outer Core after 1900. As early as the prewar period, buildings such as the Blumenthal accommodated a host of clothing and ancillary establishments. The clustering of clothing firms in multistory loft manufacturing buildings continued through the 1920s. The ten-story Gothic-style Caron Building, built in 1924 on the corner of Bleury and Concorde Streets, for example, housed thirty-four small and medium-sized clothing firms in 1929 (fig. 7.5). Built with the clothing manufacturer in mind, it came equipped with several freight and passenger elevators and had, for "firms requiring it in their manufacturing processes, a live steam and vacuum system (with outlets on each flat)." Moreover, the ten floors could "be divided to suit tenants' requirements." Several others went up, offering similar facilities as the Caron, such as the 1925 six-story Amherst Building (fig. 7.6).[29]

The concentration of the clothing industry in the uptown district and the St. Lawrence corridor was linked to the residential pattern of its

FIGURE 7.5 CARON BUILDING, 1924. Located close to the important intersection of Bleury and Ste. Catherine Streets, the Caron Building is representative of the multistory loft structures constructed for manufacturing after World War I. (*Contract Record* 38 [23 April 1924]: 393.)

workforce. The growth of the contract shop was tied to Jewish immigration beginning in the 1880s. Most Jews did not live directly in the heart of the garment district around Bleury and St. Catherine Streets but were found just north in the St. Lawrence corridor, running as far north as the Canadian Pacific Railway line in Mile End. Accounting for more

FIGURE 7.6 AMHERST BUILDING, 1925. Typical of other buildings constructed in the 1920s, the Amherst Building, with it special facilities, was designed with the needs of manufacturers in mind. (*Contract Record* 39 [7 October 1925]: 977.)

than 79 percent of the city's Jewish population in 1931, this area contained most of the group's synagogues, business and labor institutions, and retailers. This area was close to the garment district and thus it was a short journey to work for much of the industry's labor force. As well as the substantial Jewish male labor force, a significant portion of the industry's workers, especially in women's lines, were French-Canadian women. Even though these women lived in all parts of the city, they tended to reside in the East End neighborhoods adjacent to the garment manufacturing district. For the most part they were young unmarried women living in their parental homes, but in some cases they were married women bringing in wages to supplement the meager income of their husbands. As cheap, exploitable labor living close to the industry, both Jewish men and French-Canadian women were extremely useful to the subcontractor.[30]

While many large clothing firms, such as Eaton's, continued to find the core attractive, some of the large specialized, capital-intensive clothing factories began moving to suburban greenfield sites from the end of the nineteenth century. Just like the large cigarette makers, Macdonald and Imperial, specialized clothing firms such as Tooke, Peck, and Standard Shirt moved to the West, North, and East Ends (fig. 7.7). There were compelling reasons for the move to fringe sites, as the case of Tooke illustrates. Beginning as a small firm in 1870 in Old Montreal, Tooke moved to Saint-Henri in 1899, enticed by a bonus, large amounts of space, and cheap and abundant female labor. Based on specialization in one clothing line, not dependent on contract work, and with the internal economies of scale to produce effectively in a suburban site, Tooke quickly grew to be Canada's largest manufacturer of men's shirts. On the eve of World War I it had extensive national coverage: the head office in Montreal, factories in Lachine and Saint-Henri, and seven warehouses stretching from Halifax to Vancouver. By the 1920s it controlled several ancillary firms, including paper box and handkerchief makers. But even large-scale, specialized producers such as Tooke faced serious problems. In the mid-1920s the firm had to rationalize. It closed down the Lachine plant and moved its clothing and paper box subsidiaries from the central clothing cluster to its main plant at Saint-Henri. Although most clothing firms remained rooted to the central cluster and the St. Lawrence corridor, Tooke was not alone in moving to the suburbs. John Peck, the second largest clothing factory in the city, was located in the North End manufacturing district of Mile End, while Standard Shirt moved out to the East End. In their move to suburban greenfield sites Tooke, Peck, and Standard exemplify both the different spatial strategies associated with various types of garment manufacturing practice and the drive to decentralization which featured so prominently among the rest of Montreal's manufacturing sector.[31]

By the turn of the century the declining importance of Old Montreal for cigar and garment manufacture was replicated in other industries, although some firms, such as Landau and Cormack, continued to find a spatial niche in the central district. For many producers, however, the pressure of high land values, the search for larger spaces to accommodate firm expansion, and the need to restructure factory layout and to introduce new pieces of equipment forced them to seek new premises. Feeling the pressures of competition from financial and commercial activities, they found it increasingly difficult to find appropriate buildings and working conditions in the traditional manufacturing center of Old Montreal. In contrast to large, high-volume, and mechanized companies, many cigar and clothing firms did not move to locations too far from the city core. Instead, they moved to the Outer Core, especially the

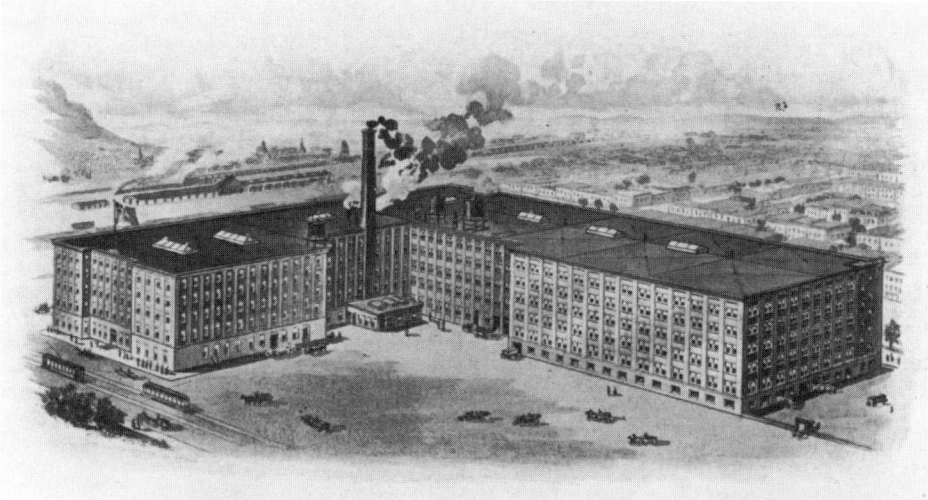

FIGURE 7.7 JOHN PECK CLOTHING FACTORY, 1915. Located in the Mile End manufacturing district, the large, vertically integrated Peck factory was one of the few clothing manufacturers to move from the central clothing cluster to a suburban location. (Lorenzo Prince, *Montreal Old and New* [Montreal: N.p., 1915], 313.)

Saint-Antoine and Saint-Lawrence districts, where cheaper land, larger amounts of floor space, and transportation facilities were available in close proximity to a large, multiskilled labor force and Old Montreal's distribution network of wholesalers and jobbers. Moreover, many of the migrating firms moved into a set of multistory buildings more appropriate for manufacturing than those found in Old Montreal. While Davis and Grothé were the pioneers in the move uptown, they were soon to be followed by other cigar firms. Similarly, clothing firms clustered along the Ste. Catherine Street cluster, following the move by several others. While some manufacturing activities remained rooted to central locations, there was another thread to the industrial geography of Montreal at the end of the century. The move of industries to locations other than the city core, noticed as early as the 1840s and firmly in place by the 1860s, had been consolidated by 1890. While the industrial districts created in the spurt of expansion at midcentury became increasingly less attractive to manufacturers, new ones, further out from the city core, developed in the early twentieth century.

CHAPTER EIGHT

"Busy Hives of Industry" of the East and North Ends

> *Acquaint yourself with the cleanly, modest homesteads that cluster around the busy hives of industry in the great manufacturing outskirts of the Canadian metropolis.*
>
> —Ernest Chambers, *Suburban Montreal as Seen from the Routes of the Park and Island Railway Co.*, 1895

By the time that Edward Chambers was writing, at the end of the nineteenth century, the East End was an important node within Montreal's spatial division of labor and continued to be so over the next thirty years. Territorially, this involved a tremendous widening of the manufacturing fringe: Sainte-Marie became more elaborate, new additions were made in Hochelaga, and a new set of industrial clusters stretched further east and to the north. Distinct in terms of their time of emergence and industrial structure, these new clusters extended and deepened Montreal's spatial division of labor and added to the East and North End's importance within the metropolitan area. The number of firms rose from 148 in 1890 to 720 in 1929, while the area's share of the city's rent grew from 15 percent to almost a third over the same period.

"DIFFERENTIATED DRIVING LOAD": THE OLD EAST END MANUFACTURING DISTRICTS

In 1907 Metal Shingle and Siding moved from Preston (Ontario) to a plant at the old Canadian Pacific Railway site in Hochelaga, where it installed "the latest types of Brown and Sharpe machine tools, such as a No. 3 surface grinder . . . and a No. 3 universal grinder and milling machine fitted with differentiated driving load." A small firm, it decided to move to the East End because a significant share of its growing business was located there. It did well at the new site, and by 1929 the company operated a much larger plant producing a variety of sheet metal products, mainly for the East End locomotive firms.[1] The Ontario firm's move characterizes the changing geography of Sainte-Marie and Ho-

Table 8.1 Manufacturing Structure of Sainte-Marie and Hochelaga, 1890 and 1929

	Sainte-Marie				Hochelaga				Sainte-Marie and Hochelaga	
	1890		1929		1890		1929		1890	1929
Industry	No. of Firms	Rent Share (%)	No. of Firms	Rent Share (%)	No. of Firms	Rent Share (%)	No. of Firms	Rent Share (%)	City Rent Share (%)	
Rubber	1	24.5	2	16.7	—	—	—	—	100.0	62.0
Food	31	21.3	32	25.9	3	10.3	19	11.8	15.4	5.5
Nonmetallic	9	9.8	5	3.5	—	—	2	0.5	55.6	3.3
Textile	4	6.9	4	21.4	2	41.8	4	29.4	45.4	30.9
Metalworking	6	6.8	11	8.0	3	2.6	15	10.1	4.5	5.7
Wood	5	4.4	6	1.4	2	2.5	3	1.1	13.4	6.9
Tobacco	3	4.2	4	1.2	1	15.7	3	19.7	46.5	28.7
Leather	10	4.2	13	4.1	2	0.2	3	1.3	4.4	7.5
Paper	1	2.7	1	0.3	1	0.3	—	—	12.7	0.9
Clothing	12	2.5	8	5.8	2	0.3	—	—	1.5	2.5
Chemical	2	1.9	6	4.2	3	0.9	4	0.7	7.0	5.9
Furniture	7	1.6	4	1.0	—	—	3	3.0	2.4	8.0
Transport	8	1.6	7	1.0	1	24.7	1	17.6	26.5	7.4
Printing	2	0.2	9	3.8	—	—	5	0.7	0.2	4.6
Total	123	100.0	131	100.0	25	100.0	75	100.0	15.1	9.5

Source: Water Tax Rolls for Montreal and surrounding municipalities, 1890 and 1929.

chelaga's manufacturing districts between 1890 and 1929. In this period they continued to rely on the same industries to form their industrial base and to have a set of linked industries. While Sainte-Marie and Hochelaga's expansion was not as significant as other districts, they did experience growth. The number of firms grew from 142 in 1890 to 206 in 1929, but their share of the city's total rent fell from 15 to 10 percent (table 8.1 and map 8.1). The differentiated load—both absolute and relative growth—in the older East End districts was common to all of the city's older manufacturing districts. Sainte-Marie and Hochelaga also continued to be characterized by a relatively polarized rent structure. Large corporations employing modern technologies and work methods, with rents of more than $10,000 in 1929, coexisted with small bakeries, carriage makers, blacksmiths, and soap makers employing less sophisticated methods operating in workshops and small factories. Several industries in 1890 retained their local importance forty years later; textile, rubber, food, and tobacco firms accounted for more than 50 percent of the districts' rent.

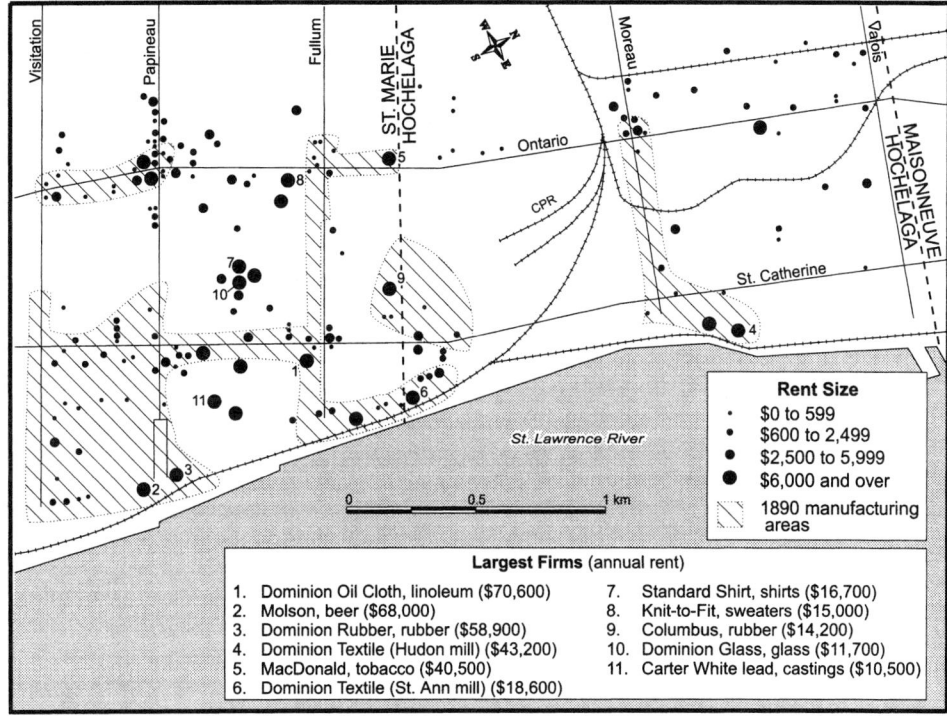

MAP 8.1 SAINTE-MARIE AND HOCHELAGA MANUFACTURING DISTRICTS, 1929. The older East End districts continued to experience manufacturing growth and formed a large factory district by 1929. (Compiled from City of Montreal, Water Tax Rolls, 1890 and 1929.)

Growth by many of the older companies was one element behind the maintenance of this structure. Along with neighboring Columbus Rubber, for example, Canadian Rubber constituted more than 60 percent of the city's rubber shoe and tire production. By 1905 its plant covered more than twelve acres and employed twenty-five hundred workers all year round. In the following year the company merged with four other Canadian rubber companies to effect changes to management, manufacturing, raw material supply, and distribution. The reorganized company grew rapidly, and by the end of the 1920s it was so large that the president, W. Eden, could state that the "main departments of tires, footwear and mechanical goods have become three great industries within one company." In 1928 the company had more than seven thousand workers across Canada and total sales of twenty-eight million dol-

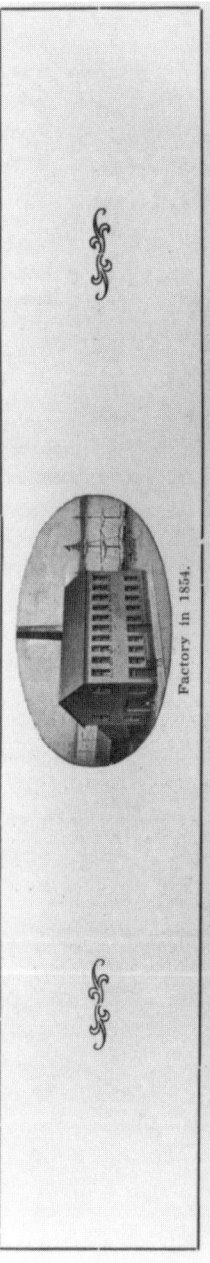

Factory in 1854.

PRESENT PLANT OF THE CANADIAN CONSOLIDATED RUBBER COMPANY, LIMITED.

Illustrative of the remarkable growth and expansion of a representative Canadian manufacturing industry we invite a comparison of the two pictures shown on this page. For upwards of fifty years this Company has been making rubber footwear for the people of Canada. It has developed from a little two-storey building, as shown in the smaller cut, to a giant corporation, with factories and main warehouses extending over thirty acres. The smaller cut shows the first factory, which was erected in the year 1854. The Company manufactures rubber footwear of every description and style, as well as a complete line of belting, hose, drug sundries, moulded goods, waterproof clothing, etc.

FIGURE 8.1 THE CANADIAN RUBBER FACTORY, 1915. After sixty years of growth the Canadian Rubber Company, in 1915, covered several city blocks and continued to be a major End East employer. (Lorenzo Prince, *Montreal Old and New* [Montreal: N.p., 1915], 310.)

lars. Despite the emergence of an extensive national system of manufacturing and distribution, the Sainte-Marie plant remained the largest plant of the company (fig. 8.1). Similarly, Hudon Cotton and Macdonald Tobacco were still two of the largest firms in the city and formed the core of Hochelaga's manufacturing cluster (fig. 8.2). An assortment of small and medium-sized new firms producing, among other things, linseed oil, biscuits, aerated water, and dairy products coexisted with these large firms.[2]

In the first decades of the twentieth century Sainte-Marie and Hochelaga attracted new firms, many practicing modern productive strategies. The case of the Standard Shirt Company is interesting in this regard; not only does it highlight some aspects of these productive practices, but it also presents a sharp contrast to the central districts' agglomeration of garment firms. In 1903 its Sainte-Marie office and factory employed twelve hundred workers. The five-story, custom-built structure had five acres of floor space and paid a rent of $7,500 in 1907, while the construction of a new building more than doubled that rent by 1912. The company operated its own power plant, with generators for lighting and electric power. Knives powered by small electric motors performed the most difficult part of the clothing production process, fabric cutting, while other departments employed various forms of specialized machinery. Particularly interesting, given the strategy of most of the city's clothing firms, is that Standard Shirt broke with the agglomeration economies of the central core and the clothing corridor along St. Lawrence Boulevard. Standard did not rely upon the intricate linkages keeping other garment makers rooted in the inner city; the profusion of interfirm links that most clothing firms depended upon were not required, as "all the products of the company are made on the premises, and they are not dependent on anyone else for any part of their manufacture." Functioning in the same manner as many other continuous-process firms from other industries, the company brought in large quantities of raw materials, manufactured a standardized product using an array of machinery, employed large numbers of unskilled workers, and sold in national markets. The East End advantages—railroad connections, large labor force, and cheap land—were too good to ignore.[3]

Another firm seeking the advantages of a peripheral location was Colin McArthur's wallpaper factory. A Scot who learned the trade in Glasgow, McArthur moved to Montreal in 1879 and entered business with J. Watson at his Montreal Wallpaper Factory in Saint-Ann. By 1890 he had established his own business in an East-End four-story plant. The basement contained the engine rooms and storage for raw

Figure 8.2 The Hochelaga Textile Mills, 1915. The old Hudon plant, now called the Hochelaga mill and one of the biggest plants of Canada's largest textile corporation, the Dominion Textile Company, remained an important component of the old suburban manufacturing district of Hochelaga. (Lorenzo Prince, *Montreal Old and New* [Montreal: N.p, 1915], 315.)

stock and manufactured color. On the first floor could be found the offices, the stock and packing rooms, and the department where skilled workers made the brass rollers for printing patterns on the wallpaper. On the second floor five printing machines manufactured staple wallpapers. While this business had in earlier days depended upon imported skilled labor, machinery had replaced hand labor on staple wallpapers. But not all of McArthur's products were made using high-volume, machine-based methods; as with firms in other industries, he combined various methods. McArthur employed skilled workers to do handwork on the most expensive wallpapers. Laboring separately from the machine minders, skilled workers produced one hundred pieces a day on the top floor, and, though they produced far less wallpaper than machines, the handworkers nevertheless fulfilled a particular market niche for the company. After the East End move the firm grew; the number of its employees increased from 50 in 1890 to 130 by 1903, and the amount of rent doubled between 1890 and 1907. The Goad Fire Insurance Atlas of 1915 indicates that new buildings had been added: two new storage sheds, a color room, and a separate two-story building that housed the company's office and warehouse. The work process of the original building had been reorganized: the first and second floors were now entirely devoted to coloring, while the upper two floors were for printing.[4]

In Sainte-Marie and Hochelaga certain firms and industries played a decisive role as magnets for further growth. Along with the creation of infrastructures and the actions of certain local alliances, they functioned as locational assets for the development of an industrial district with specific specializations. A textile node of large, mechanized firms employing thousands of workers and catering to national markets for oilcloth, thread, wool, and bedding clustered around Hochelaga's large Hudon and St. Ann cotton mills. One such firm was the Dominion Oil Cloth factory, where a mainly French-Canadian labor force produced floor, table, and carriage cloths for the Canadian market. Growing slowly at first, the firm expanded rapidly after the turn of the century and branched out into new product lines (cork carpet, and linoleum). Its land area quadrupled between 1890 and 1907, while its rent of $70,000 in 1929 was thirty times larger than its 1890 rent (fig. 8.3). Another textile firm employing a large number of workers and a large stock of machinery was Gault's Excelsior Woolen Mills. Established in 1898, by 1903 two hundred workers, using an extensive range of machinery, produced worsteds, coatings, and tweeds for the company.[5]

The single most important new manufacturing plant was the Angus

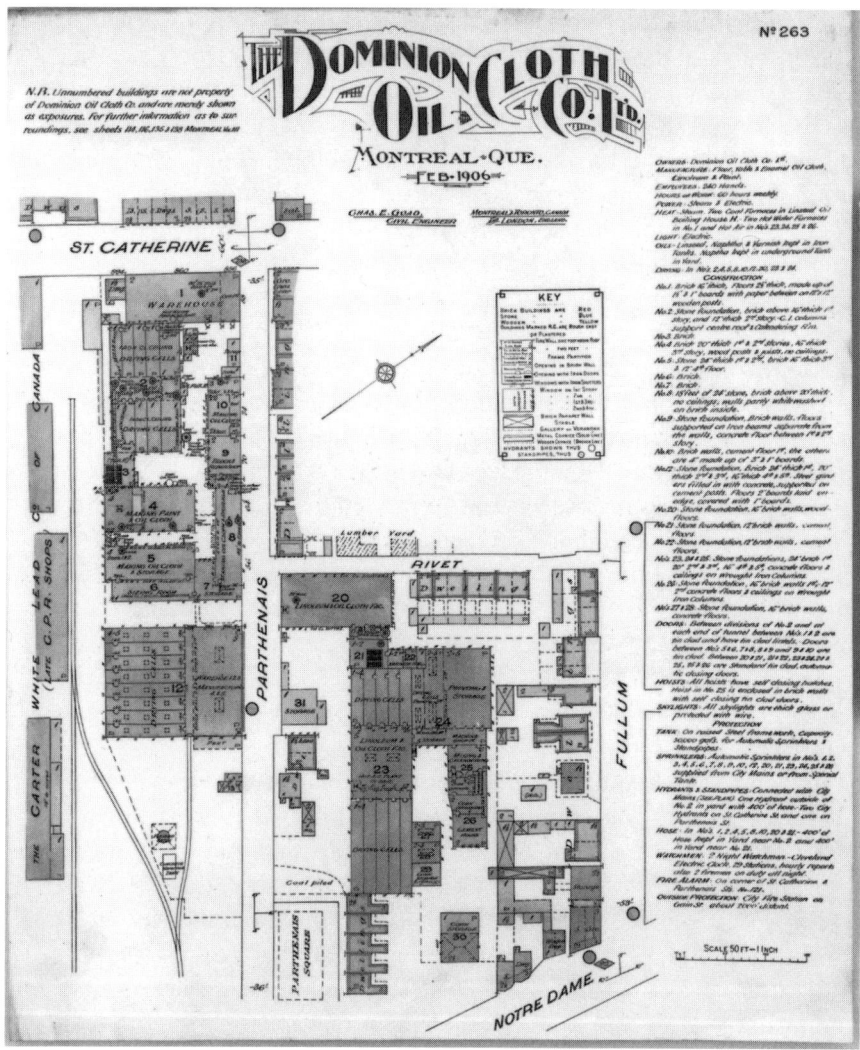

FIGURE 8.3 DOMINION OIL CLOTH COMPANY, 1906. Rapid growth after 1890 propelled Dominion Oil Cloth to the position of one of the largest firms of the Sainte-Marie manufacturing district. The 1906 insurance plan shows the company's intricate plant layout. (Charles Goad, *Fire Insurance Plan of the Dominion Oil Cloth Co. Ltd.* [Montreal: Goad, 1906]. Reproduced courtesy of the National Archives of Canada, negative no. NMC 10585.)

shops of the Canadian Pacific Railway. The creation of these locomotive manufacture works in the north section of Hochelaga in 1904 involved two intertwined projects: creation of a new industrial site and the complete renewal of a productive space. The Angus shops had a similar effect to that of the Grand Trunk shops when they built in Point St. Charles in the 1850s: in terms of size, plant layout, and the logic of the production process the shops differed from most other firms in the city. They also replaced the older shops that had been built in the southern end of Hochelaga in 1883. By the early twentieth century, with increased demand for rolling stock, the old Hochelaga site was no longer adequate, and, being surrounded by factories and housing, there was no room for expansion. With a 260-acre greenfield site the new shops were guaranteed sufficient land for long-term growth, and in this they differed from earlier firms. As one writer has noted, the shops were also "well inserted in the urban texture of Montreal through a dozen installations that responded to precise criteria of division of labour and operational efficiency." They were the epitome of modern rationalized production methods. The Angus shops were designed for the maintenance, repair, and construction of rolling stock for the Canadian Pacific national system. They also manufactured switches, frogs, and sundry other railway supplies in a separate building. In the first ten years the shops turned out more than 79,000 freight wagons, 2,000 passenger cars, and 1,300 locomotives, half of them made for the company itself. The layout on the ground reflects the vertical integration of the railroad's manufacturing: the plant consisted of three sets of workshops arranged to expedite the flow of materials. The first comprised freight and passenger car shops, a planing mill, cabinet shop, truck shops, car machine shops, wheel foundry, and dry kilns. The second set included a machine shop and erecting and blacksmith shops. The third set of buildings—a smithy and a grey iron foundry—supplied materials for all the shops.[6]

The layout of the shops and their integrated manufacture were matched with a new set of work relations, including implementation of scientific managerial methods of production. Rationalization was stimulated by the increased demand for rolling stock, the need to decrease costs, and the labor conflict culminating in the 1908 strike. To cope with these pressures, and because the older methods proved inadequate for the vast, sprawling shops containing forty departments, management sought to speed up production and ensure better labor control. After an extensive study of the shops in 1908, Henry Gantt, the American expert on factory rationalization, made two principal recommendations. The first called for greater coordination and a more efficient integration of the various parts of the plant. To that end Canadian Pacific established a central planning office to coordinate the flow of materials between and within departments. Gantt's second recommendation called for a reorganization of the production process by means of instruction cards, which described the nature, order, and required time of all individual tasks. This allowed management to assess more precisely the value of piecework and to find new ways of speeding up production. After these recommendations were implemented, productive practices were transformed; by 1913 the shops had been enlarged, labor inputs reduced, and the material flow reorganized. The acid test was, of course, whether costs were lower. In the years immediately after Gantt's recommendations were introduced, the shops could build a locomotive in fourteen days, whereas before it had taken eighteen.[7] The Angus shops generated powerful locational assets for other firms, and, as a model for companies such as Canada Cement and Imperial Oil, were a harbinger of growth farther east.

New Manufacturing Districts in the East End

Manufacturing clusters seeking special facilities farther east developed after 1890. New East End firms may have used different productive strategies than the Canadian Pacific shops, but they copied elements of the railroad company's spatial strategy. Seeking the advantages of sites farther out from nineteenth-century fringe districts, manufacturers moved their plants to Maisonneuve, Longue Pointe (or Mercier, as it was known after annexation in 1910), and Montréal Est (table 8.2). By 1929 these districts accounted for 14 percent of the metropolitan area's rent and less than 4 percent of its firms. Maisonneuve developed first. From a handful of firms in the 1890s—the large St. Lawrence Sugar and few small workshops supplying the local market—the number grew to fifty-two in 1918, when it was annexed by the city. By 1929 the district's seventy firms made up almost 5 percent of the city's total rent (map 8.2). With a sprinkling of firms from all sectors, Maisonneuve had concentrations of leather, paper, textile, metalworking, and food.[8]

In contrast, Mercier and Montréal Est's industrial base was much narrower, with almost 90 percent of its rent concentrated in a few, very large nonmetallic, transportation equipment, and chemical firms. In these districts the large capital inputs, new productive methods, and specific space requirements replicated those of the Angus shops. Montreal Locomotive, established in Longue Pointe in 1903, functioned as a magnet for later development and created the basis for the creation of mod-

Table 8.2 Manufacturing Structure of Maisonneuve, Mercier, and Montréal Est, 1929

	Maisonneuve			Mercier and Montréal Est		
Industry	No. of Firms	District Rent Share (%)	City Rent Share (%)	No. of Firms	District Rent Share (%)	City Rent Share (%)
Metalworking	7	29.2	9.0	2	5.1	3.1
Food	11	23.6	9.5	1	2.1	1.2
Textile	4	15.4	9.3	—	—	—
Leather	12	14.3	16.0	—	—	—
Paper	3	5.9	12.4	—	—	—
Chemical	4	2.0	0.4	3	7.9	14.7
Wood	3	0.9	2.2	2	2.5	12.6
Transport equipment	4	0.8	0.1	5	28.7	29.8
Printing	5	0.8	0.1	—	—	—
Clothing	2	0.7	0.1	—	—	—
Nonmetallic	1	0.1	—	4	52.8	65.8
Total	70	100	4.5	18	100.0	9.0

Source: Water Tax Rolls for Montreal and surrounding municipalities, 1890 and 1929.

ern manufacturing methods in the suburban greenfields (fig. 8.4). With capital of $1,000,000 and machine tools worth $300,000, the riverside plant manufactured locomotives, machines, and structural work—steel buildings, bridges, and roof trusses. Starting off as a Canadian financed firm, it was bought out in 1904 by the American Locomotive Company. Like other consolidated transportation equipment manufacturers, American Locomotive sought to capture the continental market, and Montreal was the best location for access to the Canadian railway companies. Being one of the largest firms in the city by 1912, Montreal Locomotive had a rent of $75,000, covered sixty-two acres, and had built its own wharf and power plant.[9] By moving to Longue Pointe, the firm extended the acceptable distance that firms could function in the wider metropolitan industrial orbit.

In the following years other large capital-intensive corporations employing new mechanical and chemical processes moved to Longue Pointe and Montréal Est. The eighteen firms in Mercier and Montréal Est in 1929 accounted for 9 percent of the total city's manufacturing rent (map 8.2). These new manufacturing districts on the extreme eastern edge of the built-up metropolitan area differed from all previous districts in the small number of firms, the size of their plants, and their industrial structure. Canadian Steel Foundries settled there in 1912 (fig. 8.5), connecting its extensive set of buildings to a Canadian National

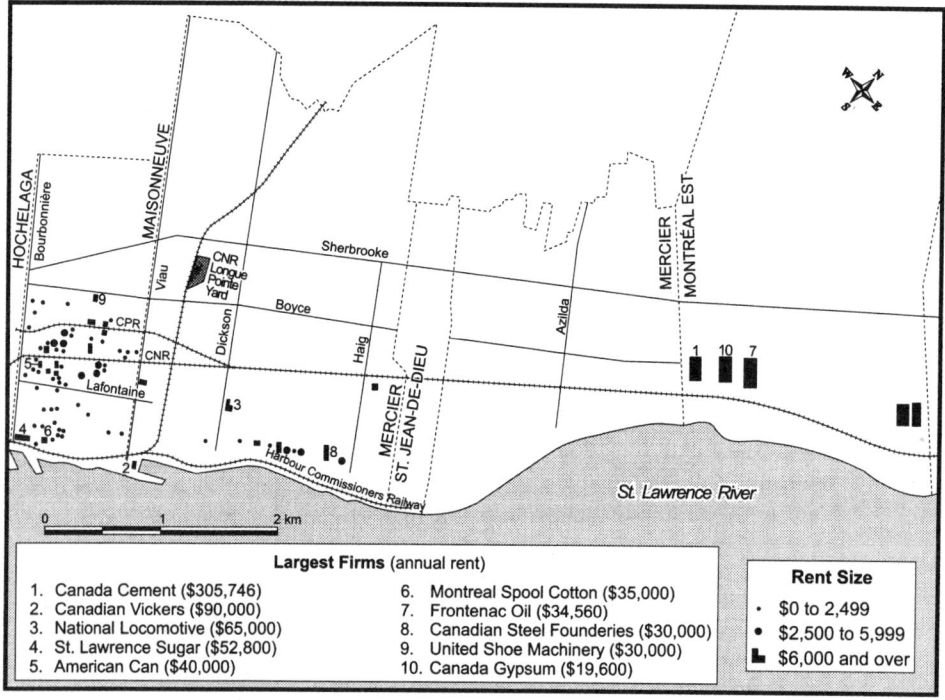

MAP 8.2 THE NEW EAST END MANUFACTURING DISTRICTS, 1929. By 1929 new manufacturing districts had emerged on rural greenfield sites, and a manufacturing belt stretched east through Maisonneuve, Longue Pointe, and Montréal Est. (Compiled from City of Montreal and Ville de Montréal Est, Water Tax Rolls, 1929.)

spur line and building its own wharf. The principal market for its steel castings was its parent company, the Canadian Car and Foundry. The Mercier plant supplied the parent firm's West End factories with locomotive frames, wheel centers, engine castings, and an assortment of other railroad products. It also made an array of steel castings for bridge, ship, and machine tool making companies. Canada's largest cement manufacturers established their major operations in this area. National Cement built on a natural cement rock outcrop in Montréal Est. When its 3,000-barrel dry-process mill situated on 102 acres opened in 1926, it was linked to the railroad and had access to river piers. Similarly, the nearby Canada Cement mill obtained material from quarries adjoining the plant and transported coal from the river to the mill by a private rail-

FIGURE 8.4 MONTREAL LOCOMOTIVE AND MACHINE COMPANY, 1905. Being one of the largest plants in Montreal when it was built, Montreal Locomotive is illustrative of an early twentieth-century propulsive firm seeking a greenfield site. It manufactured an assortment of locomotives for the national market. (Ernest Chambers, *The Book of Canada* [Montreal: Book of Canada Co., 1905], 69.)

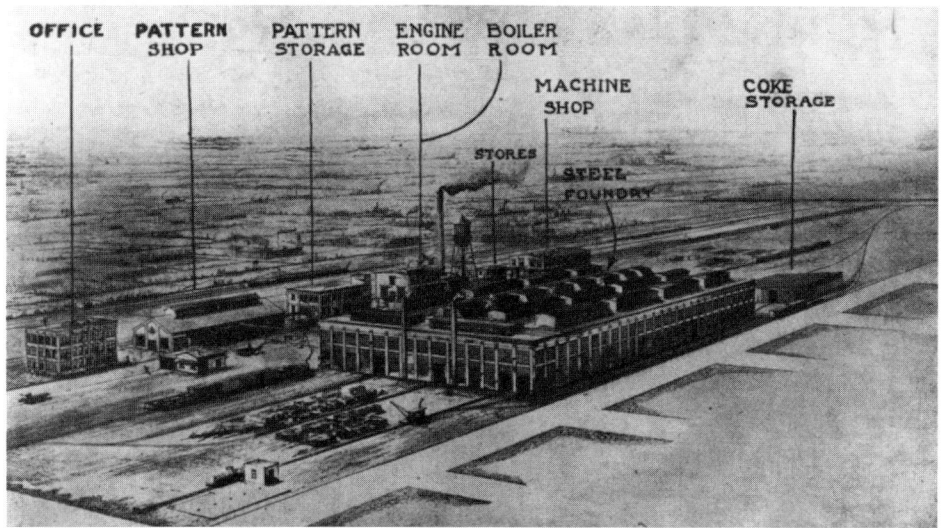

FIGURE 8.5 CANADIAN STEEL FOUNDRIES, 1912. Opened for operation in 1912, the Longue Pointe plant of the Canadian Steel Foundries company was built to manufacture castings for the Canadian railroad, ship, and machine tool industries. (*Canadian Machinery* 8 [October 1912]: 325.)

road, connected to both the Canadian National main line and to the Montreal Tramway tracks. The final additions to the cluster were the refineries of Imperial Oil, McColl-Frontenac (Texaco), and British-American.[10]

Not all of the new East End districts attracted high-volume, continuous-processing plants from a narrow range of industries. In Maisonneuve a diversified industrial base developed, built on a tight set of connections between a local political elite, property developers, utilities, and manufacturers. The development of manufacturing suburbs such as Maisonneuve is usually attributed to the availability of large tracts of land and infrastructures, loose municipal supervision of factories, and the provision of subsidies, but in this case they do not explain why firms moved at this particular time, why industrialization occurred in the suburbs in this period, and why a particular array of firms settled in these districts. Shoe factories, for example, moved from the city core to Maisonneuve, and by World War I the suburb had acquired a reputation as a major shoemaking center. Having no shoe factories before 1898, it had eleven, accounting for almost 28 percent of the city's total shoe making rent, in 1929. While the town's special facilities were obvious attractions, it is necessary to delve further to understand why shoe firms left central Montreal and moved to Maisonneuve. Industrial and metropolitan dynamics played a key role.[11]

From the 1880s Montreal shoe manufacturers faced intense competition, as cheaper production methods and price cutting destabilized the industry. After 1900 competition remained fierce, forcing shoe manufacturers to turn to new forms of cost and scientific management. Nevertheless, throughout the period costs and competition plagued the industry. In one estimate material costs represented more than 80 percent of net sales, leaving very little for overhead expenses and an adequate return on capital. In 1915 it was estimated that 94 percent of the total value of manufactured goods went to raw materials, labor, overhead, and miscellaneous costs. The squeeze on profits, cut prices, and competition were endemic to the industry.[12]

Competition was intensified by the inability of large manufacturers to deploy mechanization as a means of driving the small firms out of business and thus gaining significant control of the industry. After the 1880s manufacturers could no longer depend, as they had earlier, upon technological inventiveness or control over a patent as the means to make their way in the industry. During this period, as one writer comments, "new inventions and mechanical improvements were constantly appearing but they did not create new types of footwear or modify profoundly the division of labour and the division of operations in the manufacturing process." Even though a small number of dominant firms benefited from mechanical innovation and patent control of product and process before the 1890s, mechanization had the effect of providing new firms with a wide range of techniques to choose from. Adding to these conditions was the fact that the industry was easy to enter. As one commentator noted, "With one or two thousand dollars—enough, say, to pay wages for a week or two," the prospective manufacturer "rents ample premises, buys his machinery on credit, his leather at six months, and, with his travellers on the road, soon obtains paper enough to make him a respectable account at some bank." Despite fragmentation and the appearance of numerous small firms, there was a movement to concentrate and consolidate the industry. In 1916, while the average Canadian factory had a daily output of 350 pairs of shoes, 13 manufacturers produced more than 1,000 pairs daily, and the largest, Ames, Holden and McCready, produced upwards of 4,000. In other words, the industry was characterized by a variety of firms, of different scales, using different production methods.[13]

The small Canadian market was an obstacle for local manufacturers of all scales to achieve the types of economies common in American shoe factories. In the United States, as a leading Quebec City manufacturer stated in 1913, "business is . . . conducted on a much larger scale, several firms having each an output equal to the whole manufacturing trade of the Dominion." According to a report in 1890, Canada was a "slaughter yard for American goods," as American manufacturers could run for a long period on one style, while their Canadian counterparts "had to change round to several lines a day." Things were no better in 1929, when relatively large shoe companies such as La Parisienne still had to produce at least fifteen shoe models each month in order "to keep abreast of the times." Added to the small capacity of Canadian firms, the cost and inconvenience of making a wide range of styles—in the 1910s a retailer could carry anywhere between seven hundred to eight hundred samples—meant that firms were unable to reduce costs sufficiently to meet competition in domestic markets or from foreign manufacturers.[14]

The problem of style multiplication meant that manufacturers had to be constantly up with fashion changes, continually resetting machinery, and incurring large expenditures as lasts quickly became obsolete. Shoe manufacturers traditionally had made a wide variety of shoes, but by the 1890s, wanting to capture lower production and distribution costs, they began specializing. Caught in the bind of falling prices and increasing costs, manufacturers were forced to go into standard lines or to specialize in a particular line (men's, womens, children's, working, slippers) or a particular technique (McKay, welt, turns, stitchdown pegged, or screws). By the end of World War I, despite the small extent of the market, the Quebec industry was specialized, with only two firms producing a full range of footwear. In Montreal three factories specialized in men's goods, six were devoted to women's lines, and sixteen concentrated exclusively on children's shoes. Likewise, thirteen firms used only the McKay process, four the welt, while three used turns exclusively.[15]

The growing insurgency of shoe workers also influenced the industry. From the beginning of the transition from craft production to the factory system Montreal shoe workers had fought to maintain some control over working conditions and the labor process. In the 1860s, with skilled workers grouped in craft unions spearheading the struggles, shoemakers were most concerned with ending subcontracting. By the turn of the century the shoe industry had become more broadly based, and labor conflicts were more likely to involve workers from a spectrum of occupations. Contributing to the development of worker solidarity was the increased fragmentation of the labor process and the continued deskilling of workers: by 1900 the only skilled workers in the industry were leather cutters, who, like printers, maintained their skill and status by controlling materials and machinery. The very forces that were so essential to the competitive structure of the industry led to growing tensions within the workplace and to increased forms of struggle.[16]

Together, these changes created a situation in which manufacturers were forced to seek new strategies to increase profits and to gain a favorable share of the market. There were several options open to them. One, as we have seen, involved specialization in a particular niche of the market. But there were others: manufacturers could reduce the quality of raw materials, integrate horizontally and vertically, introduce new machinery (such as Goodyear machinery in the late 1880s), or intensify the exploitation of labor (through longer hours, decreasing piece rates, or speeding up production). Another strategy consisted of cost cutting and workplace restructuring through spatial relocation, but it was only available to a select group of Montreal shoe firms. Certain firms—including large ones with large fixed investments or large chunks of capital sunk into unvalorized machinery and small firms dependent upon centrally located jobbers and wholesalers—found it difficult to leave the tight central-city cluster. Some firms, however, were not so tied down.

The place that most of the mobile firms relocated to was Maisonneuve, in part because of the quality of Maisonneuve's special facilities. Yet other industrial suburbs had cheap land, decent utilities, and a pliable town council. One could argue, for example, that Saint-Henri made a more logical choice, as it had historically been an important leatherworking village and had an extremely industry-friendly town council. But this was exactly the point: manufacturers not only sought to find locational assets that would reduce their costs; they were also looking to escape the confines of the past by making new productive spaces. By the 1890s, as Maisonneuve began to develop as an industrial suburb, Saint-Henri had already developed an industrial heritage based on heavy industry and a strong labor presence which shoe manufacturers would not find attractive. The combination of Maisonneuve's growth after 1900 and the changes to the shoe industry promoting spatial relocation worked to make the town a suitable location. Propelled by the pressures plaguing the industry, firms sought to construct a new work environment, one tied to reducing costs and within a reasonable distance of central Montreal's shoe cluster. The conjunction of the specific character of Maisonneuve and its services and the changes occurring in the shoe industry led to the town's importance as a shoemaking center. Moreover, the move of the large United Shoe Machinery plant from Montreal to

FIGURE 8.6 UNITED SHOE MACHINERY, 1915. Moving several times since it relocated to Montreal, the United Shoe Machinery company finally settled down in the rapidly growing shoe manufacturing center of Maisonneuve in 1911. (Lorenzo Prince, *Montreal Old and New* [Montreal: N.p., 1915], 390.)

Maisonneuve in 1911 added to the towns' attractions (fig. 8.6). Once there, shoe manufacturers quickly became part of Maisonnueve's elite by laying down close links with local land developers, becoming active on the town council, and having a strong influence in the town's social and cultural affairs. With their integration into Maisonneuve society, shoe manufacturers shaped the town's political and social milieu and created the necessary mechanisms to foster a favorable environment for generating profits.[17]

This more profitable climate was made possible by the construction of modern plants, where new machinery could be installed, new lines established, and different techniques introduced. Dufresne and Locke, for example, started in a small way in the heart of the Montreal shoemaking district in 1890. After a move in 1893 to a new plant a block north, they soon felt the limitations of the central core. In 1900, after looking at several sites, they decided to move to Maisonneuve. Their new suburban factory was five times the size of their central plant and designed in the most modern manner. Concentrating on the production of McKays and turns, they grew quickly. By 1904 they employed 125 workers, and by 1912 their 350 employees worked in a building four times the size of the original one (fig. 8.7). Dupont and Frere likewise built a new factory in Maisonneuve in 1910. The old factory in Montreal had "gradually been growing more and more inadequate" and the need for new premises became "an absolute necessity." They built a typical four-story mill structure providing three times the output of the old factory. Introducing new machines in a factory better equipped for modern manufacture resulted in "better work ... being done in every department." Like other Montreal firms, Kirvan-Doig saw Maisonneuve as a place to restructure its operations. In 1912 the company moved from its old Saint-Jacques plant to a new factory on Desjardin Street. Once there it installed new and a much greater amount of machinery, arranged so "that it makes a progressive operation all around." So successful were the firms moving to Maisonneuve that, after establishing their respective specialization, they branched out into new lines: in 1910 Dufresne and Locke added Goodyears to their staple line of turns; Dupont and Frere in 1915 added men's Goodyear welts to their speciality of medium-priced women's McKays; and Rideau Shoe, after buying out the Laniel factory in 1908, converted output from cheap turns and McKays to high-grade men's shoes.[18]

Views of Factory of Dufresne & Locke, Limited, Montreal

The above illustration shows a part of the spacious Cutting Room

A corner of the Goodyear Welt Lasting Room

Exterior view of factory. Note the many large windows which make the building ideal in tne way of light and ventilation

A view of one side of the large Fitting Room

A section of the Goodyear Welt Finishing Room

FIGURE 8.7 DUFRESNE AND LOCKE, 1912. Representative of the shoe firms moving to Maisonneuve from the city core after 1900, Dufresne and Locke installed new work processes and new shoe lines in its much larger suburban factory. (*Footwear in Canada* 2 [July 1912]: 40.)

"A Conspicuous Knot": New Manufacturing Clusters in the North End

Two manufacturing districts emerged in the city's North End after 1900. The Mile End district, the largest, followed the Canadian Pacific Railway lines from the Angus shops to the northern edge of the middle-class suburb of Outremont. Plateau, the other district, formed a corridor of small firms along St. Lawrence Boulevard between Mile End and the Outer Core district of Saint-Lawrence. Before 1905 neither district had much growth. Some small firms had moved to Plateau, and several transportation and warehousing facilities settled in Mile End at the turn of the century.

The initial growth of the manufacturing cluster occurred between 1905 and World War I. In these years large factories producing, among other things, paint, clothing, electrical appliances, shoes, and auto frames were established, while firms in a variety of industries clustered nearby. Possessing firms from nearly every industrial sector, Mile End and Plateau mirrored the city's complexity, although there was a degree of specialization between them (table 8.3). Clothing and food firms accounted for more than half of Plateau's firms and two-thirds of its rent, but it was underrepresented in the traditional Montreal industries of metalworking, transportation equipment, and textiles. Plateau firms tended to be small in scale. In contrast, Mile End's firms were larger and from a wider array of industries. Clothing and food firms were important, but two types of industries dominated Mile End's cluster. The first was made up of firms from the propulsive industries of the day, including chemicals, paper, and electrical. The second consisted of stand-alone corporate firms seeking large sites close to cheap labor and transportation connections. Many were foreign firms such as the British food processors Bovril and Fry. At the same time, large warehouses and storage yards continued to locate close to the railroad yards. By World War I a substantial fringe industrial base had formed, and expansion during the war and the 1920s further extended the districts' manufacturing prowess. By 1929, according to one contemporary description, "Toward the north in the middle distance, a string of large establishments swings away from the river, skirts the more thickly settled areas, and ends in a conspicuous knot of plants at Mile End" (map 8.3).[19]

The paint industry illustrates some of the important elements of the North End's growth. Playing an important part in Montreal's older industrial districts before 1900, it had a similar role in the North End in the new century. The industry's traditional location was the bustling in-

Table 8.3 Manufacturing Structure of Mile End and Plateau, 1929

Industry	Mile End			Plateau		
	No. of Firms	District Rent Share (%)	City Rent Share (%)	No. of Firms	District Rent Share (%)	City Rent Share (%)
Food	40	26.7	10.2	21	20.6	2.7
Paper	5	14.1	40.4	2	3.1	3.1
Clothing	14	13.1	5.4	40	47.9	6.9
Chemical	15	10.8	13.7	7	2.5	1.1
Leather	10	5.1	7.9	6	3.8	2.0
Wood	14	4.1	14.2	5	1.1	1.4
Electrical	6	4.5	9.8	—	—	—
Printing	6	4.1	4.5	8	2.9	1.1
Nonmetallic	12	4.1	3.5	3	0.9	0.3
Transport equipment	6	3.9	2.8	1	0.7	0.2
Tobacco	6	2.2	5.4	4	2.0	1.7
Metalworking	11	2.1	0.9	3	0.8	0.1
Textile	6	1.2	1.0	3	2.9	0.8
Furniture	8	1.0	3.0	6	8.7	9.2
Total	176	100	6.1	116	100	2.1

Source: Water Tax Rolls for Montreal and surrounding municipalities, 1890 and 1929.

dustrial district running alongside the Lachine Canal. From the 1840s several firms produced varnish, paints, and japans for the carriage, railroad, cabinet, and construction industries. In 1890 eight of the city's ten paint firms, accounting for more than 95 percent of the industry's rent, were located there. In the early twentieth century, however, the industry experienced changes, notably the application of electricity, the shift from hand-ground to factory-made paints, and the development of new work methods. Wartime conditions induced even greater growth, as firms sought to cope with increased demand and shortages fueled by import reductions. In particular, shortages of German organic pigments forced local firms to undertake dye and other organic chemical production. After the war the automobile industry's need for quicker-drying paints to keep up with assembly line production methods led to the search for new types of paints. With its sixteen establishments producing more than 40 percent of the country's output in the late 1920s, Montreal was Canada's major paint center, in part due to the significant vertical integration within the industry. Also, replacement by large firms of wood-oil based varnishes with new chemical processes such as nitro-cellulose enamels and lacquers reduced the number of operations necessary for production and increased labor productivity.

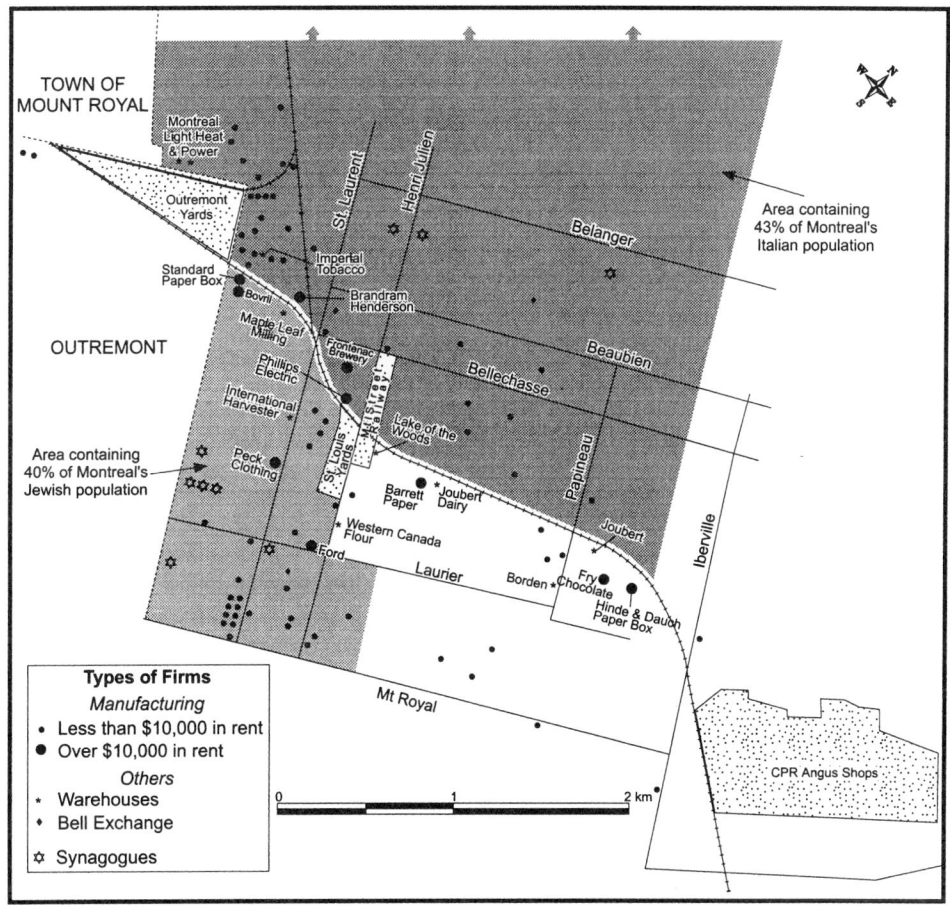

MAP 8.3 THE MILE END MANUFACTURING DISTRICT, 1929. Forming along the Canadian Pacific Railway tracks, an extensive and varied manufacturing base had developed in Mile End by the 1920s. (Compiled from City of Montreal, Water Tax Rolls, 1929.)

Although consolidation, combined with increasing demand and the introduction of new products and processes, gave rise to much larger firms, Montreal's industry continued to be characterized by a range of companies producing goods under different work conditions and at different scales. In 1929, along with some of the largest firms in the city, the industry also had several smaller, locally owned businesses producing for the local market.[20]

By the early twentieth century rapid growth and consolidation forced firms to seek out new locations for installing new work methods, factory layouts, and product innovations. For these firms large and cheap manufacturing sites in Mile End close to a growing labor force and the Canadian Pacific line were attractive. It was increasingly difficult for firms in the older manufacturing districts to expand their plant in situ. Moreover, the new chemical processes and work methods of the industry could not be effectively implemented in the older buildings, even after they had been expanded. The British firm Brandram-Henderson Ltd., like many other paint firms seeking new spaces for production, was attracted to the greenfield sites of Mile End. At the time of its merger with the American firm, Henderson and Potts, in 1906, Brandram had a small factory producing white lead and paint in the congested heart of Griffintown. After consolidating, it rationalized production and, wishing to flee the diseconomies of the older district, in 1907 built a large plant north of the Canadian Pacific lines (fig. 8.8). The new company experienced tremendous growth over the next twenty years. By the 1920s it had installed new machinery and equipment, acquired paint companies in Toronto, Medicine Hat, Vancouver, and Hamilton, and built a new varnish and enamel plant in Mile End. With a rent of sixteen thousand dollars in 1929, it was one of the largest firms in the metropolitan area.[21]

Other paint firms established their factories in Mile End for similar reasons. Martin-Senour erected a factory close to Brandram-Henderson in Outremont just north of the tracks. Another large factory, the Holland plant of International Paints (Canada) Ltd., specialized in wood, leather, and stove varnishes and small goods enamels. Finally, Barrett, the roofing paper and wallboard maker, made a specialized line of asphalt roof paints. All of these firms depended upon raw materials from all corners of the world: dry colors, colored earths, zinc oxide, flaxseed, tung oil, gums, resins, and volatiles from China, Argentina, the United States, and other parts of Canada. Some material they received directly from local suppliers: linseed oil from Dominion Linseed and Canada Linseed, white lead from Carter White Lead, and coal tar, petroleum, and containers from companies in the Mile End district and elsewhere in Montreal.[22]

Other industries flocked to Mile End after 1905 in order to deploy the latest production techniques in a greenfield site. Brewing experienced tremendous growth and consolidation after the 1880s. In order to cope better with these pressures, the Frontenac company built a large brewery in Mile End in 1911, just south of the tracks. With its refrigerat-

Figure 8.8 Brandram-Henderson Factory, 1915. Covering five acres and composed of several inter-linked buildings, the lead corroding plant of Brandram-Henderson was typical of the larger factories found in Mile End by World War I. (Lorenzo Prince, *Montreal Old and New* [Montreal: N.p., 1915], 370.)

ing ammonia compressors, horizontal Corliss engines, and Pfaudler glass-enameled tanks, it was reputedly the most modern brewery in Canada. All machinery and lighting were electrically driven, and its bottle shop had three complete bottling units with a capacity of 180,000 bottles a day. Mile End's attractiveness to paint and beer makers extended to other industries as well. One of Canada's largest cigar manufacturers, H. Simon & Son, moved its plant to Montreal from Whitby in 1902, settling in Old Montreal's cigar area. Responding to the same pressures that forced Grothé and Davis to move uptown, Simon relocated to Mile End, seeking more space and to escape the congested central districts. The company built a large, modern, daylight three-story factory with a railroad siding which, by the 1920s, produced seventeen million cigars annually and employed five hundred workers. Occupying a large site, the factory contained a hospital, cafeteria, and employee training school. And, being located in the middle of a working-class district, it had no concerns about finding cheap labor. As the company boasted, "the number of applicants for entry into its school always exceeds the company's requirements for new labour."[23]

Industrial expansion in the North End also occurred through additions, alterations, and turnover. The Perfection Glass factory, for example, was enlarged in 1926 in response to the automobile industry's demand for laminated glass. Two years later the entire layout of the plant was reconstructed to increase turnover time and reduce handling costs. In other cases vacant buildings were quickly filled. For example, the three floors and basement of A. Cimon's new Mile End shoe factory, although not large by industry standards, doubled the output of his old Sainte-Marie factory. With capital of $100,000 he began the production of high-grade welts and McKays. But disaster quickly struck. In 1913 the failure of the city's Boston's Shoe company, his biggest creditor, plunged him into financial difficulties, and he was forced to liquidate his holdings and to put the factory up for sale. At this time McFarlane Shoe, a manufacturer with a small factory in Saint-Jacques district, was looking to add a new product, children's slippers, to its line. This move would require changes, and the Cimon factory was an opportunity not to be missed. McFarlane bought the plant in 1913 and moved in a year later, after the Saint-Jacques plant's lease had expired. In the process the firm introduced some major renovations: "the latest machinery has been installed and the factory has been arranged so that every operation, from the cutting of the leather to the shipping of the finished shoe, is practically continuous, one process dove-tailing accurately with the least possible loss of time and energy." McFarlane remained a successful company and was one of the largest shoe firms in the city in 1929.[24]

Factories in Mile End were frequently part of an intra-corporate hierarchy of capital flows. Many of the firms settling along the Canadian Pacific tracks formed part of a large corporate network extending throughout the metropolitan area, Canada, and overseas. Ames-Holden-McCready, a product of some of Canada's largest and oldest footwear manufacturers, was one such firm. In 1911 the company was created from a merger of two firms: Ames-Holden and James McCready, whose original plants were located in Old Montreal's shoemaking district. Facing the same types of pressures that forced shoe firms to move to Maisonneuve, they reorganized their work and spatial relations. By the time of the incorporation in 1911 the company had three factories: the McCready plant located in the North End; an Ames-Holden factory in the Saint-Antoine district; and another Ames-Holden factory in the small industrial town of St-Hyacinthe, thirty miles from Montreal. Each of the plants had a specific role in the corporate specialization: women and child fine-grade McKays in the North End, fine-grade men and boys' Goodyear welts in Saint-Antoine, and low-grade peg and standard screw shoes at St-Hyacinthe. According to company president D. Lorne McGibbon, the company had expended considerable energy and resources "with a view to effecting economies in operation, production, and distribution, and at the same time improving the quality and style of the goods." Along with specialization the company also undertook a policy of vertically integration. In 1912 the company established a separate cut-sole factory with up-to-date machinery in its North End plant and bought a heel factory equipped with a Grover heel-building machine and presses, and the two plants provided soles and heels for the North End and Inspector plants.[25]

"An Excellent Enterprise": Housing and Infrastructure in the East and North Ends

Real estate developers such as Maisonneuve's Mendoza Langlois contributed to the expansion of existing and the formation of new manufacturing districts in the East and North Ends after 1890. As he told the newspaper *La Patrie*, "My experience as a real estate agent demonstrates that land, when situated in an industrial district and bought with the object of subdividing into lots and then selling, constitutes an excellent enterprise." Sensing the great profits to be made from opening up the area outside the built-up part of town, Langlois bought farmland, subdivided it into lots, and sold it to builders. Districts such as Maisonneuve were attractive to entrepreneurs because industry, harbor facilities, railroads, and the town's spirit of progress ensured residential growth. This process of converting city fringe land was repeated by numerous others, resulting in new belts of working-class homes clustered around the manufacturing nodes on the edge of the metropolis. Although no overall plan of metropolitan development existed, shifting alliances of developers, utility owners, local politicians, and manufacturers practiced a range of strategies in individual areas, from ad hoc and uncontrolled approaches to systematic and well-organized plans. Unable and unwilling to install a coherent Progressive plan of citywide growth, local alliances forged a territorial link between working-class residences and factories. In the process they established a patchwork of spaces conducive to suburban industrial development.[26]

Infrastructure developments, especially Israel Tarte's large-scale port redevelopment after 1900, formed a critical building block of the new districts. While overall metropolitan planning continued to have little importance to decision makers, new harbor construction reinforced the links of the eastern districts with the rest of the city and established the basis for industrial expansion by providing services to East End firms. An elevated railroad running the length of the waterfront, from the central harbor to Hochelaga and Maisonneuve, improved the circulation of goods and people between Montreal and the eastern districts. A new dry dock straddling Maisonneuve and Longue Pointe's boundary extended the area's docking facilities. As a site, it had few infrastructural impediments and was well linked to the Harbour Commission's railway tracks, which were connected to the city's railroads. The protected basin, with its wharves and thirty acres of developed land, was supplied to Canadian Vickers Ltd., a subsidiary of the British firm of Vickers. Working under contract with the federal government and the Harbour Commission, the company constructed a floating dock, a shipyard, and a repair plant. The dock, along with other modern port facilities after 1900, presented potential East End manufacturers with important locational assets.[27]

Developers were quick to realize the advantages of transportation development for fringe and suburban residential development. In conjunction with the new port extensions to the east and the various railroad corridors of the area, real estate companies interested in the profits to be made from simple land conversion initiated extensive working-class residential growth. Land speculation and working-class housing construction in Maisonneuve and Longue Pointe was spurred by the extension of port facilities. Similarly, the Canadian Pacific's Outremont yards stimulated residential growth, both in Outremont and the adjacent districts of Mile End and Park Extension. The yards, as one writer stated, "will give employment to a good number of men and will create a demand for houses in the locality, which, at present, is occupied as farm land." Local entrepreneurs, such as Mile End's Joseph-Octave Villenueve, took advantage of these developments and constructed extensive working-class housing close to the Canadian Pacific Railway yards and tracks. Along with this, he was also instrumental in the extension of the city electric streetcar into Mile End. Continuing after the war and extending through the 1920s, other entrepreneurs constructed a belt of duplex and triplex housing fanning north to the Rivière Prairie along the major north-south streets and east-west railroad lines and strung out to the east along the harbor and railroad facilities.[28]

During the building cycles of the turn of the century and the 1920s, the large-scale process of farmland conversion into working-class residential districts occurred in the East and North Ends. For land development companies such as Charrune and Daoust, according to a report in the *Montreal Star* in March 1906, "the greatest demand by far [is] from the east end of the city. The workingmen seem to realise the importance and advantages of having their own homes and consequently are buying up a great deal of property with the idea of building just as soon as their means allow." The first to be settled was Maisonneuve in the 1890s, followed by extensive construction in Longue Pointe and Pointe-aux-Trembles. According to one real estate agent in 1906, growth had been so quick that "very few of the citizens of Montreal have any idea of the wonderful expansion now going on to the north and east especially, extending for miles over land which only a few years ago was lying idle or used for farming purposes." With these housing developments the population of the North and East Ends increased dramatically in the first three decades of the twentieth century. To the north Rosemont, with just 315 people in 1901, grew to more than 45,000 in 1931, while over the same period the corridor running along the Canadian Pacific railway line grew from less than 11,000 to almost 70,000. A similar spurt occurred in the East End. The Maisonneuve-Mercier district, for example, grew sevenfold, from 7,000 to more than 55,000 over the same period.[29]

Employment opportunities in manufacturing, transportation, and warehousing contributed to rapid working-class residential growth. Although there are few studies outlining the direct links between jobs and workers' residence, there is sufficient evidence to suggest the importance of a strong geographic link between working-class housing and factory construction. In some cases manufacturing led development and attracted house builders and workers, while in others a fringe blue-collar residential population stimulated manufacturing decentralization. Frequently, the two processes occurred together. The Amherst Park subdivision north of the Canadian Pacific Railway lines provided lots for the workers employed in the Mile End factories (fig. 8.9), and the construction of the new Longue Pointe plant of the National Bridge Company before World War I, for example, led to houses "being built adjoining the works for employees." Even earlier, ripples of housing construction occurred in the vicinity of the Angus shops, as railroad employees sought accommodations close to their suburban place of work. Their demand for new living quarters immediately initiated a spate of housing projects on the city fringe and in the new northern suburban areas of Rosemont, Alexandra Park, Fairmount, Hastings farm, and Propriéte Préfontaine. According to U. Dundurand, of the Rosemont Land Improvement

FIGURE 8.9 AMHERST PARK, 1905. The trolley and an extensive advertising campaign were part of the strategy for selling small lots of suburban land to the working class. Close to the Mile End factories, working-class Amherst Park was settled by a mix of British, French Canadians, and Italians. (Ernest Chambers, *The Book of Canada* [Montreal: Book of Canada Co., 1905], 108.)

Company, the "boom stage" in these developments had passed by 1906, and lot buying and house building had "settled down to a regular steady demand." Of the 2,362 lots originally put up for sale in 1903 in Rosemont, only 900 remained in 1906. Even though Rosemont was a mainly residential suburb, it does not fit the mold of a stereotypical commuter bedroom community; while some residents found work in the Angus shops, many others walked to jobs in the nearby Mile End, Hochelaga, and Maisonneuve factories. Acknowledging this quality, Dundurand noted that "the growth of industries in the East End of the city assured a continued good demand for lots in these sections."[30]

The North and East Ends had a mixed-class composition. In a few cases areas of high-class housing appeared, the most important being the suburb of Outremont. As a product of two land developers who had bought land from the ubiquitous Sulpicians, by World War I it was one of Montreal's elite districts. Outside of Outremont and a few scattered blocks, however, most of the North and East End residential districts contained a mix of blue-collar, lower-white collar, and self-employed workers. In a sample of 1,878 Mile End residents taken from the 1931 street directory, for example, almost two-thirds were blue-collar workers, another 15 percent were clerical workers, and 12 percent were small business people servicing local needs. In other words, more than nine out of ten Mile End residents were from the lower classes or the small proprietor class. The attractiveness of Mile End was related to its better housing and environmental conditions. Along with the pull of suburban jobs, the deteriorating conditions of the central districts forced those who were able to afford the commute to seek housing on the periphery. Those with high wages and stable employment could afford the better class of suburban housing, which tended to be triplex flats with four to six rooms. Most of the suburban and fringe-area working class could not afford the trolley fare; residents either had to walk long distances to work, or, more likely, they had to live close to their suburban place of work. For those at the lower end of the wage scale or working in suburban manufacturing establishments, the densely built rows of duplexes and triplexes on narrow blocks provided relatively low-rent housing close to the burgeoning suburban factory districts. But not all suburban housing was in good shape; there were numerous blighted areas in the North and East Ends.[31]

People from an assortment of ethnic backgrounds lived in the North and East Ends. Even though ethnic diversity had been an element of the suburban districts' social composition before 1900, an increasing number of eastern and southern European immigrants appeared in the North and East Ends in the early twentieth century. By World War I, as one Montreal City Mission worker reported, "the downtown streets have fewer inhabitants, and the foreigners are spread out into the north and east where they secure more room and for less money." This was evident in the declining importance of the Jewish concentration in the central manufacturing districts. By World War I the Mile End area just south of the Canadian Pacific lines had become the major Jewish residential, religious, cultural, and commercial district. Likewise, just north of the tracks, in Saint-Jean ward, were large numbers of Italians. In the first decade of the twentieth century the Italian population's center of gravity moved from the inner city to the northern reaches of the built-up metropolis, as signified by the establishment of the Madonna della Difesa parish in Mile End in 1910. By the 1920s Italian residential and institutional clusters were found farther north and to the east. Similarly, concentrations of British immigrants and French Canadians formed in the northern parts of Villeray, Rosemont, and Delorimier. Finally, by the late 1920s a major Ukrainian settlement was located north of the group's major employer, the Angus shops.[32]

In many cases the extension of manufacturing establishments and working-class homes into the eastern and northern fringes after 1900 was a chaotic process, featuring little planning, an absence of zoning, and unregulated property development. But this was not always the case. In some cases suburban manufacturing districts were created with a loose assemblage of institutional actors. In others, however, community formation was tightly controlled by alliances of local politicians, land developers, and manufacturers. One such organized suburb, Pointe-aux-Trembles, emerged in the far eastern tip of the Island of Montreal at the end of World War I. Conceived as a working-class garden city, Pointe-aux-Trembles was also to be a "centre of manufacturing jobs and factories." Trusting that Montreal's harbor commissioners would continue to extend the wharves farther eastward and using a special charter based on the Quebec Housing Act, the town developer and mayor, Rosaire Prieur, believed that factories would seek out the "most logical spot on the Island of Montreal for manufacturing." To add to its geographical attractiveness, the town council laid down asphalt streets, provided lighting, sewerage, and waterworks, and established bylaws defining the types of housing that could be built. The overriding concern of council members was, according to Prieur, that "however many factories or houses are erected, there can never be overcrowding of the workers— one of the curses of industrialism." But Pointe-Aux-Trembles never became a model industrial workers' suburb. Despite the rhetoric, factories did not see the logic of the site for manufacturing. Moreover, the distance and cost of commuting to Montreal or other employment nodes were too great. Housing costs also proved to be out of reach of all but the highest paid blue-collar worker (fig. 8.10). Consequently, the town's population only reached 2,970 in 1931. Likewise, the industrial suburb of Montréal Est developed in the same period. Originally promoted as a middle-class garden suburb, the land developer and mayor, Joseph Versailles, turned to industrial development as a way to squeeze out a profit after the initial plan fell through. Unlike Pointe-Aux-Trembles, Montréal Est was never planned as a model working-class town. Instead, it became the home of a few very large noxious industries and, with just over two thousand people in 1931, a very small working-class population.[33]

A more successful attempt at creating a planned industrial suburb combining manufacturing and the working class occurred closer to the

FIGURE 8.10 WORKERS' COTTAGES AT POINTE-AUX-TREMBLES, 1918. Conceived as a working-class and manufacturing suburb, Pointe-aux-Trembles' modest houses remained out of reach of most blue-collar workers, while very few manufacturing plants found the advertised advantages of the suburb attractive enough to move there. (*Canadian Municipal Journal* 13 [August 1918]: 239.)

city limits. Maisonneuve, the "Pittsburgh of Canada," represents a well-documented case of the intersecting interests of local state, developers, and manufacturers. Taking advantage of growing demand for suburban manufacturing sites from the late 1890s, a local elite integrated land and industrial capital and used its control over the local state to establish a growth program based on two pillars. The first was the creation of a systematic industrial policy to attract firms. Maisonneuve's policy of industrial growth consisted of municipal subsidies in the form of tax concessions, cash grants, and infrastructures. From the town's incorporation in 1883, the municipal council attracted firms, many of which came from Montreal. The first was St. Lawrence Sugar, which moved to the town after a fire had destroyed its West End refinery. Locating close to the river, the Sutherland Pier, and the harbor commissioners' railroad line, it received a twenty-year tax exemption and a railroad right-of-way. By 1900 it covered more than two blocks, with the main refinery

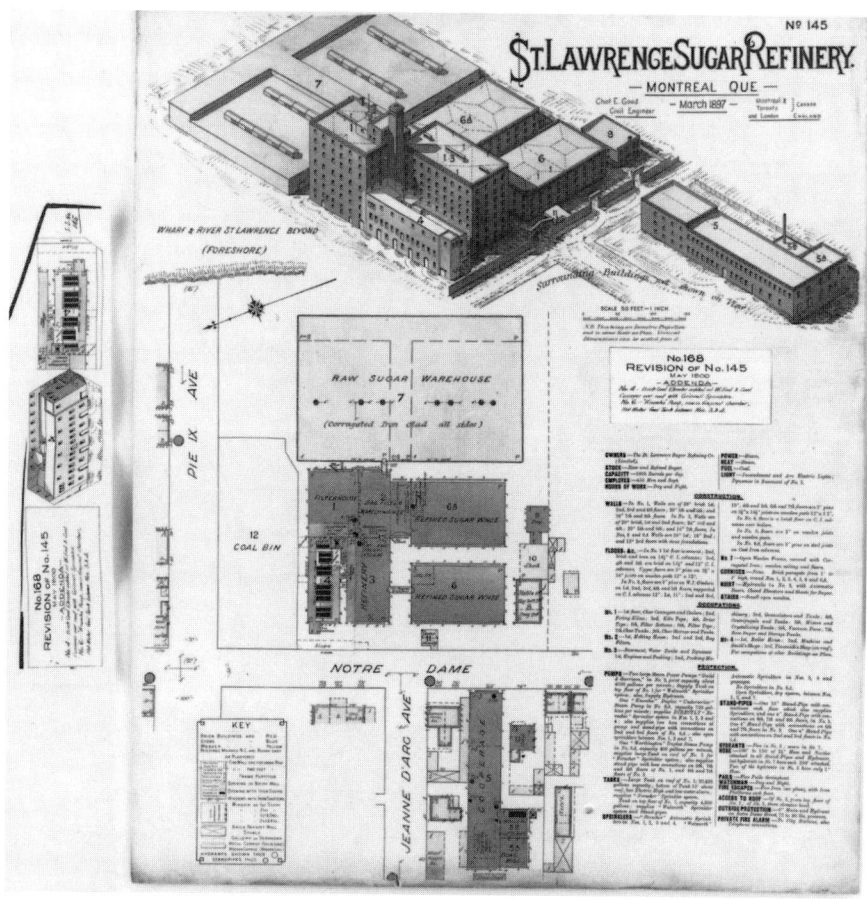

FIGURE 8.11 ST. LAWRENCE SUGAR CO., 1900. Moving to Maisonneuve from Griffintown in 1887, St. Lawrence Sugar played an important part in the building up of the industrial suburb's manufacturing base. (Charles Goad, *Fire Insurance Plan of the St. Lawrence Sugar Refinery* [Montreal: Goad, 1900]. Reproduced courtsey of the National Archives of Canada, negative no. NMC 10604.)

soaring eight storys into the air (fig. 8.11). Following St. Lawrence, a wave of growth occurred between 1894 and 1900, when the town provided ten firms with exemptions worth more than $83,000. After a quiet period in the first years of the new century, another boom took place, with eighteen companies moving to Maisonneuve with subsidies. Along with tax reductions and grants, manufacturers were also attracted by the town's extensive set of infrastructures—the trolley, railways, and harbor facilities, electric street lighting, and a water system.[34]

The second element of the town's industrial policy was the building of a model working-class town. Aware that the success of the industrial suburb depended upon a resident labor force, the town's elite undertook a program of house building for workers and underwrote the systematic creation of a working-class district. Even before the incorporation of the town large landowners, many of whom were manufacturers and were extremely influential on the city council, controlled expansive tracts of land. According to the first assessment in 1884, eight families were assessed for almost three-quarters of the town's property value. Although this concentration was to diminish over time as owners sold their land, a significant portion of the district's land remained in the hands of a small landowning elite. These families, almost without exception, had a representative on the town council. Land was subdivided by the landowners and sold to small builders and building companies that, in the Montreal style, built relatively cheap, two- and three-story row housing. In 1911, for example, the vast majority of the suburb's population worked in factories and construction sites or as domestic servants in the homes of Montreal's wealthy; less than 6 percent were in professional or managerial occupations. While they were not up to the standards of Montreal's wealthy suburbs, Maisonneuve's services were of high quality compared to what was found in most of working-class Montreal. Landowners controlled the direction of Maisonneuve's growth by promoting industry and would also reap the profits from the full-scale development of working-class housing. The incentives granted to manufacturers thus were a way for land capital to cream substantial benefits.[35]

The combined dynamics of manufacturing growth, property development, and the involvement of a local alliance produced an organized industrial suburb in Maisonneuve. The same set of dynamics was responsible for the formation of new, and the expansion of existing, manufacturing districts throughout the East and North Ends. Housing developments in Mile End, Rosemont, Montréal Est, and elsewhere formed an extensive industrial working-class residential belt encircling the late nineteenth-century built-up city. This belt in the eastern and northern sections of the expanding metropolis was fragmented into a series of distinct manufacturing districts: the original Sainte-Marie section and the industrial suburb of Hochelaga had expanded considerably; the large Angus shops generated an important cluster of firms traversing the railway lines in Mile End; Maisonneuve, before annexation, had become one of the most important industrial suburbs in Canada; and Mercier and Montréal Est were home to large, capital-intensive corporations. Each district had several defining features: a specialized set of industries which made them distinct from one another; the early establishment of a few propulsive firms from different industries which established the base for future growth; the coalescing of an assortment of other firms around the large firms; and a complex pattern of intra-district and intra-metropolitan linkages. The expansion of old manufacturing nodes and the appearance of new ones was taking place in the metropolitan area's East and North Ends. In the West End new manufacturing districts also appeared.

CHAPTER NINE

"Expanded in All Directions"

The Western Manufacturing Districts

> To-day, in 1913, the city has accelerated its already rapid growth, expanded in all directions, reached out to the shores of the Back River, extended many miles up and down the St. Lawrence, surrounded Mount Royal, thrown up a stream of residents across the St. Lawrence, and become contiguous to Lachine, once nine miles to the west.
> —William Lighthall, mayor of Westmount, 1914

The process of industrial and working-class decentralization continued in the western part of the city after 1890. As the midcentury fringe manufacturing districts of Griffintown and Canal declined in importance, districts farther out became the focus of growth. Building on their early manufacturing base, Saint-Henri and Sainte-Cunégonde experienced substantial growth after the late 1880s, while the two industrial satellites of Lachine and Saint-Pierre added to the manufacturing cluster along the Lachine Canal. The West End's share of total manufacturing rent increased from 30 percent in 1890 to more than 33 percent in 1929, and, as a result, it remained metropolitan Montreal's most important manufacturing corridor. It also continued to possess the highest rents of the city; its median rent of $1,100 and mean rent of $6,221 in 1929 were well above the city equivalents of $800 and $2,866, respectively. Despite the large scale of the area's firms, its rent distribution remained diverse, as large corporations coexisted with smaller firms from a range of industries. Finally, it maintained its specialized manufacturing base: metalworking, food, transportation equipment, and textiles accounted for three-quarters of the West End's rent in 1929.

It is generally accepted that early-twentieth-century North American manufacturing suburbs were characterized by stand-alone corporate firms with few linkages to one another, and this has certainly been the view of Montreal's western districts. In his study of Saint-Henri's industrial structure Gilles Lauzon states that on the eve of World War I the suburb "appeared as a patchwork, with little logic."[1] In fact, a strong logic underpinned the structure and practices of West End industries. In many cases the local linkages were indirect, typically manifested in a set of local locational assets created by land developers, builders, municipal elites, transportation companies, and manufacturers established in an earlier period. There were also direct linkages, as firms employing a variety of manufacturing pathways spatially coalesced to take advantage of flows of goods and information between one another.

"Alterations and Additions" to Canal and Griffintown

By the late 1880s, as with many other firms in Canal and Griffintown, Fergusson, Alexander was experiencing problems brought about by the changing dynamics of the continental economy. Firms were forced to adjust and recalibrate their strategies in response to increasing competition, product line profusion, growing consolidation, more distant markets, and the introduction of new technologies. A major problem confronting firms was the changing requirements of work space. With the addition of more machinery, equipment, product lines, and workers, the question of changing space needs was pressing. In 1889, for example, Fergusson, Alexander and Company were "making extensive alterations and additions to their central lead, color and varnish factory on College street. Every inch of space is being utilized to make room for additional plant and machinery. As an 'overflow,' the firm have leased buildings and a plot of ground near St. Gabriel Locks in the suburbs which will assist, when necessary, in relieving the congestion at their city works." Adding to these difficulties was the installation of new manufacturing methods. Technological and organizational changes not only contributed to the growth of firms' physical size but also involved the rationalization of factory work processes. All too often, older firm layouts were unable to accommodate manufacturing change. These problems affected Griffintown and Canal firms of all descriptions. The vintage of many factories inhibited rationalization along modern lines, while these districts' lack of space, especially compared to other districts, fueled the problem of increasing demand for more and new work space. In the face of these pressures firms could choose from several options. One was to build modern factories in greenfield sites in other parts of the city, as the paint and steel companies did in Mile End, Mercier, and Montréal Est after 1900. Other strategies involved searching for new manufacturing sites in the immediate western districts or, as in the case of Fergusson, Alexander, undertaking extensive alterations and additions. All of these options affected Canal and Griffintown between 1890 and 1929.[2]

The two nineteenth-century fringe districts remained the home of some of the city's most important sectors and firms (figs. 9.1, 9.2, and 9.3). Canal and Griffintown's leading sectors after 1900 were similar to those of the nineteenth century (table 9.1). Metal, food, chemical, and wood products remained important, and sawmills, foundries, machine shops, and food processors continued to cluster along the Lachine Canal and in Griffintown. Canal remained a node of large, high-volume, capital-intensive plants producing metal products, flour, and sugar, while Griffintown's smaller-scale metal fabricating, wood, and food enterprises continued to produce an assortment of products (map 9.1). With their concentration of metal, transportation equipment, food, textile, electrical, and chemical firms, Canal and Griffintown's industrial structure was quite different from other parts of Montreal. Some of the largest firms in the city were found here, many tracing their origins back to the mid-nineteenth century. The Grand Trunk shops, Redpath Sugar, and James Robertson remained in the older western districts and grew in situ by extensive alterations and additions. After 1890 firms from the same industries coalesced around these old plants, adding to the geographical specialization of manufacturing. The antecedents of Canal and Griffintown's manufacturing on the eve of the Great Depression were firmly rooted in its nineteenth-century historical geography.

While these developments made possible an absolute increase in the numbers of firms, the relative share of the two districts declined significantly, indicating that, with the development of new productive spaces in other parts of the city, the strength and vitality of the older districts would be slowly undermined. By 1929 Canal and Griffintown accounted for just under a tenth of Montreal's rent, while in 1861 they accounted for more than a third. The process of relative capital disinvestment in the older districts is illustrated by the historical geography of the city's flour milling industry. Before the 1840s flour milling remained concentrated in numerous small rural mills scattered throughout Lower and Upper Canada. The enlargement of the Lachine Canal in the 1840s provided nascent urban flour millers with access to the wheat passing along the canal and to hydraulic power needed to operate the mills. Construction of four big flour mills in the 1840s and 1850s centralized Canadian milling along the canal, and by 1860 it accounted for more than half of all Canadian exports (map 9.2). Along with the sugar mills and metal factories, these capital-intensive, high-volume flour mills formed a dominant presence along the Lachine Canal. As one contemporary noted, they towered "into the air on the canal bank" and were surrounded by a "stream of loaded cars pouring along the wharf front . . . ; the floating elevators steaming about the harbour . . . ; the bagging

FIGURE 9.1 MONTREAL FROM STREET RAILWAY POWER HOUSE CHIMNEY, QC, 1896. The bird's-eye views of the Lachine Canal shows the string of factories, mills, refineries, storage spaces, and harbor facilities vital to the Canal District's industrial prominence. (Notman Photographic Archives, McCord Museum of Canadian History, Montreal, accession no. view-2943.)

FIGURE 9.2 MONTREAL FROM STREET RAILWAY POWER HOUSE CHIMNEY, QC, 1896. View from St. Railway 1896. (Notman Photographic Archives, McCord Museum of Canadian History, Montreal, accession no. view-2944.)

FIGURE 9.3 THE VALLEY OF THE LACHINE CANAL, QC, C. 1910. This 1910 view of lower Griffintown presents a detailed picture of the mass of factories, railroads, and docking facilities that hugged the banks of the Lachine Canal. (Notman Photographic Archives, McCord Museum of Canadian History, Montreal, accession no. MP.000.879.17.)

barges with their crowds of busy baggers, and the endless processions of heavily laden lorries with their loads of bagged grain." As the Canadian wheat economy moved further west to the Prairies, however, the city's share of the market tapered off. Montreal millers faced a saturated domestic market and heavily competitive foreign markets by the 1870s. The declining hegemony of Montreal's flour milling quickly became apparent and showed up in the restructuring of local firms and the rationalization of their canal manufacturing space.[3]

The flour companies responded in a variety of ways, as illustrated by Ogilvie, a Montreal milling company. One option involved expansion outside Montreal. As early as 1888, the Ogilvies had three mills and twenty elevators in Ontario and Manitoba. Another response was to be open to product and process innovations. In the mid-1870s Ogilvie introduced the Hungarian steel rollers into his mill, a process allowing for

Table 9.1 Manufacturing Structure of Griffintown and Canal, 1890 and 1929

	Griffintown				Canal				Griffintown and Canal	
	1890		1929		1890		1929		1890	1929
Industry	No. of Firms	Rent Share (%)	No. of Firms	Rent Share (%)	No. of Firms	Rent Share (%)	No. of Firms	Rent Share (%)	Share of City Rent (%)	
Metalworking	20	40.1	28	45.1	15	20.8	13	9.5	29.0	12.6
Food	7	16.0	9	11.2	9	24.1	10	30.3	32.4	16.8
Wood	11	9.5	3	3.4	6	7.2	4	3.6	34.6	21.0
Tobacco	4	9.0	—	—	—	—	—	—	13.6	0
Electrical	1	7.8	1	1.1	1	1.9	1	19.5	80.0	57.3
Chemical	3	4.3	10	8.5	6	5.9	11	7.0	31.5	15.9
Furniture	3	3.9	1	0.7	1	0.1	—	—	44.4	0.9
Leather	1	2.6	1	3.6	4	0.4	2	0.5	2.5	3.1
Textile	3	1.4	5	9.1	4	9.3	4	7.4	14.2	11.0
Transport	5	1.0	4	9.0	4	26.6	5	19.7	41.7	20.9
Total	67	100.0	73	100.0	68	100	59	100.0	17.3	10.4

Source: Water Tax Rolls for Montreal and surrounding municipalities, 1890 and 1929.

a third more work with half as much power and less maintenance. By the 1880s it was the dominant process in Canadian flour mills. Finally, given the demand placed on capital reserves from new technology and expanding markets, firms could consolidate. Between 1865 and 1902 the Ogilvie family through its various corporate forms acquired all of the city's flour mills. At first these mills were remodeled and equipped with the latest machinery and equipment, but by the end of the century all but one mill were closed down. In 1905 the remaining company under Ogilvie control purchased the entire site along Mill Street and made extensive additions and installed new machinery to the Royal mill (map 9.2).[4]

The consolidation of capital as represented by the flour industry formed one aspect of the geography of Canal and Griffintown. Firms were also moving lock, stock, and barrel to other locations. A case in point is the electrical apparatus industry, which historically had found Canal and Griffintown attractive locations. At the end of the century Royal Electric and General Electric made electrical products in Griffintown, but their manufacturing operations were quickly closed down. Royal was bought out by American interests and stripped of its assets in 1900, while General Electric transferred its Montreal lines to its Ontario factories in 1906. Their sites were taken over by warehouses and broken

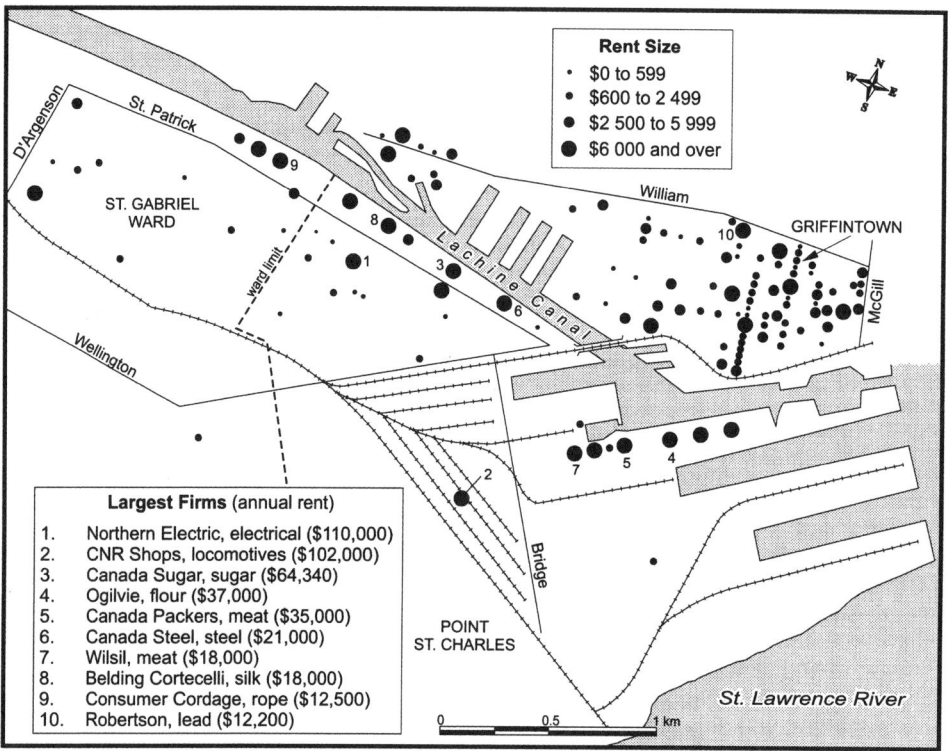

MAP 9.1 GRIFFINTOWN AND CANAL MANUFACTURING DISTRICTS, 1929. Despite their waning importance, Canal and Griffintown continued to be home to a wide assortment of large firms from the metal, locomotive, and food industries. (Compiled from City of Montreal, Water Tax Rolls, 1929.)

up into smaller workshops. In other cases disinvestment benefited other metropolitan locations. Phillips Electrical, Canada's only maker of copper rods, insulated wire and cables for communications equipment, moved from Canal to greener pastures in 1905. The original factory was located alongside the Lachine Canal, and, despite large-scale additions in 1901, the old factory could not keep up with demand nor provide the type of layout that the company needed. In its new Mile End factory the company had more than three times the floor space of the old one and came equipped with a spur line, well-lighted buildings, and a modern

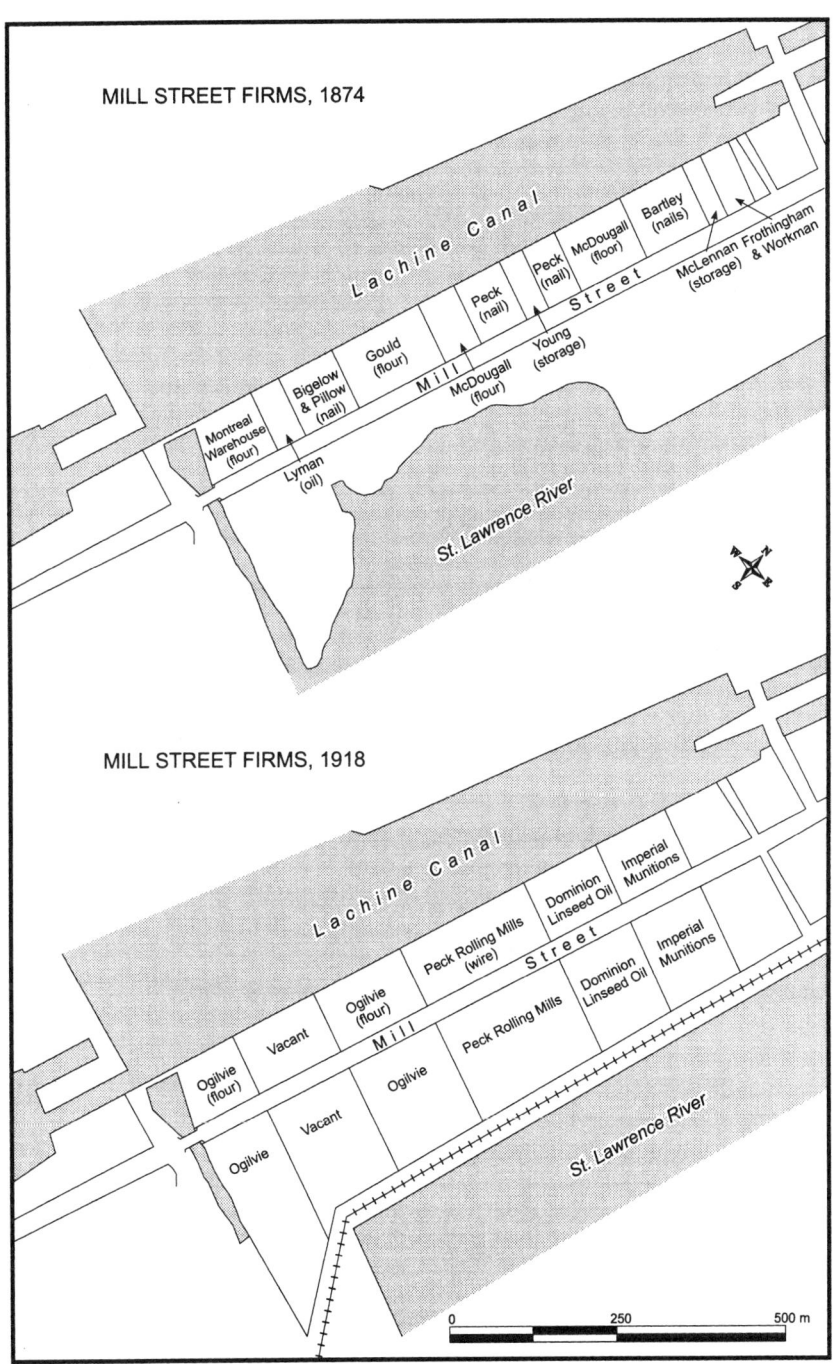

MAP 9.2 MILL STREET, 1874 AND 1918. From the 1840s Mill Street had been home to some of the largest firms in the city. Its changing land-use reflected the changing conditions of the Canal district. (Compiled from "Report of the Commissioners of Public Works," Sessional Papers, 1874-75; City of Montreal, Water Tax Rolls, 1918; Henry Hopkins, *Atlas of the City and Island of Montreal* [Montreal, 1879]; Charles Goad, *Atlas of the City of Montreal and Vicinity* [Montreal: Goad, 1918–51].)

electrical generating system. At its new location it implemented changes that would have been onerous and costly, if not impossible, on the Lachine Canal. Two years later the company added a new building and increased output by 50 percent, and in 1920 it built a new plant to accommodate the manufacture of a new line, enameled wire.[5]

Firms from other industries were also fleeing Canal and Griffintown for greenfield sites. In 1880 the wallpaper manufacturer J. C. Watson converted two stores in Griffintown into a small factory. Before long, however, Watson found the intrusion of central manufacturing districts' functions into the area to be problematic and sought, like his old partner, Colin McArthur, the advantages of a peripheral location. Enticed by a $9,000 cash grant and a twenty-year tax exemption, the company moved to Maisonneuve in 1897. By World War I the greatly expanded plant employed 250 workers and produced 60,000 rolls of wallpaper daily. Meanwhile, the old Watson factory was parceled out to several smaller firms, yet by 1929 it was no longer occupied by manufacturers. Similarly, in his sprawling Griffintown foundry, covering three acres and built over forty years, William Clendinneng manufactured an assortment of stoves, bedsteads, ornamental castings, and furnaces (fig. 9.4). In 1894 Clendinneng moved to Saint-Henri, attracted by a ten-acre site and a twenty-year tax exemption. The suburb's leaders argued that the firm would provide employment for 450 workers and increase property values. Despite the vocal opposition of some of Saint-Henri's important property owners, who thought the terms too high, the bonus was passed, and Clendinneng constructed a building valued at $100,000 in the core of Saint-Henri's manufacturing area.[6]

These examples demonstrate the beginning of industrial sclerosis in the old western districts. Nonetheless, Canal and Griffintown continued to experience growth, as many old firms and some new ones found reasons to stay there. In some cases new firms were attracted by the locational assets that had made these important manufacturing districts since the mid-nineteenth century: proximity to transportation facilities, a multiskilled labor force, and some remaining large lots of land. In 1890, for example, Dominion Glass opened a large factory in Point Saint-Charles, close to the Grand Trunk Railway yards. Over the next two decades, in response to the changing demands of the national market and increasing competition, the company pursued a policy of buying out competitors and capturing a large share of Canada's glassware markets. Twenty years later it reorganized, absorbing many companies and rationalizing individual plants. Even though it had captured 95 percent of the Canadian glassware market by the mid-1920s, the company

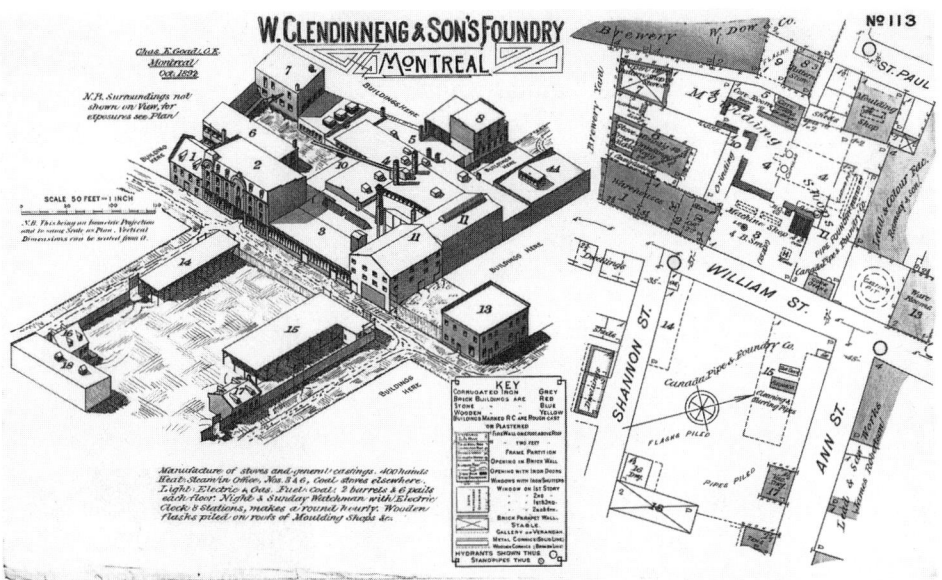

FIGURE 9.4 THE CLENDINNENG FOUNDRY, 1892. The fire insurance plan of the Clendinneng Foundry clearly shows how industrial plants were built up incrementally over the years and how they were tightly interspersed with other manufacturing establishments. (Charles Goad, *Fire Insurance Plan of W. Clendinneng and Son's Foundry* [Montreal, 1892]. Reproduced courtesy of the National Archives of Canada, negative no. NMC 16283.)

had overextended itself. Declining profits, the demand for greater product variety, and general market uncertainty led to product and plant concentration. The company instituted a policy concentrating on traditional lines (glass bottles and jars), shutting down the illuminating glassware lines and disposing of its Hamilton and Toronto properties. The plant at Point Saint-Charles, however, was considered to be a viable one, partly because of the capital investments made in the plant over the years and partly because of the locational assets of the district. In 1925 Dominion Glass extended and made improvements to the plant, almost doubled its floor space, and bought four acres adjacent to the factory.[7]

Griffintown and Canal were home to other firms that were content to alter and add to their existing plants. Many established firms, as the electrical manufacturing industry once again illustrates, were compelled to remain in situ because of the frozen character of large chunks of fixed capital. Even though the industry was generally undergoing expansion in this period, its development in Montreal was not impressive. Its twenty-four firms in 1929 accounted for less than 3 percent of Montreal's rent. While manufacturing a range of products, the local industry grew slowly and in some cases, such as General Electric, lost plants to other cities. Dominated by small American branch plants, including Victor Talking Machine and Black and Decker, the local industry functioned as a simple assembler of American technology.

An exception to this was Northern Electric. From a small telephone assembly shop in 1884, the company grew to be one of the largest firms in Montreal. The drive to reach such a scale centered on the company's ability to manufacture switchboards and telephones for a growing market, the introduction of workplace innovations, and mergers with competitors. In most cities electrical apparatus firms sought out greenfield sites in the suburbs, but this was not the case for Northern Electric. Although on several occasions the need for larger quarters forced it to seek new premises, the company continued to find the older western districts satisfactory for its needs. It made several moves to different Griffintown sites between 1882 and 1906 before finally settling down, just before World War I, south of the Lachine Canal in a multistory plant with more than 500,000 square feet of floor space (fig. 9.5). Even at the new location it made alterations and additions as demand and changing operations necessitated. In 1926 the company added four stories to the plant's central wing, and the resulting manufacturing plant covered four city blocks and had more than one million square feet of floor space.[8]

In situ growth was evident in other industries as well. The Grand Trunk Railway's mass of railway lines and assortment of buildings, constructed at different times since 1856, compelled it to remain at its Point Saint-Charles location. Frequently, however, it was forced to update its works; in 1892 it built a new rolling mill, and in 1929 it opened a new motive power shop building. Likewise, while the R. C. Jamieson paint and varnish works established in 1858 grew by buying out local competitors, its main plant remained the original one located close to the Lachine Canal (fig. 9.6). Another example is Darling Brothers, makers of heating equipment and elevators. Established in 1888, the firm grew incrementally at the same location over the next forty years. In 1899 it added a two-story building on what had formerly been the yard room next to the original premises. During World War I it made three additions to its plant and in 1918 built a new one-story concrete foundry opposite the old one to produce new lines such as centrifugal pumps. The company still considered Griffintown—with its access to the waterfront and railroads, a well-trained labor force, and Old Montreal wholesalers—a viable loca-

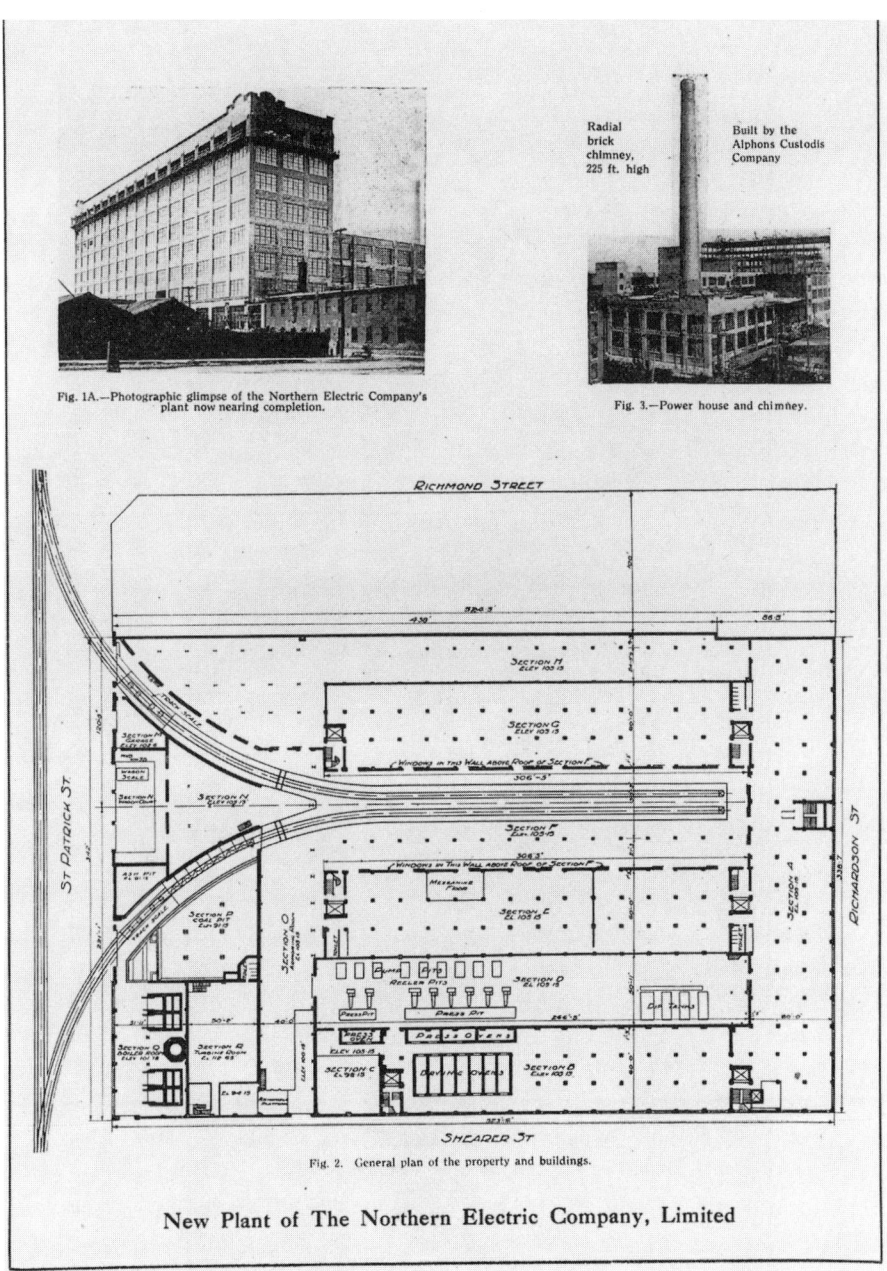

FIGURE 9.5 NORTHERN ELECTRIC, 1914. Built on the eve of World War I, the Northern Electric factory was one of the largest manufacturing establishments in Montreal. The plant's scale, modern factory design, and relationship to railroads can be ascertained from the photograph and plan. (*Contract Record* 28 [May 1914]: 685.)

FIGURE 9.6 R. C. JAMIESON PAINT FACTORY, 1915. While modern paint factories were being built in Mile End, the Jamieson company decided to remain located along the Lachine Canal. The 1915 photograph shows the process of in situ growth that had occurred since it was built in the 1850s. (Lorenzo Prince, *Montreal Old and New* [Montreal: N.p., 1915], 381)

tion. Unlike an increasing number of firms, the benefits of the suburbs were not an option. There was still enough land available in Griffintown, especially given the demolition of housing and the removal of firms to the suburbs, for Darling Brothers to build a large modern foundry, featuring new design elements and a layout permitting more efficient through-flow.[9]

The importance of new factory space and work processes for the locational decisions of firms was evident in other industries. Established in the 1870s, Canada Paint was small and produced a limited range of lines (fig. 9.7). Consolidating in 1894, the new firm sought large economies, reduced handling costs, increased interplant specialization, and greater control over raw material supplies. By 1910 its three factories—in Montreal, Toronto, and Winnipeg—specialized in different aspects of the trade, while the company owned and operated oxide and graphite mills. With the boom of the early twentieth century the company experienced, according to an observer in the *Canadian Engineer,* a "gratifying increase of business which ... necessitates large premises." Even though paint firms would shortly begin their move to Mile End, Canada Paint found that Canal had the "merits" of being central and close to the main plant. Building the new premises not only doubled the company's color-making capacity and enlarged all other departments; it also led to the reorganization of departments and work flows within the new and extended plant.

To make this transition successfully, Canada Paint had to have an integrated works. The company's managers believed that moving to the suburbs, even with all of their advantages, would sever the linkages among the different parts of the plant and undermine the company's ability to compete. They were also concerned that the fixed capital invested in the original factory inhibited the firm's ability to move to a greenfield site in Mile End or elsewhere. By 1929, as a result of several rounds of additions and alterations, Canada Paint had a sprawling factory covering many blocks and producing a wide range of paints, varnishes, colors, and stains.[10]

The different elements of the older manufacturing districts were crystallized in a few firms. One example comes, once again, from the electrical industry. Manufacturing at Marconi Radio began in a small machine shop in Glace Bay, Cape Breton, servicing its wireless telegraphy operations. In 1910 the company moved its manufacturing operations to Montreal's East End to get a larger plant and to be close to the nation's engineering center. Once in Montreal the company branched out into the manufacture of high-speed wireless transmission equip-

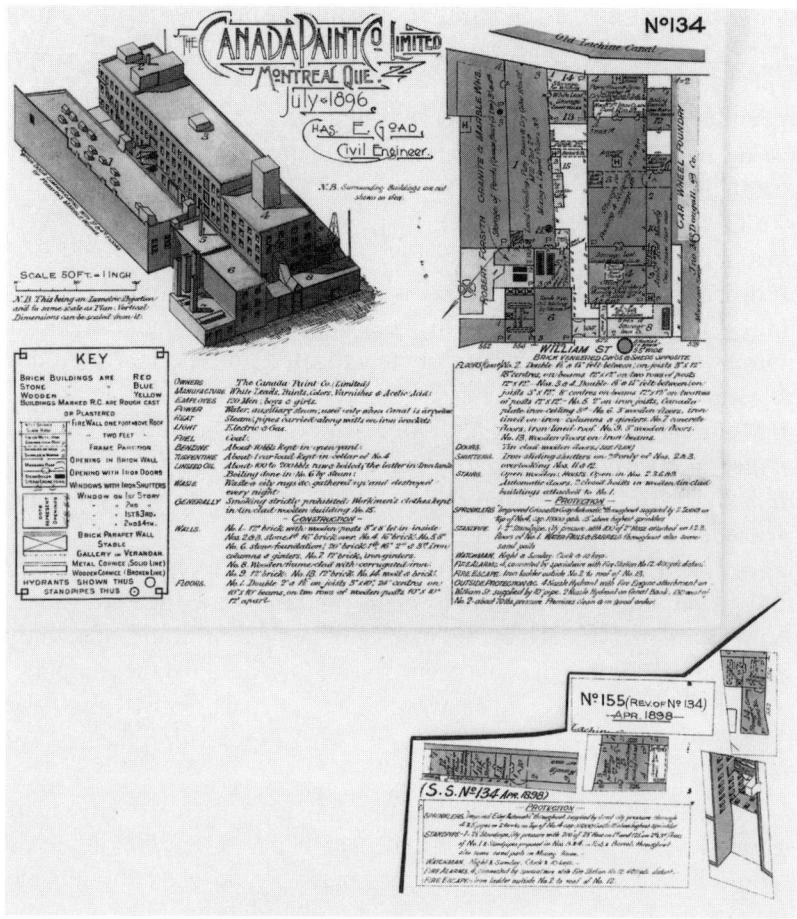

FIGURE 9.7 PLAN OF CANADA PAINT, 1896. The 1896 fire insurance plan of Canada Paint shows the plant before it was converted from a multi-product establishment to one in which it manufactured a much smaller range of products. (Charles Goad, *Fire Insurance Plan of the Canada Paint Co. Ltd.* [Montreal, 1896]. Reproduced courtesy of the National Archives of Canada, negative no. NMC 10758.)

ment, a product previously bought from other companies. During World War I growing demand, coupled with the difficulty of finding skilled labor, forced the firm to move to a large building in the heart of Griffintown, close to other electrical firms. Finding that the recycled building met their needs and being in close proximity to similar firms and a large multiskilled labor force, the firm remained there for several years. In 1928, however, it decided to move to the North End, where it built a plant designed to meet the specific nature of its work. Marconi illustrates the dual character of the older western districts in this period: on the one hand, networks of expertise, labor, and built environment attracted firms; on the other, the older district could not compete with greenfield sites, where the factory, services, and labor force could be shaped more to a company's liking.[11]

Locating Industry: A New Round of Industrial Suburban Growth

While Canal and Griffintown experienced strong currents of disinvestment after 1890, the western suburbs, with their favorable conditions, underwent pronounced growth. Building on their nineteenth-century suburban origins, Saint-Henri and Sainte-Cunégonde grew rapidly, while a minor industrial resurgence occurred in Côte Saint-Paul (table 9.2, map 9.3). The emergence of the industrial towns of Lachine and Saint-Pierre stretching along the Lachine Canal further fueled the expansion of the manufacturing complex. From a small, preindus-

Table 9.2 Manufacturing Structure of the Western Suburbs, 1890 and 1929

	1890			1929		
Industry	No. of Firms	Share of District Rent (%)	Share of City Rent (%)	No. of Firms	Share of District Rent (%)	Share of City Rent (%)
Metalworking	22	40.9	33.5	38	37.7	60.1
Textile	4	21.0	32.1	21	13.2	41.6
Food	15	17.8	14.9	54	11.1	16.2
Transport Equipment	7	4.8	7.6	10	12.6	34.2
Wood	8	4.8	6.9	18	4.0	24.7
Leather	11	2.7	4.8	9	1.3	7.4
Chemical	4	2.6	11.1	20	5.2	25.2
Clothing	13	2.1	1.8	3	1.1	1.8
Electrical	1	0.5	7.1	3	1.7	13.8
Printing	2	0.2	0.8	25	0.7	3.0
Nonmetallic	1	0.1	0.6	20	5.3	17.4
Tobacco	—	—	—	2	4.4	41.3
Other	19	3.2	3.5	24	1.1	3.1
Total	104	100	12.7	247	100	23.2

Sources: Water Tax Rolls for Montreal and surrounding municipalities, 1890 and 1929.
Note: Western suburbs includes Saint-Henri, Sainte-Cunégonde, Côte Saint-Paul, Verdun, Nôtre Dame de Grace, Lachine, and Ville Saint-Pierre.

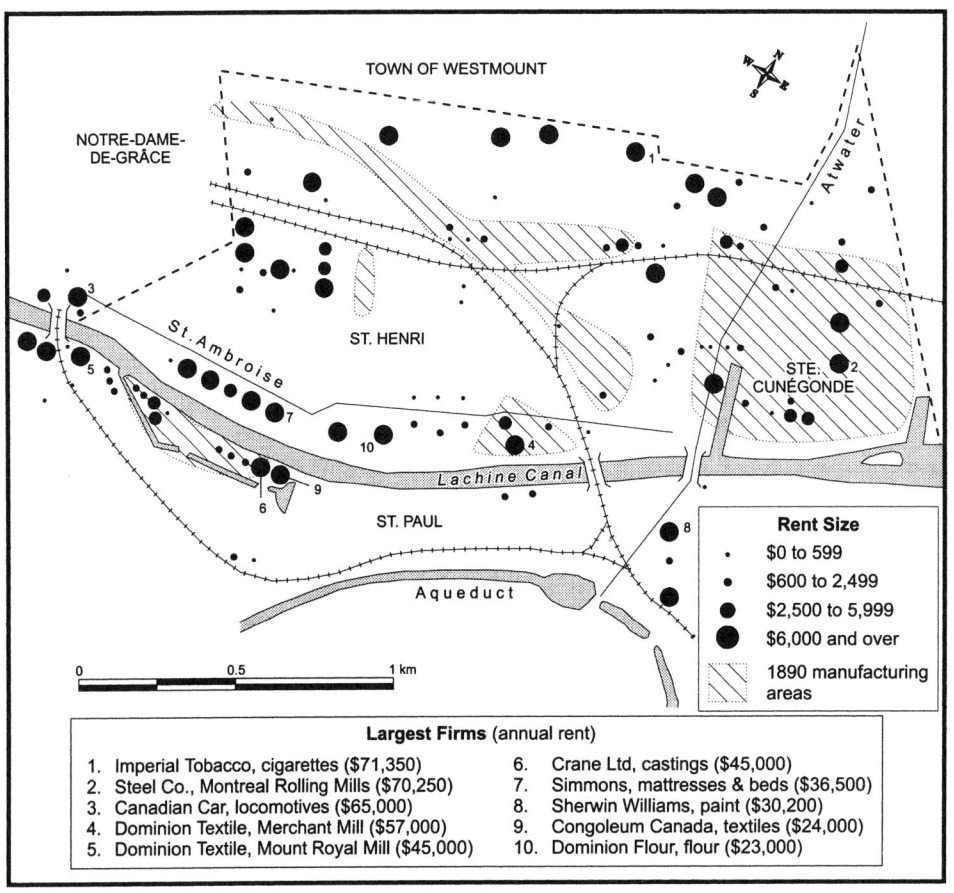

MAP 9.3 THE OLDER WESTERN SUBURBAN MANUFACTURING DISTRICT, 1929. Over the course of seventy years of industrial development the old western suburbs, part of the city since annexation in the early twentieth century, had acquired an extensive and diversified industrial base. (Compiled from City of Montreal, Water Tax Rolls, 1890 and 1929.)

trial village Lachine grew into a satellite town dominated by a small number of large corporate steel and machine-making firms. Adjacent to Lachine but differing in its scale, Saint-Pierre developed after 1900 (map 9.4). Even though each district had its own specific character, the western suburbs' industrial structure exhibited similarities and developed out of nineteenth-century antecedents. The metal, textile, and food-processing sectors continued to be important, even though their

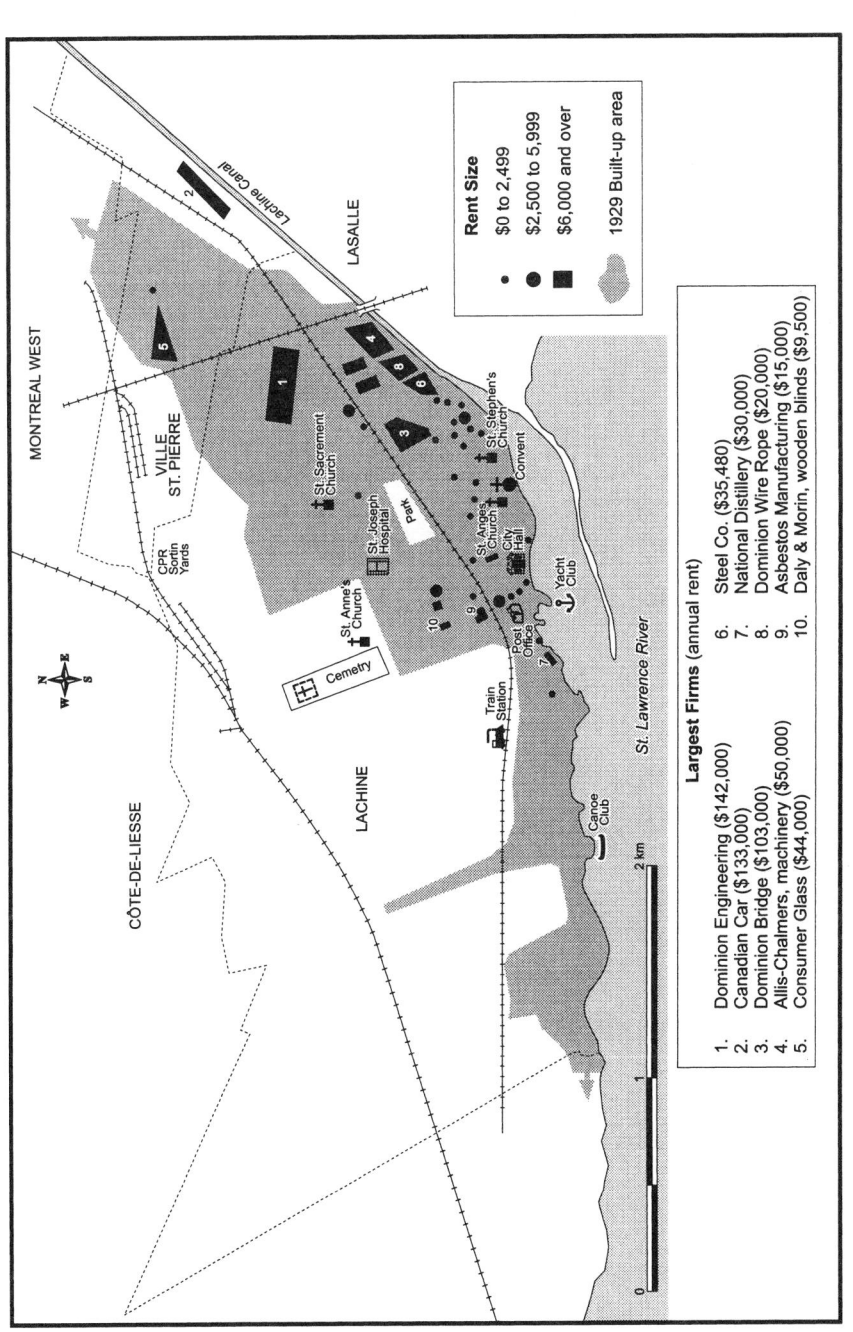

Map 9.4 The Lachine Manufacturing District, 1929. By the early twentieth century Montreal was ringed by several satellite cities. In the west end the mainly working-class towns of Lachine and Ville Saint-Pierre were the home of large steel, machinery, and glass plants. (Compiled from Water Tax Rolls of the Towns of Lachine and Ville Saint-Pierre for 1929.)

share of the aggregate rent declined from 80 percent in 1890 to 62 percent in 1929. At the same time, new sectors (chemicals, locomotives, and nonmetallic) became important components of the western suburbs.[12]

The western suburbs functioned as magnets for both large corporate firms and a variety of smaller firms operating within specialized niches. The classic interpretation of industrial suburbanization stresses the movement of stand-alone, corporate firms employing high-volume, capital-intensive methods. By implementing internal economies of scale and new machinery, these firms freed themselves from the city core and settled on the urban fringe—one important element of the western suburbs' growth. But the development of Lachine and Saint-Henri was also predicated on the agglomeration of firms from a variety of industries, deploying an array of manufacturing strategies, and having linkages to other firms in the districts. After 1890 the West End's attractiveness for firms able to decentralize lay in their ability to build factories of all scales in districts containing a large labor force, excellent transportation facilities, a wide selection of existing firms, and a successful industrial heritage. Manufacturers also moved there with the knowledge that the local elite was ideologically aligned with a policy that put the interests of firms before those of the local working class.

The result, as illustrated by the metal sector, was that quite different firms, ranging from the large corporation to small proprietary establishments, coexisted. The origins of the western suburban metal complex lay in the establishment of the Montreal Rolling Mill in the late 1860s. Over the next forty years the firm grew rapidly and remained an anchor of industrial development, and by the early 1900s it had annual sales of $4.5 million and produced 25,000 tons of rolled products, 12,000 tons of skelp, and an assortment of nails, wire, and other finished metal products. By the early twentieth century, however, the mill, like other metal firms, faced several problems, notably overproduction and competition from Ontario and large integrated American mills. It was forced to implement a new manufacturing strategy. Rather than following the British model, in which hundreds of small-scale proprietary firms dotted the industrial landscape, the company imitated the American, selecting a corporate model of consolidation and applying new processes. Montreal Rolling Mill bought out local competitors (Pillow and Hersey; and Hodgson Iron) and vertically and horizontally integrated with Ontario firms to form the Steel Company of Canada in 1910. Over the next twenty years the company instituted policies of plant expansion, product diversification, and cost reduction by replacing labor with machines. In 1929 the plant—with twenty-two buildings containing rolling mills;

pipe, tack, wire, nail, and horseshoe works; warehouses; engine rooms; offices; and cooper and carpenter shops—sprawled for more than ten acres along the Lachine Canal.[13]

The transformation of the Montreal Rolling Mills from a relatively small nineteenth-century firm to a highly capitalized, integrated corporation had several implications for the West End's metal complex. Most important, it was instrumental in the creation of local locational assets that drew other metal firms to the suburbs. One firm that recognized the attractiveness of the West End was Canadian Car and Foundry, the nation's largest maker of railroad rolling stock. Established by a merger in 1909, it grew rapidly. By 1914 it owned five steel foundries, large timber holdings, and four car works, two of which were located along the Lachine Canal. The two West End plants were affected by a policy of greater plant specialization introduced after the merger: Dominion focused on railroad equipment manufacture (brake beams, bolsters), while Turcot built steel freight and passenger cars. To facilitate the links between the two works, the company built a two-mile railway track in 1913. Similarly, other large structural steel companies constructed plants in Lachine and Saint-Pierre. The first was Dominion Bridge in 1882, followed by Allis-Chalmers, St. Lawrence Bridge, and Dominion Wire. These firms formed part of integrated Canadian and American corporations with extensive marketing and production abilities and a high degree of plant specialization. The West End's metal complex also consisted of smaller niche firms, manufacturing a range of products, including machinery, metal lamps, regulators, valves, and wire meshing. In 1889, for example, Robert Mitchell moved his brass foundry from Old Montreal to a large four-story building in Sainte-Cunégonde. A large site permitting construction of a finely segmented work space and with access to a specialized labor force was a powerful incentive for a firm selling architectural bronze products in national markets.[14]

Other industries sought out the West End manufacturing suburbs. One of the attractions was the particular type of manufacturing space found in the West End. Most important, many came seeking greenfield sites where a brand-new factory could be constructed and other locational assets tapped. An important aspect of the western suburbs was that firms could build factories that suited their particular needs in terms of the internal layout of the firm and the provision of land that could be built upon in times of expansion. Following its move from a cramped Old Montreal workshop to a Saint-Henri factory, Alaska Feather and Down expanded greatly. In 1903 Alaska occupied almost ten times more work space than at the earlier site, while by 1918 the factory covered almost four acres. Similarly, Sherwin-Williams built a modern paint factory on a large Lachine Canal site in Côte Saint-Paul in 1903, where its paint, varnish, linseed oil, can, and box departments were "all connected and operated by one central power plant, insuring economy and efficiency in operating." Obtaining advice from an agent with years of experience in Montreal, the company knew that a western suburban site offered substantial benefits, especially for firms employing a high-volume pathway. Moreover, the western districts had a stock of buildings that could be profitably recycled. When Everleigh, the leather trunk maker, moved from Montreal to an old leather factory in Saint-Henri in 1910, it had not only enough space to expand its premises and install more machinery but also an existing building already suited to the company's manufacturing process.[15]

The agglomeration of firms in the West End was rooted in linkages operating at several geographic scales. West End firms were plugged into national and international raw material markets. Montreal Rolling Mill obtained iron ore from Nova Scotia and Britain and coal from the United States. Colonial Bleaching and Printing imported cotton for its bleaching and dyeing rooms from Britain, Europe, and the United States. Sherwin-Williams received the oxide and graphite essential for paint making from its mines, and the Dominion and St. Lawrence flour mills brought in wheat from western Canada. These distant raw materials were shipped to the western suburban plants via the Great Lakes ports, along the Lachine Canal, or by railroad. Suburban firms also established linkages with local producers, in some cases resulting from vertical integration. This could occur within a single plant: the Montreal Rolling Mill plant had primary steel and finishing facilities as well as cooper and carpentry shops. It could also occur in different local plants of the same firm: the structural steel producer Dominion Bridge transferred its machine-making activities to its adjacent St. Lawrence Bridge subsidiary in 1920; a locomotive maker obtained steel castings from its canal foundry and steel wheels from another local subsidiary. The four western suburbs' Dominion Textile mills each had a specialized position within the corporate hierarchy and functioned within intrafirm flows of skill, information, and materials. There were also local interfirm linkages: local structural metal plants, machinery shops, and foundries obtained rolled metal from the Steel Company; local wadding, bag, and garment manufacturers received finished cotton from Dominion Textile; and Canadian Car provided considerable business for the district's rolling mills, hardware factories, paint producers, sawmills, and brass foundries. Firms in the West End also had metropolitan-wide links.

Andrew Gault's Mount-Royal Spinning Wool supplied a variety of textiles to shirt factories in the city core. Dominion Bridge opened up its East End National Steel plant in the late 1920s because, according to Meredith Burrill, it was "best suited for a light type of work that cannot be profitably fabricated at Lachine."[16]

In some cases firms were drawn by locational assets that did not involve direct linkages to local firms. Imperial Tobacco, Canada's leading cigarette manufacturer, moved to Saint-Henri seeking a particular array of western district assets. Imperial was distinguished by a set of productive and spatial strategies which, with few exceptions, set it apart from other tobacco firms. In 1895, with its acquisition of D. Ritchie, Montreal's first cigarette maker, the company began a policy of consolidation. In the following years it acquired snuff, tobacco, and cigar firms, giving the company an important entry into all branches of the tobacco industry, except cigars. Related to this was Imperial's strategy of making a small number of standard lines of Virginia cigarettes, the primary differentiating factor being competition through brand names backed up with an extensive advertising campaign and distribution network. Associated with the building of a Canadian tobacco empire was the implementation of a strategy based on large-capital investments. The introduction of Bonsack machines enabled Imperial, a highly mechanized firm from the very beginning, to pursue a policy of high-volume production using continuous-processing methods and a largely unskilled work force. As a result, after the late 1880s, according to a commentator, in 1915, in the *Canadian Cigar and Tobacco Journal,* the machine "revolutionized the cigarette industry and when run at top speed it turns out over a hundred thousand cigarettes daily, whereas but a thousand or so were previously made by hand."[17]

The spatial strategy of Imperial emerged out of these productive strategies. First established in Old Montreal, it moved three times before building Canada's largest tobacco factory in Saint-Henri in 1907, "the most perfect example of monolithic construction in the world," according to a piece in the *Canadian Cigar and Tobacco Journal* that year. Imperial's move to Saint-Henri was not directly motivated by the introduction of new technologies or internal economies of scale. Nor were linkages to local firms important; Imperial was basically a stand- alone corporation. Several other factors motivated the move: the need to expand its premises to capture more of the rapidly expanding cigarette consumption; a desire to consolidate its position as the largest tobacco manufacturer in Canada through its factory showcase; and the need to create a more efficient plant for mechanized, high-volume cigarette production. As one writer stated, cigarette machines "are now complicated and ingenious, and are tended generally by a man operator and by several girls [*sic*], whose duty is chiefly to remove cigarettes, which are constantly streaming out."[18]

Saint-Henri offered significant advantages for firms like Imperial. The new factory differed from other buildings it had previously occupied. In 1918 the new plant was four times the size of its last factory. The layout was different: the Saint-Henri plant was much longer, to allow for the flow-through of material demanded by the Bonsack machines. At the same time, Saint-Henri's transportation facilities allowed for easy distribution of the finished products and receiving raw materials. Finally, it contained a large working-class population from which to draw. In the case of Imperial, with its need for a large female workforce, Saint-Henri had a large pool of women needing to supplement the low wages that husbands or fathers brought home.

In the 1860s the Montreal Rolling Mill was a harbinger of the coming of industrial capitalism to an artisanal corner of the Montreal region. In successive waves of growth over the next sixty years new manufacturing clusters developed in the western district. Many of the firms established there sought new productive spaces in which a high-volume, capital-intensive manufacturing strategy could be installed. For the stand-alone firms deploying this strategy, the West End suburbs offered an alluring combination of advantages: excellent transportation facilities, accessibility to a large and varied labor force, proximity to existing large industrial districts, and large sites where large buildings could be constructed and eventually expanded. The western manufacturing districts also consisted of an assortment of firms employing different manufacturing strategies. The Montreal Rolling Mills, unlike Imperial Tobacco, despite its scale and capital inputs, remained committed to the coexistence of elaborate technologies, skilled labor, and a range of products. Similarly, foundries, chemical factories, textile mills, and other firms ranging in scale and manufacturing type forged the link underpinning the development of agglomeration economies.

"Great Activity": Locational Assets of the West End

Extensive manufacturing districts in the western suburbs could not have emerged as they did without the formation of large working-class residential districts. Like what had taken place in Saint-Ann a generation before, the land development industry underwrote the building of working-class districts on the western fringe of the expanding metropolis. A new surge of growth occurred with the building booms of 1896–1912 and the 1920s. At the height of the building cycle in 1906, one oberver noted that "there has been great activity out by Turcot during the past year. Railway activity has chiefly been the cause; besides which a great many factories have recently located in the vicinity. Among these are Tooke's, the Lang Manufacturing Co., Colonial Bleaching Co., Alaska Feather and Down Co., and Canada Malting Co. Further out are the Canada Car Works, Simplex Railway Appliance and others." In 1906, according to the real estate agent, E. Haugh, very few of the six hundred lots that he had for sale in Saint-Henri were "unsold." They were not being held for speculation; most had been built upon. New factories springing up on the metropolitan fringe were attracting workers. In the more westerly Woodlands Park subdivision, for example, Haugh was happy to report that lots were selling fast, "owing to [their] proximity to the Canada Car Company's and other works." But his joy was tinged with concern. Increasing working-class demand for housing close to the West End's factories and railway shops could not be met, he said, as "there is very little unsold land in that section of the city, anywhere, and no more can be brought into the market." He was not alone in feeling despair. Another real estate agent specializing in western property, L. Deneau, pointed to the "shortage of buildings" for workers. Haugh and Deneau were exaggerating the situation in order to boost the demand for land in the western part of the city. Nevertheless, their hyperbole does capture the basic fact that suburban factory building was intimately tied to extensive suburban working-class housing construction during the opening decades of the twentieth century.[19]

An assortment of working-class districts defined the social geography of the western corridor running along the Lachine Canal and the railroad tracks leading out of Montreal to Lachine, ten miles away. Building on its nineteenth-century heritage, the residential-manufacturing corridor running from Saint-Ann through the old suburbs of Saint-Henri and Sainte-Cunégonde continued to be almost entirely working class. While Saint-Ann experienced population decline over the period, losing more than seventy-five hundred people between 1891 and 1931, growth occurred in the western suburbs. Together, the working-class industrial suburbs of Saint-Henri, Sainte-Cunégonde, and Côte Saint-Paul and the residential districts of Saint-Gabriel and Emard grew from thirty-four thousand in 1891 to more than ninety-seven thousand in 1931. Aside from a small number of local businesspeople, Saint-Ann, Saint-Henri, and Sainte-Cunégonde remained entirely working class.

Contemporaries such as the reformer Herbert Ames, in his survey of the "city below the hill," and the Chicago-trained sociologist Carl Dawson, in his studies of Montreal's social ecology, pointed to the working-class character of the area. The early origins of these districts shaped their class composition in the first decades of the twentieth century.[20]

Even further west the industrial towns of Lachine and Saint-Pierre were drawn into the widening metropolitan orbit after 1880. As the older working-class districts filled up, new suburbs developed as nuclei of industrial and working-class settlement. Originally an outpost of the fur trade, Lachine experienced little growth for most of the nineteenth century. With only 1,306 people in 1861 it grew slowly, reaching 3,761 in 1891. By the 1880s, however, industrial growth and local land development lay the basis for Lachine's rise as a manufacturing center on the metropolitan fringe. Over the next forty years the population crept up, reaching 18,630 by 1931. Established a generation later, Saint-Pierre grew slowly. Along with Lachine, even though it only had a population of just over 4,000 in 1931, it constituted an important working-class node in the West End.

In contrast to the older western manufacturing suburbs, these newer industrial satellites had a mixed class structure. In 1921, for example, Lachine and Saint-Pierre had, respectively, 29 percent and 18 percent of their household heads working in managerial, professional, and self-employed occupations. These shares were much higher than in Griffintown or Canal. The towns, especially Lachine, had a small number of professionals and managers providing local health, legal, and religious services or working in the steel corporations. Also serving the local economy were the 10 percent working in the construction trades. Most of the workforce, however, worked in blue-collar occupations. Although a large share of the 60 and 70 percent of the two towns' respective household heads were laborers, there was a range of skilled workers. Concentrations were to be found in the metal (machinists, fitters, molders) and the railway (brakemen, engineers, baggagemen) trades. The French were the largest ethnic group in Lachine (54 percent) and Saint-Pierre (64 percent), but both towns had a substantial English population (38 and 29 percent, respectively), while southern and eastern Europeans made up the rest of the population. Lachine and Saint-Pierre may not have been one-class towns, but by far the largest share of their population was working class.[21]

The Lachine Canal corridor had another type of working-class district. Suburbs such as Côte Saint-Paul (Saint-Paul, after annexation in 1910), despite its early industrial development and the small burst of growth at the turn of the century, were almost entirely residential. An outstanding example was the working-class town of Verdun. Before 1900 Verdun was a small country village comprised of a few houses and a church. By World War I it was the largest suburb in the metropolitan area, and by 1931 its population was more than sixty thousand. Yet population growth did not mean industrial growth—in 1931 the town had only three factories—but it did mean that Verdun became a residential dormitory. This was not, however, a middle-class dormitory but one almost entirely populated by workers employed in the adjacent West End manufacturing districts. Throughout the period more than 90 percent of the workforce were laborers and semiskilled and skilled workers. The remainder were mainly low-paid clerks, with a sprinkle of local businesspeople.[22]

Housing in the various western industrial working-class districts after 1890 was typically built in the speculative land development pattern of the day. A broad range of developers, ranging from large companies with a reach across the entire metropolitan area to small local firms, and an assortment of builders actively constructed residential and manufacturing space. Most of the housing built in the western working-class corridor was of poor quality. In the older districts such as Point Saint-Charles housing stock was "composed of long rows of tenement houses, strongly built, and situated against the street." The problem was that, since their construction in the 1860s and 1870s, many had degenerated into slums as a result of workers' precarious labor conditions, in particular low wages and high unemployment, and the failure of absentee landlords to reinvest their profits into upkeep. Although better housing conditions existed in the newer sections, quality still left a lot to be desired. Verdun's housing was poorly designed and cramped and had little sunlight, while in Saint-Paul, "a low lying, unhealthy section" of the city, housing was "given over to a cheap type of dwelling." With few exceptions, buildings in the western districts were two- and three-story row houses; there were few detached homes, and row housing tended to be small in size, leading to very high population densities. But speculative housing was not the only type to be built. In Saint-Pierre, for example, a number of homes were built by the owners with "their own hands and in their spare time," while scattered throughout the districts some manufacturers, such as Lachine's Dominion Bridge, built "neat little brick cottages" for their workers. Despite evidence of owner- and company-built housing, it appears that most dwellings in the West End's working-class districts were built by the city's construction industry.[23]

The electric streetcar connected these working-class residences to the city. Although Montreal had a relatively small streetcar system, and many suburbs had difficulty obtaining lines, the suburbs eventually did get service. In 1896, for example, Lachine received a line from the Montreal Park and Island Railway Company, but only after receiving sole privileges and a thirty- year tax exemption. Despite the apparent necessity of the streetcar, it was not of great importance for many of the districts' inhabitants. Most workers, either in residential suburbs such as Verdun or satellite towns such as Lachine, were within walking distance of their places of work. In some cases the streetcar brought in workers from the city. Labor for the factories of Saint-Pierre, according to one commentator, came "from the huge labour market of Montreal which for some time has been tapped by the street railway system, bringing hundreds of men each morning to work, and taking them home in the evening." Similarly, some workers commuted from the East End to work in the Point Saint-Charles Grand Trunk shops. In most cases, however, workers were unable to commute, as residential settlement preceded the trolley. Even when it did not and even if suburban workers wished to spend their time commuting, many could not afford to do so. The major reason was, as Lloyd Reynolds notes in his study of British immigrants in Montreal, "the labouring family lives relatively close to the subsistence level," and only skilled blue-collar and white-collar workers were consistently above the poverty line. As the large majority of the suburbs' population did not fit into either of these occupational categories, a large share of suburban dwellers could not afford to commute to and from the city. The only group to use the trolley to a significant extent to get to work was the small clerical workforce who commuted downtown. Even though progressively higher wages and a more accessible public transit system made a longer journey to work possible, most workers walked to work at the local factories.[24]

In many of the western districts a mill-type community developed, each "impregnated with small town attitudes." A strong work-home relationship in Point Saint-Charles dating from the 1850s was found between the Grand Trunk shops and the residential area. As late as the 1930s, the Point continued "to house a wage-earning population which would find work in the neighbouring industries," especially the Grand Trunk's repair shops, yards, and locomotive manufacturing works. This was reflected in the long tradition of sons following fathers into the company's workforce. In 1929, for example, almost two hundred workers in the motive power shops were second, third, and fourth generation. Extensive commercial and recreational facilities supported the mill quality of many working-class districts and added to the importance of a

Figure 9.8 Wellington Street, Pointe St. Charles, Montreal, QC, c. 1910. As the main street of the mainly British, working-class area of Pointe Saint-Charles, Wellington Street catered to the everyday shopping, institutional, and social needs of the manufacturing district south of the Lachine Canal. (Notman Photographic Archives, McCord Museum of Canadian History, Montreal, accession no. MP.000.879.3.)

short journey to work. Verdun's wide range of social and cultural institutions and retail shops meant that residents only went downtown for specialized commodities. The Point had baths, a Grand Trunk–funded recreational ground, and saloons, where, according to Mary Davidson, "there is hell to pay every Saturday night" (fig. 9.8). Saint-Pierre had twenty-five stores, two churches, two schools, a fire brigade, a police force (with four officers), and a town hall. These facilities, combined with a local labor shed and local environmental barriers restricting movement, produced a mosaic of separate communities. As Davidson said of Verdun and the Point in 1933, their "physical isolation from the larger city in the past tended to make these two areas self-sufficient in their own social life, even today."[25]

Orchestrating the flow of industrial and property capital into the western suburbs after 1890 were politicians as generous and malleable as their predecessors. Western municipalities had long been active in shaping local policy for the benefit of manufacturers. By the 1870s local machines helped create the nucleus of an important manufacturing and working-class complex in Saint-Henri and Sainte-Cunégonde. In all of the suburbs between 1890 and 1929 the activities of speculators and developers, as in all of Montreal's districts, were closely tied to that of local governments. In Verdun, for example, zoning and building by-laws were "manipulated for private ends by speculators" such as John Crawford and U. H. Dandurand. Along with the extensive land development undertaken by the likes of Haugh, Deneau, Crawford, and Dandurand, the provision of bonuses and the neglect of any systematic planning laid the basis for a working-class residential and manufacturing corridor along the Lachine Canal between Griffintown and Lachine. These policies, formulated at the expense of the local working class, added to the suburbs' economic competitive edge. With each new growth cycle a district's place in the city's spatial division of manufacturing revolved around the ability of local boosters to capture industrial expansion and to build working-class housing.[26]

One locational asset determining the nature of the western fringe's industrial base were industrial bonuses. Saint-Henri and Sainte-Cunégonde, in particular, became the home of firms no longer finding Montreal a fertile place for business. Many large firms settling in the industrial suburbs were escaping problems associated with their location in Montreal. The city believed that "if this great city, the chief seat of Canadian manufacturing enterprise, cannot get along without giving annual bonuses to factories, it had better lower its flag to the level of the small municipalities that compete with each other for a blacksmith's shop.... Montreal is not a hospital for decrepit industries."[27]

While Montreal may not have wished to lower its flag, the western suburbs did. In the late nineteenth and early twentieth centuries municipalities devised a similar and highly competitive industrial policy centered on the provision of tax exemptions, free land, and bonuses. The 1890s were particularly busy years. Tooke, the large Montreal shirt firm, for example, received a $35,000 bonus in 1899 to move to Saint-Henri. In 1892 Côte Saint-Paul granted a $10,000 bonus and a twenty-year tax exemption to the Canada Axe and Harvest Tool Manufacturing Company. Lachine attracted the Simplex Railway Company, a producer of railway supplies, from Saint-Henri. Moreover, firms already established in a district were provided with subsidies to induce them to remain or expand in situ. In the 1890s Saint-Henri extended a $20,000 bonus and a twenty-year tax exemption to Merchant Manufacturing for an extension to its mill (fig. 9.9). Suburban councils also presented other types

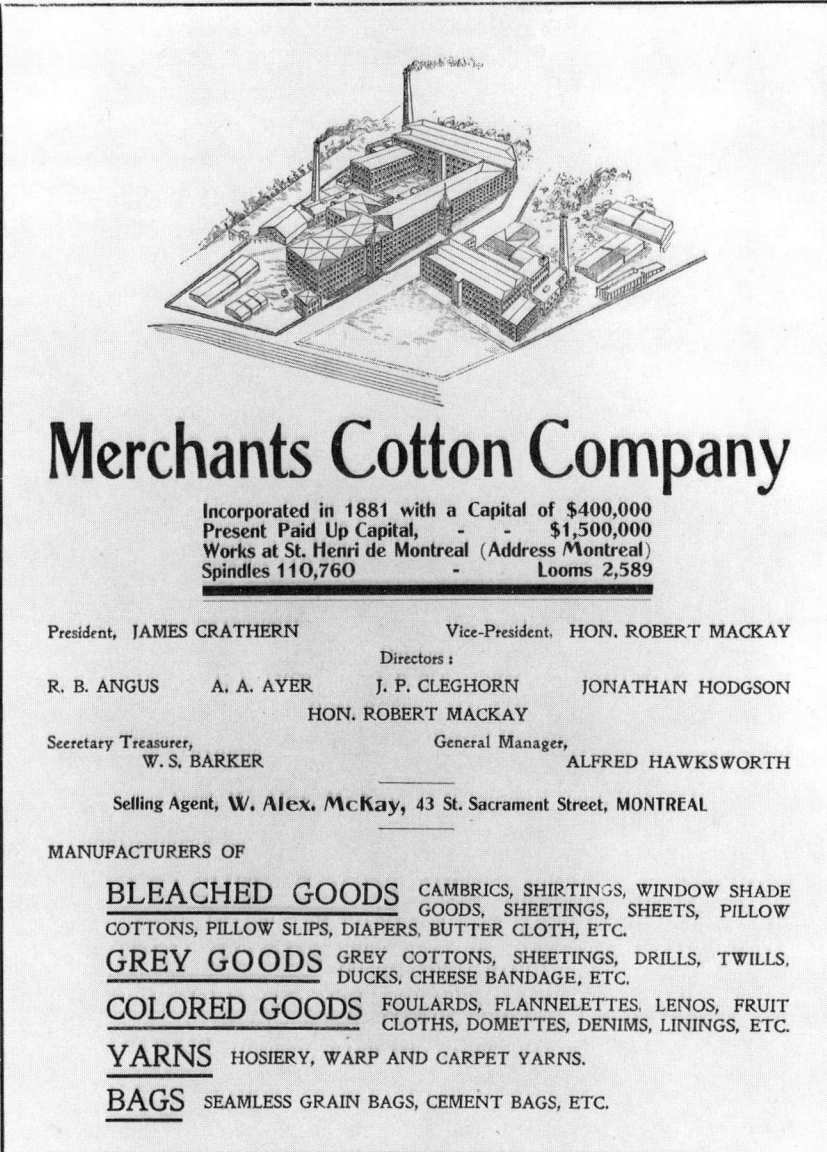

FIGURE 9.9 MERCHANT COTTON ADVERTISEMENT, 1905. Located on the banks of the Lachine Canal in Saint-Henri, Merchant Manufacturing was a typical large, high-volume plant producing staple goods for the national market settling in Montreal's manufacturing suburbs. (Ernest Chambers, *The Book of Canada* [Montreal: Book of Canada Co., 1905], 334.)

of economic activities with bonuses. Côte Saint-Paul gave the Grand Trunk Railway a $15,000 bonus and a twenty-year tax exemption so that "the factory town ... [could] have railway communication with the city." The suburbs, however, were not always successful. In 1898 Lachine granted a $15,000 bonus to the Montreal leather trunk manufacturer, George Barrington. Yet the company did not move to Lachine. Instead, it opted to remain in Montreal, perhaps because the conditions under which the bonus was granted (two buildings to be constructed with a value of more than $25,000, permanent employment of eighty workers, and proper fire insurance) could not outweigh the attractions of the central factory district. Despite examples such as Barrington, bonuses and tax exemptions pulled firms seeking a suburban site to the western suburbs.[28]

Despite the form of control exerted by local alliances of developers, manufacturers, and municipal governments, the tremendous industrial and working-class residential growth taking place in the Lachine Canal corridor was almost entirely uncoordinated at the metropolitan level. Other than the provision of bonuses, there was little formal regulation of urban form and social conditions. Suburban expansion was undertaken without any overall pattern or plan, as in the case of Lachine, where, according to the Montreal Junior Board of Trade, "there is not much attempt at zoning or planning and the chief business of the City seems to be persuading firms to locate there and allowing them to do so in an unplanned manner." One consequence of this policy of uncoordinated growth was the rapid deterioration of many West End areas. By the 1930s the districts that had experienced the great boom that Haugh and Deneau took part in at the turn of the century showed all the signs of the classic industrial slum: poverty, high infant mortality rates, few services, terrible housing conditions, and the intermingling of factories and homes. As Gabrielle Roy describes in her novel of life in Saint-Henri, "streets with low houses descended in two directions toward the areas of greater poverty, on this side to Workman Street and St. Antoine, and on the other, down to Lachine Canal where St. Henri stuffs its mattresses, spins it thread of silk or cotton, runs it looms, reels off its spools, while the earth trembles at the rushing trains, and the foghorns blast, and the ships, engines, screws, rails and whistles spell out the adventure of the world."[29]

By the turn of the century the West End's manufacturing and working-class districts were firmly caught in the orbit of Montreal's space economy. Even though these districts are considered to have developed without a logic to their industrial structure, this was not evident between

1890 and the Great Depression. There was a logic to the actions of local municipalities. The particular strategies they invoked underpinned the making of manufacturing spaces attractive to manufacturers. The local bourgeoisie, in conjunction with actors at a metropolitan scale, created malleable spaces regardless of the long-term costs—heavy debts, environmental degradation, lack of essential services, and poor housing conditions. This situation stands in sharp contrast to the few non-working-class West End suburbs. In Westmount, Notre Dame de Grâce, and Montreal West the establishment of restrictive bylaws created quite different environments than those found in the industrial suburbs and city manufacturing districts. The suburb of Westmount, located on the slopes of Mont Royal, to the north of Saint-Henri, for example, developed from the very beginning as an exclusive enclave for Montreal's English-speaking bourgeoisie. Here, as in Notre Dame de Grâce and Montreal West, the establishment of a set of land-use and urban design restrictions shaped its social composition. Westmount prohibited factories in most parts of the suburb, established a strict set of guidelines for housing design and lot size, and installed an elaborate set of infrastructures.[30]

There was also a logic to the linkages established in the districts. These were not necessarily direct linkages between firms of the same industry, which was common, but also the linkages created by firms having similar productive pathways and the ability of these firms to carve out or take advantage of new productive spaces. They were districts built upon an assortment of firms pursuing a range of manufacturing pathways. Some, like the cotton and flour mills, operated a highly mechanized, high-volume production process manufacturing basic lines for a national market and employing a large, unskilled workforce. Others had more specialized markets and depended on a greater range of skills.

Despite the differences, firms settled in a manufacturing corridor with a long and successful history. The new districts after 1890 were built upon earlier waves of growth, were specific to the western part of the city alongside the canal, and were linked to one another and other parts of the city. Once set in motion, these suburbs were home to particular industries and a large working-class population living in separate communities. The western suburbs formed a specialized district within Montreal's spatial division of labor and depended on several forms of capital acting together to create new manufacturing strategies and manufacturing spaces.

Conclusion

Montreal's Manufacturing Districts, 1850 to 1929

> We already find that, to the east, the north, the west, mills, factories, railway freight terminals and warehouses are pushing forward indiscriminately, and becoming unpleasantly intermingled with many of our most desirable residence districts, making it difficult to design and control a proper scheme of highways, boulevards, parks, business thoroughfares, and local transportation systems.
>
> —Arsene Lavallé, mayor of Montreal,
> *Proceedings of the Fifth National Conference on City Planning*, 1913

Since the mid-nineteenth century functional differentiation, residential decentralization, and ethnic and class segregation have been fundamental features of the North American city. Building on these themes, generation of writers, from the Chicago School to contemporary urban historians, have pointed to the spatial polarization of the metropolitan area. Urban research for the most part has followed this broad picture. One important strand of work has been to trace the building of the densely populated working-class and immigrant neighborhoods and factory districts surrounding the functionally specialized central business district; another has been to document the emergence of the bourgeois utopia in the middle-class suburbs before World War II. The result of this separation of the metropolitan area into a working-class central city and middle-class suburbs has ensured that the importance of manufacturing and working-class decentralization has been largely ignored. This is true for Montreal as well as other North American cities. Writers such as Gérald Martin, Jean Delage, and David Slater have constructed an image of Montreal in which manufacturing and working-class suburbanization was relatively unimportant before the end of the 1930s.[1]

While this urban structure may have some relevance to contemporary American metropolitan areas, the dual structure bears little semblance to how North American cities developed before World War II. Certainly, the picture I have drawn here of Montreal's manufacturing geography between 1850 and 1929 fits neither the description nor the interpretation of manufacturing and working-class decentralization given by successive cohorts of urbanists. In this study I have shown that a succession of manufacturing districts pushed their way out of Montreal's built-up area and settled on greenfield sites from the 1840s. Two broad cycles of growth have been identified. In the first, a set of manufacturing districts appeared on the fringe areas of the West and East Ends: the midcentury districts of Griffintown, Canal, and Sainte-Marie were followed a generation later by Hochelaga, Saint-Henri, Sainte-Cunégonde, and Côte Saint-Paul. In the second round of growth after 1890 some of these older districts declined, in some cases absolutely, in others relatively. Yet this decline was more than balanced by the continued expansion of other older districts and the ongoing march of manufacturing to more distant sites, notably Maisonneuve, Mercier, Montréal Est, Mile End, Lachine, and Saint-Pierre.[2]

Central to this historical geography of Montreal's manufacturing was the need to manage the tensions associated with the need for industrial capital to be both place bound and geographically mobile. Manufacturing capital had to be operationalized at specific sites. From the placing of an individual machine in a factory to the distribution of labor through a plant and the laying down of a spur line, all firms had to coordinate and regulate their industrial organization in place. The character of the spaces constructed by manufacturers varied according to the demands of particular manufacturing strategies; owners and managers of small machines shops, large steel mills, garment lofts, and cement factories had to create place-specific sites for their different productive facilities and labor forces. Firms also had to ensure that their location was directly linked to surrounding firms; they had to tie their place-bound organization to other geographically proximate firms. Despite long-distance relationships with suppliers, markets, inventions, and financial institutions, manufacturers forged short-distance links to nearby suppliers, customers, credit, and workers. Even though these links were more important for vertically disintegrated firms, all firms, even those with developed internal economies of scale, had to bind themselves to a specific site with an array of services, infrastructures, workers, and locational assets such as transportation facilities, roads, and sewers. Over time these place-bound links also stretched, generally in one of two ways. First, over a generation, sometimes two, manufacturing districts attracted new manufacturing investment in the form of factories, machinery, and material-handling equipment and new social capital in the form of housing, sewers, and transportation facilities. Second, manufacturers depended on the renewal of existing place-bound locational assets in order to accommodate changing production needs. In the process firms became embedded in geographically organized manufacturing complexes stretching over generations.[3]

Manufacturing capital also constantly sought out new territory, as districts created in one generation became a problem for the next. Spurred by competitive conditions operating at regional, national, and international scales, firms had to adapt continually if they were to be successful. In their search for profits manufacturers introduced new technologies, intensified the division of labor, expanded the plant, changed product lines, and sought out new markets while keeping production at the same site. But there was another option available to manufacturers; they could shift capital from old industrial sites to greenfield ones. Three principal sets of dynamics induced manufacturers to shift capital investment to different parts of the expanding metropolitan area. Changes internal to manufacturing, such as the introduction of new machinery, new work processes, and increased firm size, generated demand for new forms of factory layout and scale, which in turn forced firms to seek land where modern and space-extensive plants could be located. Firms serving growing national and international markets required modern and more efficient transportation facilities, many of which were built on the city fringe. Finally, firms seeking to escape the problems of high rents, congestion, and rundown sites shifted investment to new manufacturing spaces on the metropolitan periphery.

Leading the march out of the built-up area were large, propulsive firms. In each succeeding growth phase they tended to be larger and more technically sophisticated, to demand different and larger work spaces, to deploy new work processes and a much finer division of labor, and to serve wider markets. By today's standards many of these firms would be considered relatively small and technically "primitive." At the time, however, their new productive relations and technological attributes modified existing productive pathways, challenged existing locational options, and created new place-bound opportunities. By taking advantage of their widening scope of technological and organizational change, propulsive firms were able to construct a spatial flexibility centered on relocation to fringe areas. This was the case in Montreal, as firms from several growth industries formed the core of various manufacturing districts: iron and steel mills, locomotive shops, and textile mills in the West End; large paint and paper factories in Mile End; and cement works and petroleum refineries in Montréal Est.

But large firms were not the only ones moving to new manufacturing districts. It is a generally accepted tenet of urban industrial decentralization theory that central factory districts were the repository of agglomeration economies and the home of small, labor-intensive firms. In contrast, the few industrial suburbs that are considered to have developed before World War II are viewed as the recipients of large, capital-intensive corporations. Evidence for this can be found in Montreal. Montréal Est's space-extensive petroleum refineries and cement mills, employing high-volume production methods, are one example of industrial suburbanization centered on a small number of vertically integrated, stand-alone firms with relatively few linkages at the district level. But this was the exception. More typically, firms from a variety of industries employing a range of productive strategies and scales settled around propulsive firms located in metropolitan greenfield sites. Some sought the same sort of advantages as large firms, while others clustered to supply or receive semifinished products from firms already in place. From the 1840s West End foundries, metal finishing shops, sawmills, cooperages, and textile firms of all scales and organizational structures developed close to large rolling mills, flour and sugar refineries, and cotton mills. Similarly, Mile End paint factories after 1900 were well connected to a set of local suppliers and customers. What the Montreal evidence indicates—even though the volume of interfirm links may have diminished with the growth of the large, vertically integrated firm after 1900—is that an extensive linkage network was forged between large and medium semi-integrated firms and niche producers.

Strong local linkages, however, can be identified even in those manufacturing districts in which large, stand-alone firms likely dominated. In his description of Montreal's manufacturing Jean Delage views Canadian Copper's Montréal Est plant as representative of the interwar stand-alone, vertically integrated, capital-intensive suburban firm. He argued that the suburbs were enticing because firms such as Canadian Copper could efficiently distribute their product to national and international markets and could install an advanced form of high-volume, continuous production on a large greenfield site with excellent power and shipping facilities. There is some truth to this account. Land was cheap and plentiful in Montréal Est, and some of Canadian Copper's electrolytically refined copper went to export and national markets. The major customer, however, was not the nonlocal market. Most of the refined copper found its way to the newly constructed and adjacent rod mill and cable plant of Canada Wire and Cable.[4]

In many ways Montréal Est, with its small number of large, capital-intensive firms, narrow industrial structure, and small working-class population, was not representative of other manufacturing districts. But it was typical in two important regards. First, the symbiotic manufacturing and geographic relationship between Canadian Copper and

Canada Wire illustrates the close ties existing between corporate-owned plants on an industrial site on the metropolitan periphery. Despite their apparent lack of local linkages, many Montréal Est firms were not autonomous, stand-alone entities. Along with Canadian Copper and Canada Wire, the suburb's cement factories, for example, relied on locally produced raw materials. Moreover, all firms depended on localized labor markets and had strong links to the city core, either because their head office was located there or because of their dependence on the core's financial, information, and distribution networks. These types of linkages were common throughout the metropolitan area. Second, the Montréal Est case highlights how important the balancing of the tension between immobile and mobile capital was for Montreal's manufacturing geography. From the development of the Lachine Canal and Sainte-Marie districts in the 1840s to the establishment of Montréal Est in the interwar period, manufacturing districts represented the fusing of place-bound links with geographically shifting capital. In each cycle of growth new territorial nodes consisting of new forms of modern plant design, technological innovation, and factory methods were superimposed upon the metropolitan landscape.

Our understanding of metropolitan manufacturing geography is further complicated by how historians and historical geographers have defined the suburbs. For most scholars the political fragmentation of the metropolis into city and suburb is believed to have had an important impact on industrial decentralization. For social historians, such as Robert Fishman and Kenneth Jackson, the suburbs before World War II are by definition, middle class, white, and politically separate from the city. While the same view is held by economic geographers such as Jean Delage, David Slater, and Allen Scott, they also point to the beginnings of manufacturing suburbanization from the turn of the century in independent municipalities surrounding the central city, such as Chicago's Pullman and Gary. Just as with those who portray the central city–suburban divide as a decisive element of the social geography of metropolitan areas, the boundary dividing city from suburb is equally critical for writers who point to industrial suburbanization before World War II. Their argument is that the bounded nature of social and political life allowed small suburban municipalities to offer firms a set of direct and indirect assets not obtainable in the politically fractious central city. In particular, by moving to the suburbs, large corporations could establish suburban mill fiefdoms and gain extensive economic and political control.[5]

The argument that municipal boundaries had a decisive impact on the geography of manufacturing by enabling large firms to move to suburban municipalities, however, is questionable. Despite the manufacturing suburbs controlled by a U.S. Steel or Pullman Palace Car, there are very few examples of manufacturing suburban fiefdoms. This is not to gainsay the power that firms could exert, but their strength usually resulted from the accumulated influence of numerous firms and the industrial growth ethic of local government, rather than the overriding power of one large corporation. The existence of a multitude of firms in Montreal's suburbs ensured that manufacturers could not exercise monopoly power, yet several Montreal industrial suburbs, including Saint-Henri, Sainte-Cunégonde, Lachine, Maisonneuve, and Montréal Est, did establish industrial and social policies that placed manufacturers' interests ahead of the local working class. From the 1840s steel, textile, locomotive, petroleum, and cement corporations, among others, took advantage of the search by suburban elites to attract manufacturing. In the process manufacturers were able to construct a set of power relations in the suburbs which were less opaque than those in the central city. The ability of firms to control the destiny of industrial suburbs nevertheless only partially accounts for the tremendous volume of suburbanization occurring after 1850.

The geography of metropolitan manufacturing is also problematic because of the method typically used to measure the timing and degree of suburbanization. Almost without exception, studies have taken the post-annexation definition as the sine qua non of *central* and *suburban*. Although this makes data collection a simple task, it is weak methodological practice. Cities experienced massive territorial growth before 1920 as suburbs of all types were enveloped within the judicial boundaries of aggressively expanding cities. While central-city aggrandizement was a response to a variety of city-suburban issues, the point to note here is that a substantial part of what is considered central-city manufacturing growth occurred through the annexation of industrial suburbs. This was the case in Chicago, New York, and Philadelphia. It was also the case in Montreal, where the absorption of suburban Hochelaga, Saint-Henri, Sainte-Cunégonde, Longue Pointe, and Maisonneuve between 1883 and 1918 added considerably to the city's manufacturing base. Given that suburbs were defined as part of the central city once they were annexed, it is not surprising that writers found extensive centralization of industry in Montreal before 1920. But what they fail to acknowledge is that their analysis was based on a history of manufacturing decentralization dating back to the 1840s.[6]

For these reasons, the importance of political boundaries separating the different parts of the metropolitan area has to be shown, not assumed. In Montreal from an early date manufacturers sought noncentral locations, whether or not in separate municipalities. Sometimes this was because nuisance industries were pushed by bylaw or public pressure to the edge of an urban settlement. In most cases, however, it was because firms sought out locational assets being created on the city's unbuilt fringe as well as in the suburbs. Independent suburbs did not have a monopoly on these assets; city fringe districts had them in equal measure. While suburban bonuses attracted firms, we have to be clear how they affected the firm's locational decision-making process. What was important was that subsidies were only decisive once a company had already decided to move to a suburban location. Bonuses by themselves could not entice a firm to settle on the fringe; they certainly played a part in locational sorting by suburb but not in the decision to decentralize. For a firm the decision to move to the suburbs lay not in the savings induced from a bonus or a tax exemption but in how its manufacturing strategy fit into the broader competitive dynamics of the industry. In other words, decentralization had little to do with the number of firms on either side of what was often an arbitrary line separating the central city from the suburbs. Much more decisive was the manner in which the metropolitan built-up area was stretched to accommodate the changing requirements of industrial firms.

The stretching of Montreal's urban form to accommodate capital's spatial mobility and immobility led to the formation of a distinct spatial division of manufacturing between 1850 and 1929. The Canal district, for example, with its iron and flour mills, differed from Sainte-Marie, with its brewing and rubber factories; from the central manufacturing districts, with their garment and printing plants; and from Mile End, with its paint, chemical, and paper mills. The linkages spun off from the need to bind manufacturing capital to place strongly influenced the geography of new investment and shaped the specialization of Montreal's districts. Similarly, both the timing of an industry's growth and the history of particular manufacturing pathways produced a spectrum of locational choices that took on specific geographies. For example, the growth of the textile industry after 1870, coupled with its large scale and its specific structure, established the basis for the industry's geography. It was no coincidence that the industrial suburbs emerging in this period—Hochelaga and Saint-Henri—were home to Montreal's large cotton manufacturers, with their high-volume, mass-market strategy.

Similarly, a generation later new industrial suburbs became the home of rapidly expanding modern paint, locomotive, petroleum, and steel plants.

Suburban manufacturing districts were not alone in their changing geographic character. Central manufacturing districts were also transformed. Clothing, printing, and carriage-making plants of all descriptions—from large, technologically innovative firms introducing extensive work rationalization to the small workshop functioning as a niche producer—had to contend with changing industrial dynamics. In adapting to these changes, centrally rooted firms refashioned the geography of the central manufacturing districts. In Montreal the most obvious change was the move of the city's classic centralizing industries from Old Montreal to the Outer Core, especially after the late nineteenth century. For tobacco and clothing firms the shift in manufacturing investment involved moving uptown, where they built an array of plants and sought out the new locational assets created there.

Again, the pattern and processes of Montreal's manufacturing geography were not unique. As a few studies suggest, the dynamics driving manufacturing district formation in Montreal were replicated in cities as diverse as New York, Baltimore, Philadelphia, Chicago, Pittsburgh, Detroit, St. Louis, and San Francisco. In all of these cities an array of specialized manufacturing clusters with local linkages formed on the moving metropolitan fringe, both within the city itself and adjacent areas. Despite these commonalities, Montreal differed from these cities in several ways. The social and political legacy of British colonialism and the "two solitudes" of the French and British shaped Montreal in a manner not replicated anywhere else on the continent and imposed a unique imprint on the city's geography. Another notable difference lay in the structure of its manufacturing pathways. In the second half of the nineteenth century, for example, Montreal's balance of milling, mechanical, technical, and craft pathways differed from that of other cities. Accordingly, the demands on Montreal's manufacturing space were unlike Pittsburgh and Chicago, with their concentrations of heavy industry, and New York and Toronto, with their large share of labor-intensive producers. These differences translated into different space requirements and different manufacturing geographies.[7]

There were also industrial differences between Canada and the United States. Firms north of the 49th Parallel had a more restricted industrial structure than in the United States. This partly had to do with the resource-based economy bequeathed by British colonialism and the development of Canada's industrialization within the shadow of America. It also had to do with the small scale of the Canadian market and the early growth of powerful multiplant corporations in the United States. Canadian firms were unable to reach the same scale as American firms and thus were unable to construct the same economies of scale and scope. In addition, the move of American corporations to Canada from the 1870s led to substantial control over Canadian manufacturing and arrested technical innovativeness and investment in growth industries. Thus, even though cities had a common set of dynamics, a different set of locational parameters emerged in Montreal.[8]

"Capitalists and Manipulators": Building Manufacturing and Working-Class Districts

Along with its industrial dynamics, another critical element shaping Montreal's geography was the power that business and political elites had to redraw the industrial landscape. This point was acknowledged by an anonymous writer in the *Canadian Engineer,* who noted in 1905 that a handful of capitalists wielded considerable sway over the everyday life of Montreal's citizens: "In a vain endeavor to get effective competition and cheap rates, encouragement has been given in the past to private companies, but one another after opponents have been absorbed by capitalists and manipulators who influence the legislation of the province as well as the city, and the result is that Montreal is a city of 300,000 where it might be a city of 900,000 if natural advantages were administered for the benefit of its citizens and manufacturers instead of for the benefit of a handful of capitalists." Echoing this sentiment, the story told here of Montreal's manufacturing districts has emphasized the power of manufacturers, merchants, land developers, and politicians to shape urban form. Through their control over capital, land, and power, they, sometimes as individuals, sometimes as members of local coalitions, decided where and when factory sites were to be opened up, working-class housing constructed, and urban infrastructures laid down.[9]

These elites, however, were not omnipresent, nor did they always operate cooperatively. Although they made many of the decisions leading to the development of manufacturing districts, they functioned as individuals bounded by their class, the competitive conditions they faced in their industry or business, and the obstacles and opportunities of Montreal's political economy. Nor did they construct a coordinated plan at the city and metropolitan scales. Functioning in an ad hoc manner at the city neighborhood and suburban scales, individuals and elite alliances went to considerable trouble to pursue their own ends without the benefit of formal regulation at the metropolitan level. Reacting to the wider structures of capitalist social relations but coming together as individuals and coalitions operating within a shifting set of institutional arrangements, local elites created individual landscapes that bore little relationship with those built by others. This was not for want of trying. In Montreal a variety of reform governments, commissions, committees, and leagues were patched together in an attempt to generate the appropriate mechanisms for overseeing urban growth in an efficient and orderly manner. The solutions, however, did very little to deal with the structural roots of the fragmented and unequal city. Nevertheless, spurred by the twin needs to increase the volume of capital circulating through the expanding metropolitan area continually and to promote nodes of investment on the urban landscape, local coalitions established the place-bound locational assets underpinning manufacturing and working-class districts.

The success of these manufacturing districts depended on the ability of urban elites to construct a viable intrametropolitan transportation network. The extension of East End harbor facilities laid one foundation for the formation of manufacturing districts in Hochelaga, Maisonneuve, Longue Pointe, and Montréal Est. The cyclical redevelopment of the Lachine Canal underwrote the move of manufacturing from Saint-Ann Ward to the western suburbs as far as the town of Lachine. The Grand Trunk and Canadian Pacific Railway lines formed manufacturing corridors in all directions. While laying the basis for individual manufacturing districts, port, and railroad investments also linked firms to one another and integrated individual sites, districts, and the entire metropolitan area into one functioning entity. These transportation facilities, however, did not just appear on the urban landscape. Each round of transportation development was built by a spatially nested coalition composed of local and national business elites and international capitalists who channeled huge amounts of public funds into a private landscape of infrastructure. In some cases they were built with the needs of individual firms in mind, notably the building of railroad spurs and shipping piers. In other cases they were constructed for other reasons: competing with other North American cities' harbors, capturing long-distance rail freight, and bypassing the physical constraints of the river. Despite disagreement over details—the different aims of promoters and the occasional failure to achieve the required ends—the outcome was the same: a set of networked locational assets that attracted, sometimes very quickly, sometimes over a much longer period, factories and ancillary activities.

Equally important was the redevelopment of Montreal's existing built environment by local business and political elites. Refitting the landscape to manufacturing needs took many forms. A common one was the razing of buildings and the construction of new ones. The redevelopment of Old Montreal and the Outer Core's commercial buildings over several building cycles, for example, contributed to the development of a central factory area. In the second half of the nineteenth century much of this activity occurred in Old Montreal. The old Grey Nuns and Hôtel Dieu Buildings, for example, were torn down and replaced by warehouses that provided space for manufacturers and an assortment of mercantile functions essential to the formation of central agglomeration economies. By the end of the nineteenth century the conversion process had moved north. Parts of the Outer Core underwent demolition and were replaced by multistory loft buildings housing hundreds of small and medium-sized firms dependent on communication and subcontracting networks. Along with conversion of the central factory districts, the manufacturing built environment was redeveloped in other ways. Rejigging the street system by laying down new streets and widening and modernizing existing ones refashioned the locational assets of various districts. Waves of investment in the harbor and canal involved not only the building of new docks but also equipment renewal and the redesign of facilities. None of this occurred within a vacuum. The ability of local alliances to channel chunks of public and private capital into the (re)construction of the built environment offered manufacturers a new set of locational assets and options.

Development of the local property market also widened the range of locational sites. Speculators, developers, and builders responding to the profits to be made from urban development created locational assets in the form of metropolitan-wide working-class residential districts. Typically, the hitching of working-class residential areas to factory districts before World War II is seen as taking place in the central city. This was the case in Montreal. Adjacent to middle-class and elite residential areas, the central factory districts of Saint-Antoine, Saint-Lawrence, and Saint-Jacques contained a very large working-class and immigrant population living in close proximity to centrally located manufacturers. But this symbiotic relationship between central employment and worker housing was also replicated on the city fringe after the mid-nineteenth century, as shifting coalitions of builders and developers marched out of the built-up city. In the process the land development industry, in conjunction with local suburban politicians, financiers, utility companies, merchants, and manufacturers, produced an assortment of residential areas on the urban fringe.[10]

It is difficult to determine whether industrial or working-class suburbanization was the leading factor in the spread of the city. The Montreal evidence suggests that both were important. On the one hand, the building industry constructed a suburban quilt of cheap working-class housing not directly connected to manufacturing employment. For example, the populating of the northern working-class suburbs of Saint-Jean-Baptiste and Côte Saint-Louis in the second half of the nineteenth century was neither the product of the establishment of factories within their boundaries nor the large-scale shift of manufacturing northward. On the other hand, suburban builders followed the exodus of employment to the city fringe. Existing manufacturing clusters spread into adjacent areas and drew residential areas to them. Greenfield sites separated from existing districts attracted manufacturing and formed the core of new working-class districts. The development of midcentury residential development south of the Lachine Canal and in Sainte-Marie, for example, coalesced around iron works, foundries, flour mills, breweries, and rubber works.

The picture, however, was not so neat and tidy. In nearly all cases working-class residential and industrial suburbanization functioned simultaneously. In many of the city's manufacturing districts the establishment of a few initial firms generated working-class housing construction. The ensuing move of workers to these fledgling areas frequently led to the arrival of more factories, which in turn attracted more housing development. In other cases relatively autonomous satellite towns and suburban villages with an existing industrial and population base were incorporated into the wider metropolitan fabric. Large steel manufacturers were attracted to Lachine because of the presence of a small working-class population, infrastructures, and services, while Saint-Henri and Sainte-Cunégonde's growth centered on the spiraling effect of population and the expansion of manufacturing. Similarly, Saint-Jean-Baptiste and Côte Saint-Louis, despite their origins as residential districts, provided a labor pool for the North End manufacturing districts of Mile End and Plateau.

Also contributing to the formation of metropolitan manufacturing and working-class districts was the creation of manufacturing sites throughout the city. Motivated by the same factors underpinning residential land development, local alliances subdivided land for manufacturing sites. In each cycle of growth the majority of these lots were opened on the expanding urban fringe. The succession of sites along the Lachine Canal, for example, resulted from shifting alliances of merchants, manufacturers, landowners, the Sulpicians, and local and federal governments. To the east and the north developers, in conjunction with railroad companies, harbor authorities, and the city and suburban governments, laid out sites close to the harbor and railroads. In some cases, such as Maisonneuve, a tightly focused bourgeoisie carved out manufacturing sites while building adjacent working-class housing. In other cases, as in Canal, Saint-Henri, Lachine, and Mile End, developers sold land on an ad hoc basis or, as in the Outer Core, converted residential sites into manufacturing ones. Regardless of the manner in which Montreal's working-class and manufacturing districts came about, a common thread was the power that local alliances had in the acquisition, subdivision, and (re)development of land for manufacturers and workers.

The forging of the appropriate infrastructures and land development process by local coalitions which underpinned the making of Montreal's manufacturing districts was replicated in other North American cities. Chicago's extensive railroad system connected the industrial suburbs to the central city, to one another, and to the agricultural hinterland. Development of Nashville's railroad and fringe manufacturing areas was simultaneous: firms clustered around the lines and subsequently laid down new lines and spurs, which activated even more manufacturing growth. The building of the San Pedro Harbor linked industrial activity in that area to the wider Los Angeles economy, spurring industrial and working-class development. The industrial suburbs of Chicago, Pittsburgh, Toronto, and Milwaukee were built from some combination of owner-building, informal suburban land development directed by local property developers and a more planned process orchestrated by manufacturers and other decision makers.[11]

Emphasizing the overwhelming power of the city's elite to redraw the urban landscape downplays the role played by workers in the making of the city's manufacturing districts. Writers such as David Gordon, on the other hand, argue that workers from the 1880s had a central role in manufacturing decentralization. According to Gordon, firms fled the growing labor conflict of the central city for independent suburban municipalities, where they could exercise considerable local control. There is no doubt that labor and capital relations played a role in decentralization. The broad thrusts of capital accumulation and class conflict manifested in the struggle over profits, conditions, and workplace control formed and directed urban-industrial growth. While these dynamics structured class relations, the notion that manufacturers were able to escape tensions with labor by deploying capital outside the central city is questionable. There is little evidence to support the idea that firms actually uprooted their plants from central Montreal as a direct response to labor conflict. Rather, the role of labor in the relationship between manufacturers' spatial strategy and manufacturing districts was more opaque. Firms had a variety of ways, other than moving, to undermine workers' power. In particular, reacting to the imperatives of labor-capital relations, they squeezed profits and controlled labor by introducing bigger individual machines, larger ensembles of machinery, more space-extensive work processes, occupational and ethnic divisions in the workplace, and greater amounts of material-handling equipment. Rather than fleeing central sites of labor conflict, it was the opportunity presented by workplace changes, expanding markets, and new place-bound locational assets which enabled firms to suburbanize.[12]

Moreover, it was only firms with the ability to move which could uproot themselves from the central factory districts. This did not always include those facing intense labor conflict, though in some cases it did: some firms, such as Pullman and Apollo Iron and Steel, consciously distanced themselves from Chicago and Pittsburgh's central city in order to avoid the labor problems of the heavily congested central factory districts. But Pullman and Vandergrift were the exception rather than the rule. Montreal firms moving to the fringe relied on an assortment of other factors: transportation facilities, a well-stocked labor pool, and sites where they could build large, well-designed factories and install new work processes, machinery, and equipment. Furthermore, many central-city firms with the apparent ability to decentralize faced the problem of large-scale investment already sunk in place. Many firms had built their plants over a generation or more. Starting off in a small building, they eventually created a sprawling plant composed of investments in buildings, machinery, railroad spurs, and equipment which, depreciating over different time horizons, could not be easily and profitably dismantled.[13]

While workers may have had little direct control over the major decisions responsible for the location of manufacturing and housing districts, they did create strong neighborhood ties. The link between relatively self-contained nineteenth- and early-twentieth-century working-class neighborhoods and a factory district has been pointed out by various writers, from the social ecologists of the Chicago School to present-day historians. Even though most writers emphasize the central location of this relationship, others have pointed to its existence throughout the metropolitan district. Roderick McKenzie in 1921, for example, stated that around the fringe factory districts of Columbus, Ohio, "independent communities develop which have a life of their own distinct from the rest of the city." Historians studying Philadelphia, Pittsburgh, Chicago, and other cities have also shown how working-class and immigrant homes were interspersed with or adjacent to factories, warehouses, and railroads. With their places of worship, union halls, taverns, credit unions, and ethnic and employment networks, neighborhoods developed relatively separate from the currents of other metropolitan neighborhoods.[14]

Montreal's working-class manufacturing neighborhoods also centered around localized labor markets and social networks. In her study of nineteenth-century Montreal families Bettina Bradbury shows the importance of ethnic-based local employment and social interaction to the neighborhood fabric: "for much of the working class the texture of daily life revolved ... around the sociabilities built up within families, between neighbours and friends, on the streets, or in taverns and shops." Similarly, I have shown how close-knit, self-contained working-class neighborhoods developed around manufacturing areas. The nineteenth-century West End was composed of several neighborhoods distinguished by ethnicity and occupation centered on tight local interaction. The vast majority of the city's Jewish population before 1930 lived in the St. Lawrence Boulevard corridor, running from the city core to the North End. Living within walking distance of their major workplace—the garment factories and lofts—they built an impressive range of retail stores, institutions, and synagogues that catered to most of their everyday needs. In the face of intense discrimination Montreal's Italian and Ukrainians forged strong geographic links between the location of jobs and their cultural networks, institutional facilities, and residence. Other areas, ranging from residential Verdun and Rosemont to the satellite manufacturing towns and suburbs of Maisonneuve, Lachine, and Saint-Pierre, consisted of blue-collar neighborhoods featuring local working-class and ethnic institutions.[15]

The existence of localized labor markets and social networks rested on the sporadic interaction that workers had with the central city and a short journey to work. While many suburban workers may have occasionally taken the trolley to visit downtown stores or theaters and some may have taken it daily to get to work at downtown or crosstown locations, Montreal's working class remained rooted to neighborhoods. Just what proportion of the population of one district commuted to other metropolitan districts is not known. Certainly, before the trolley most workers walked to work. Beginning in the 1890s and the 1920s, respectively, however, the electric trolley and the automobile promoted a greater range of housing options for the working class. The provision of cheap mass transit and the advent of the automobile not only accelerated manufacturing dispersal to the suburbs but also freed workers from living close to their place of work downtown. By taking the trolley or owning cars, workers could commute from their central-city homes to suburban factories, and vice versa.[16]

The Montreal case is a lot more complicated than this, however. There is little doubt that white-collar and well-paid, skilled blue-collar workers could afford to take the trolley after the 1890s and to own a car by the 1920s. Because they were no longer tied to living within walking distance of central workplaces, these workers had access to a greater range of housing and employment options than previously. There is little evidence, however, that most industrial workers could afford the trolley carfare or that blue-collar workers used the trolley to any significant degree to get to work before 1930. A study of the Grand Trunk locomotive shops in Point Saint-Charles between 1880 and 1917, for example, shows that the large majority of workers lived within walking distance of the shops. Despite a scattering of some French-Canadian workers throughout other parts of the city, Point Saint-Charles was "an area which combined manufacturing activity and workingmen's residences.... The community that developed was very stable and had a closely-knit population." Similarly, despite the dramatic increase in automobile registration after 1920, most Montreal workers did not have access to a car. Montreal was not unique in this regard. As studies of Philadelphia and Milwaukee suggest, even though the distance traveled between home and work was greater in the 1920s than in the nineteenth century, a large share of the manufacturing labor force walked to work. Similarly, a study of early-twentieth-century Toronto points to a shorter journey to work for those employed in suburban factories.[17]

In the eighty years after 1850 Montreal's transformation from a small mercantile city to a large industrial metropolis involved the reshaping of its economic and social landscapes. From the mid-nineteenth-century manufacturing areas of Griffintown, Canal, and Sainte-Marie hugging the perimeter of the mercantile city to the twentieth-century satellite towns of Lachine and industrial suburbs of Montréal Est fifteen miles from the city core, a succession of manufacturing districts and close-knit working-class residential neighborhoods were built on the expanding metropolitan fringe. Throughout the city working-class homes, institutions, and stores were interspersed with factories, sewers, railroad lines, and piers. Two principal sets of dynamics fueled the building of this metropolitan landscape. First, urban elites driven by a search for the profits to be gained from the city-building process constructed a suburban landscape. This post-1850 suburban landscape was not just a middle-class one; it also belonged to manufacturing and the working class. Second, Montreal's manufacturing and working-class districts were built as a response by the city's business elites to the needs of manufacturers to implement new technologies, new layouts, and new labor processes in new spaces. The incessant imperative of industrial restructuring combined with the ability of the elite to construct new industrial and residential districts over several generations produced a new landscape for the city. Between 1850 and 1930 a rationalized metropolitan landscape consisting of linked and specialized manufacturing and working-class districts held together by elite-controlled networks of capital developed in Montreal.

APPENDIX

Sources, Case Studies, and Scale

Urban historians and historical geographers have traditionally used three primary sources to describe a city's manufacturing stock. In the case of Montreal the census manufacturing manuscripts are the most common source, but they do not provide a long-term picture of the city's manufacturing structure and geography. The census before 1871 does not provide a comprehensive description of the city's manufactures. According to William Hutton, the chief census taker, the manufacturing section of the 1851–52 census was "undefined and so incorrectly stated that little reliable information . . . can be gleaned from the Returns." The same problem exists for 1861. The 1871 census, with more inclusive definitions, is the first one to provide an adequate appraisal of Canada's industry. Henceforth, manufacturing would "apply to all industries of *any importance* conducted in *separate* establishments or workshops" (my emph.). The 1871 manufacturing manuscripts are extant, and, for the first time, it is possible to describe the character and location of Montreal's industry. Since the 1881, 1891, and 1901 manuscripts have been destroyed, the census is of limited value for later years. Although the published census material can be useful, it is impossible to extract information with a geographical precision finer than the census district. Not only did the definitions of these districts change over time, but each district was very large. The published census reports of 1881, for example, distribute the entire city's manufacturing stock among three census districts. Furthermore, the extensive and differentiated suburban belt surrounding the city was broken into only two divisions, the counties of Hochelaga and Jacques Cartier. Thus, few meaningful statements can be made about the internal structure and location of industry in the city or the suburbs. Another weakness of the census is the 1901 redefinition of *manufacturing establishments.* Beginning in 1871, the term included all firms, regardless of the number of workers.

In 1901, however, the definition was changed to include only firms with at least five employees. The reason was that "the anomaly of treating as a manufacturing establishment every room or workshop in which one or two persons are employed should not be entertained when the whole tendency of industries is towards the concentration of capital, management, skill and labour." The effect is to make meaningful comparisons between different years impossible.[1]

Another source is the city directory. While it is useful for cross-referencing with data taken from other sources, the directory does not provide the basis for an empirically detailed historical geography of manufacturing. The street and alphabetical listings provide no information aside from address, name, and sector. The absence of any indication or surrogate of such important variables as employment, capital, or product lines reduces its usefulness as a source. Nor does it consistently distinguish manufacturing from commercial or residential entries. While the "Business Classified Directory" is helpful, it is incomplete and biased toward larger firms.

An alternative source is the water tax assessment (*le rôle d'évaluation*). Beginning in 1847 Montreal undertook an annual levy, along with its property assessment, of the rent paid by each household, business, and organization for the purpose of rasing taxes to cover the costs of the newly acquired municipal water system. For any year beginning in 1847, therefore, it is possible to construct a listing of all manufacturing, commercial, and financial enterprises in the city. Along with giving the exact location of every business firm in the city, the annual rents taken from the rolls can be used as an estimate for the size of the enterprise, capital value, or employment. All businesses were required to pay a 7.5 percent rate on the rent they paid or, if they owned the premises, the market rent equivalent. Although suburban water tax rolls sometimes differed from Montreal's, it was possible to identify the location and rent of suburban manufacturing firms.

Despite the obvious biases and discrepancies that creep into any enumeration system, the rents paid by firms provide an excellent picture of the scale of operations. Statistical tests show a high degree of correspondence between annual rent and the variables taken from the census and a list of manufacturers' capital investment. I have tested this assertion for two years. In 1871 the rents from firms in Saint-Ann ward were compared with the census values for capital invested and number of employees in the same firms. The degree of linear association is 0.95 between rent and capital and 0.79 between rent and number of employees. A similar test was undertaken for the relationship between rent and capital investment of a random set of the city's firms in 1907. The linear relationship in this case is 0.79.[2]

The tax rolls provide a more complete listing of firms than either the census or the city directory. According to the census, there were 69 manufacturing firms in 1851 and 144 in 1861. Yet in both years the tax rolls list nearly that many in just one ward (Saint-Ann). Similarly, the directory contained only a small number of firms compared to the tax rolls. Moreover, directory firms were generally the city's largest. The fact that small firms were frequently missing can be ascertained from one example. According to the 1890 directory, the city had 15 baking firms, while the 1890 assessment listed 124 firms. The median business rents of these two sets of baking firms also differed greatly: the median for the tax rolls was $267, while it was $446 for the firms collected from the directory. In other words, rent data taken from the water tax rolls have two primary strengths: they contain the most complete listing and the exact location of all manufacturing firms in the city, and they provide an extremely useful and usable measure of enterprise physical and employment scale.

Differentiating manufacturing firms from retailing and wholesaling establishments was a problem, as the rolls do not indicate a firm's function. The strategy employed to solve this problem was to include the ambiguous firms in the initial stage of data collection. They were then cross-referenced with the classified business sections of the annual Lovell city directory and other business directories. In some cases I was able to identify the firm's function easily. In some cases, however, the distinction between manufacturing and nonmanufacturing was blurred because so many firms performed multiple functions. This was especially true in the nineteenth-century clothing, baking, carriage making, saddlery, and jewelry industries. Even the labels are not trustworthy and straightforward. The term *watchmaker,* for example, suggests someone in the process of making a watch or clock. In many cases, however, I found that a watchmaker was simply a retailer, who may or may not have had a watch repair shop at the back of the store. In the case of the shoe industry it was difficult to differentiate repair shops from full-fledged manufacturers. The same set of problems existed for blacksmithing, furniture, pharmaceutical, hat, and fur firms. A related difficulty was with manufacturing firms that had an office or warehouse in Montreal but whose manufacturing operations were located in another city. These firms called themselves manufacturers in the listings but did not perform any manufacturing activity in Montreal. Only long searches through an assortment of business directories solved these problems.

A case study approach provides the basis of this study. A first set of case studies provides broad empirical generalizations at the urban level through the generation of data sets consisting of annual rents for all metropolitan Montreal manufacturing establishments in selected years. Three snapshots of Montreal's manufacturing structure taken from the city and the surrounding suburbs' water tax rolls of 1861, 1890, and 1929 provide the basis for this study. In each year every manufacturing establishment in the city and the suburbs was collected. After completing the sorting and cross-referencing of the water tax data the number of manufacturing firms in metropolitan Montreal was 635 in 1861, 1,365 in 1890, and 2,444 in 1929. These cross-sections were used to describe the location of industry on these dates and to make comparisons over time.[3]

While these cross-sections provide an important cut into metropolitan Montreal's industrial geography at the three dates, they provide a static picture. The processes, events, firms, and people involved in the formation of manufacturing districts are missing; the rolls do not allow for an interpretation of the dynamics underlying the patterns. To examine the processes and changes generating the representations of Montreal's industrial geography in 1861, 1890, and 1929, case studies of the tobacco, clothing, metalworking, carriage, cotton, printing, and paint industries and several manufacturing districts are presented. This intensive research complements the broad generalizations gained from extensive research. The case studies of districts, firms, and industries provide insights into the actual mechanisms identified by theoretical observations and highlight how particular combinations of events produced specific spatial outcomes. In order to do this the water tax rolls need to be complemented by other sources. Biographies of firms, industries, and districts are constructed from an assortment of sources, the most important of which are city directories, industrial and trade journals, booster pamphlets, fire insurance atlases, government reports, newspapers, and company and neighborhood histories. These sources permit me to draw a richer picture of the changing historical geography of Montreal's manufacturing. They illuminate the geographic processes that are the concern of this study.[4]

One of the pressing problems of the examination of metropolitan industrial geography is that of spatial scale. This problem has always bedeviled geographers. How historians and historical geographers have grappled with this question has had an enormous impact upon both the generation of urban geographic theory and the development of empirical findings. Although I do not propose any solutions to this problem, I do examine the historical geography of Montreal's manufacturing through a range of scales. The decision to undertake a study of urban industrial geography based on broad spatial limits is fraught with dangers. In particular, the reliance upon ward or other administrative boundaries can, and often does, hide more than it reveals. Very few urban phenomena follow political and administrative boundaries. This is the case, for example, with industrial suburbs. They are administratively separate from, yet integrated in and dependent upon, the central city. In some cases these industrial suburbs are outgrowths of industrial districts within the city, while in others they are separate satellites with no direct connection to city manufacturing districts. Moreover, those areas within the jurisdictional boundaries of the city which are at the edge of urban development can be viewed as "suburban" in the sense of their relationship to the rest of the city. Even though they do not have separate political institutions, these areas function in many other ways as suburban areas. In order to avoid these problems, the method utilized in this study is the construction of several spatial scales.[5]

Manufacturing districts form the smallest geographic scale and are distinguished from one another by their industrial organization, their timing of development, and their social geography. They are built up from data for individual firms, which are assigned to cells making up a grid placed over a map of the city. Once the industrial content of each cell was known, the cells were then allocated to a particular district based on the three factors mentioned earlier: there were seven districts in 1861, eleven in 1890, and fourteen in 1929. The number of cells making up a district ranged from one in Griffintown to nine in the districts of Maisonneuve and the North End, and the number of firms in a cell ranged from zero in all years to 256 in 1929. At the other end of the spatial spectrum, and in order to make broader statements about the geography of Montreal's industry, districts that had similar industrial properties and were spatially contiguous were collapsed into four (1861 and 1890) and five zones (1929). There are two central zones: Old Montreal only contains the Old Montreal district, while Outer Core is made up of the Saint-Antoine, Saint-Lawrence, and Saint-Jacques districts. The West End zone consists of the Griffintown, Canal, Saint-Henri, and Lachine districts. On the other side of the city, the East End zone is made up of Sainte-Marie, Hochealga, Maisonneuve, and Montréal Est districts. Finally, in the northern part of Montreal is the North End, containing the Mile End and Plateau districts.[6]

The largest scale at which analysis is undertaken is that of the urban district of Montreal, defined as the municipal corporation and its contiguous urban territory at the end of the study period. But why stop at the city and its immediate surrounding areas when Montreal's economic reach extended far beyond these boundaries? One reason is that a large proportion of Quebec's industry was located in Montreal and its adjacent suburban districts. Another is that the difficulty of finding assessment data and other information matching what exists for Montreal compelled me, along with the restrictions of time, to focus on the city and its immediate area.

The question of scale is also relevant to industry. There is no objective, simple, or incontestable definition of what constitutes an industry. To get around the problem of the conceptual baggage that comes with breaking up the industrial economy into separate parts, industry is defined by several scales. I will use industrial sectors such as textiles (made up of, among other things, cotton, blankets, and silk) and metals (primary metal, metal fabrication, and machinery) as one entry into the larger-scale movements of industrial capital over time. This, of course, follows the normal census Standard Industrial Classification system, which usually classifies firms by their relationship to the material requirements of firms. This is not always such a successful system, however, and can cause a great deal of confusion. Industries, a subset of sectors, are defined as a set of firms that are more closely related than those making up a sector because of their specific array of technologies, labor processes, markets, and interfirm linkages. To grasp the varied productive strategies undertaken by individual industries such as machinery, baking, cotton, tobacco, and shoes, it was necessary to make a more elaborate classification of manufacturing firms. By examining individual industries and, in some cases, individual firms and branches, it is possible to delve deeper into manufacturing dynamics.[7]

Notes

CHAPTER ONE: "LIVING TENDRILS"

1. Gérald Martin, "Études des facteurs qui ont déterminé la localisation de l'industrie à Montréal et des les banlieues," *Revue Trimestrielle Canadienne* 34 (1934): 297–335; Jean Delage, "L'industrie manufacturière," in *Montréal économique,* ed. Edras Minville (Montréal: Fides, 1943), 183–241.

2. For quotes, see Leo Schnore, "Metropolitan Growth and Decentralization," *American Journal of Sociology* 63 (1957): 173; Robert Fishman, "Megalopolis Unbound," in *Metropolis: Center and Symbol of Our Times,* ed. Philip Kasinitz (New York: New York University Press, 1995), 396; Brian Berry and Yehoshua Cohen, "Decentralization of Commerce and Industry: The Restructuring of Metropolitan America," in *The Urbanization of the Suburbs,* ed. Louis Masotti and Jeffrey Haddon (Beverly Hills, Ca.: Sage, 1973), 451; Allen Scott, "Production System Dynamics and Metropolitan Development," *Annals of the Association of American Geographers* 72 (1982): 188; David Slater, "Decentralization of Urban Peoples and Manufacturing Activity in Canada," *Canadian Journal of Economics and Political Science* 27 (1961): 76, 83–84; Sam Bass Warner, *The Urban Wilderness: A History of the American City* (Berkeley: University of California Press, 1972), 118.

3. For quotes, see Fishman, "Megalopolis Unbound," 402–3; Schnore, "Metropolitan Development," 175; Warner, *Urban Wilderness,* 118; Berry and Cohen, "Decentralization of Commerce and Industry," 451; David Goldfield and Blaine Brownell, *Urban America: From Downtown to No Town* (Boston: Houghton Mifflin, 1979), 352. Also see Allen Scott, "Locational Patterns and Dynamics of Industrial Activity in the Modern Metropolis," *Urban Studies* 19 (1982): 111–42; and "Production System Dynamics."

4. For quotes, see Fishman, "Megalopolis Unbound," 396, 402. Also see Ernest Burgess, "The Growth of the City," in *The City,* ed. Robert Park, Ernest Burgess, and Roderick McKenzie (Chicago: University of Chicago Press, 1925), 47–62; Robert Fishman, *Bourgeois Utopias: The Rise and Fall of Suburbia* (New York: Basic Books, 1987); Kenneth Jackson, *Crabgrass Frontier: The Suburbanization of the United States* (New York: Oxford University Press, 1985); John Stilgoe, *Borderland: Origins of the American Suburb, 1820–1939* (New Haven: Yale University Press, 1988); David Ward, *Cities and Immigrants: A Geography of Change in Nineteenth-Century America* (New York: Oxford University Press, 1971); Sam Bass Warner, *Streetcar Suburbs: The Process of Growth in Boston, 1870–1900* (Cambridge: Harvard University Press, 1962). For a different view, see Richard Harris and Robert Lewis, "Constructing a Fault(y) Zone: Misrepresentations of American Cities and Suburbs, 1900–1950," *Annals of the Association of American Geographers* 88 (1998): 622–39.

5. Richard Stott, *Workers in the Metropolis: Class, Ethnicity and Youth in Antebellum New York City* (Ithaca: Cornell University Press, 1990), 42–48; Alexander von Hoffman, *Local Attachments: the Making of an American Urban Neighborhood, 1850–1920* (Baltimore: Johns Hopkins University Press, 1994); Henry Binford *The First Suburbs: Residential Communities on the Boston Periphery, 1815–1860* (Chicago: University of Chicago Press, 1985); Philip Scranton, *Proprietary Capitalism: The Textile Manufacture at Philadelphia, 1800–1885* (New York: Cambridge University Press, 1983); Edward Muller and Paul Groves, "The Emergence of Industrial Districts in Mid-Nineteenth Century Baltimore," *Geographical Review* 69 (1979): 159–78.

6. Graham Taylor, *Satellite Cities: A Study of Industrial Suburbs* (New York: Appleton, 1915); Glenn McLaughlin, *Growth of American Manufacturing Areas: A Comparative Analysis with Special Emphasis on Trends in the Pittsburgh District* (Pittsburgh: Bureau of Business Research, University of Pittsburgh, 1938), 127. For overviews of industrial suburbanization, see Robert Lewis, "Running Rings around the City: North American Industrial Suburbs, 1850–1950," in *Changing Suburbs,* ed. Richard Harris and Philip Larkham (London: E and FN Spon, 1999), 146–67; Richard Walker and Robert Lewis, "Beyond the Crabgrass Frontier: Industry and the Spread of North American Cities, 1850–1950," *Journal of Historical Geography* (in press);

7. Regional Plan of New York and Its Environs, *Regional Survey of New York and Its Environs,* vols. 1A and 1B (1927; rpt., New York: Arno Press, 1974); Richard Harris, "Industry and Residence: The Decentralization of New York City, 1900–1940," *Journal of Historical Geography* 19 (1993): 169–76; James Kenyon, *Industrial Localization and Metropolitan Growth: The Paterson-Passaic District* (Chicago: Department of Geography, Research Paper No. 67, 1960), 24–66; McClellan and Junkersfeld, Inc., *Report on Transportation in the Milwaukee*

Metropolitan District (New York: McClellan and Junkersfeld, 1928), 2 vols.; Gerald Bloomfield, "Shaping the Character of the City: the Automobile Industry and Detroit, 1900–1920," *Michigan Quarterly Review* 25 (1986): 167–81; Olivier Zunz, *The Changing Face of Inequality: Urbanization, Industrial Development, and Immigrants in Detroit, 1880–1920* (Chicago: University of Chicago Press, 1982), 97–101, 292–309; Greg Hise, "Nature's Workshop: Industry and Urban Expansion in Southern California, 1900–1950," *Journal of Historical Geography* (in press); Richard Walker, "Industry Builds the City: The Suburbanization of Manufacturing in the San Francisco Bay Area, 1850–1945," *Journal of Historical Geography* (in press).

8. Kevin Kane and Thomas Bell, "Suburbs for a Labor Elite," *Geographical Review* 75 (1985): 319–34; Henry McKiven, *Iron and Steel: Class, Race, and Community in Birmingham, Alabama, 1875–1920* (Chapel Hill: University of North Carolina Press, 1995); John Marshall, "Railroads and Urban Growth," in *Growing Metropolis: Aspects of Development in Nashville,* ed. James Blumstein and Benjamin Walter (Nashville: Vanderbilt University Press, 1975), 65–80; Edward Muller, "Industrial Suburbs and the Growth of Metropolitan Pittsburgh, 1870–1920," *Journal of Historical Geography* (in press); Anthony Orum, *City-Building in America* (Boulder: Westview Press, 1995); Roger Simon, "Housing and Services in an Immigrant Neighborhood: Milwaukee's Ward 14," *Journal of Urban History* 2 (1976): 435–58; Robert Cramer, *Manufacturing Structure of the Cicero District, Metropolitan Chicago* (Chicago: University of Chicago, Geography Department Research Paper No. 27, 1952); Hise, "Nature's Workshop"; Richard Walker, "The Transformation of Urban Structure in the Nineteenth Century and the Beginnings of Suburbanization," in *Urbanization and Conflict in Market Societies,* ed. Kevin Cox (Chicago: Maaroufa Press, 1979), 165–212; Scranton, *Proprietary Capitalism,* 177–271; Caroline Golab, *Immigrant Destinations* (Philadelphia: Temple University Press, 1977), 111–56.

9. Also see Patricia Thornton, Martha Langford, and Brian Slack, "Etude historique du patrimoine industriel de Montréal" (report submitted to Ministère des Affaires Culturelles, Gouvernment du Québec and Service de l'Habitation et du Développment Urbain, Ville de Montréal, 1991).

10. John Dickinson and Brian Young, "Periodization in Quebec History: A Reevaluation," *Québec Studies* 12 (1991): 1–10.

11. "Elements in the Selection of a Business Location," *Canadian Engineer* (18 August 1910), 208.

12. Alfred Chandler, *The Visible Hand: The Managerial Revolution in American Business* (Cambridge: Belknap, 1977); Naomi Lamoreaux, *The Great Merger Movement in American Business, 1895–1904* (New York: Cambridge University Press, 1985); Daniel Nelson, *Managers and Workers: Origins of the New Factory System in the United States, 1880–1920* (Madison: University of Wisconsin Press,

1975); Harry Braverman, *Labor and Monopoly Capital: The Degradation of Work in the Twentieth Century* (New York: Monthly Review Press, 1974); Richard Edwards, *Contested Terrain: The Transformation of the Workplace in the Twentieth Century* (New York: Basic Books, 1979); David Hounshell, *From the American System to Mass Production, 1880–1932: The Development of Manufacturing Technology in the United States* (Baltimore: Johns Hopkins University Press, 1984).

13. Philip Scranton, *Proprietary Capitalism, Figured Tapestry: Production, Markets and Power in Philadelphia Textiles, 1885–1941* (New York: Cambridge University Press, 1989); and *Endless Novelty: Speciality Production and American Industrialization, 1865–1925* (Princeton: Princeton University Press, 1997); John Brown, *The Baldwin Locomotive Works* (Baltimore: Johns Hopkins University Press, 1995); Regina Blaszczyk, "'Reign of the Robots': The Homer Laughlin China Company and Flexible Mass Production," *Technology and Culture* 36 (1995): 863–911; John Ingham, *Making Iron and Steel: Independent Mills in Pittsburgh, 1820–1920* (Columbus: Ohio State University Press, 1991); Stephen Meyer, "Technology and the Workplace: Skilled and Production Workers at Allis-Chalmers, 1900–1941," *Technology and Culture* 29 (1988): 839–64; Charles Sabel and Jonathan Zeitlin, "Historical Alternatives to Mass Production: Politics, Markets, and Technology in Nineteenth-Century Industrialization," *Past and Present* 108 (1985): 133–76.

14. Richard Walker, "The Geographical Organization of Production-Systems," *Environment and Planning D* 6 (1988): 381–82.

15. Scott, "Locational Patterns"; Raymond Fales and Leon Moses, "Land-Use Theory and the Spatial Structure of the Nineteenth-Century City," *Papers of the Regional Science Association* 28 (1972): 49–80. For a similar but more historically nuanced account, see Walker, "Transformation of Urban Structure"; and "A Theory of Suburbanization: Capitalism and the Construction of Urban Space in the United States," in *Urbanization and Urban Planning in Capitalist Society*, ed. Michael Dear and Allen Scott (New York: Methuen, 1981), 383–429; Allan Pred, "The Intrametropolitan Location of American Manufacturing," *Annals of the Assocation of American Geographers* 54 (1964): 165–89; Ward, *Cities and Immigrants*. On the growth of other central economic functions, see Mona Domosh, *Invented Cities: The Creation of Landscape in Nineteenth-Century New York and Boston* (New Haven: Yale University Press, 1996); Gunter Gad and Deryck Holdsworth, "Corporate Capitalism and the Emergence of the High Rise Office Building," *Urban Geography* 8 (1987): 212–31, and "Building for City, Region, and Nation: Office Development in Toronto, 1834–1984," in *Forging a Consensus: Historical Essays on Toronto*, ed. Victor Russell (Toronto: University of Toronto Press, 1984), 273–319; Christine Rosen, *The Limits of Power: Great*

Fires and the Process of City Growth in America (Cambridge: Cambridge University Press, 1986).

16. G. Ferguson, "Decentralization of Industry," *Journal of the Town Planning Institute* (April 1924): 13.

17. Canada, *Census of Canada, 1890–1891* (Ottawa, 1897), 4:283. Also see Jeremy Atack, "Firm Size and Industrial Structure in the United States during the Nineteenth Century," *Journal of Economic History* 44 (1986): 463–75; Chandler, *Visible Hand;* Nelson, *Managers and Workers;* Anthony O'Brien, "Factory Size, Economies of Scale, and the Great Merger Wave of 1898–1902," *Journal of Economic History* 48 (1988): 639–49. For new factory form, see Reyner Banham, *A Concrete Atlantis: U.S. Industrial Building and European Modern Architecture* (Cambridge: MIT Press, 1986); Lindy Biggs, *The Rational Factory: Architecture, Technology, and Work in America's Age of Mass Production* (Baltimore: Johns Hopkins University Press, 1996); Grant Hildebrand, *Designing for Industry: The Architecture of Albert Kahn* (Cambridge: Harvard University Press, 1974).

18. For quote, see Michael Storper, "Technology and New Regional Growth Complexes: The Economics of Discontinous Spatial Development," in *Technological Change, Employment and Spatial Dynamics,* ed. Peter Nijkamp (Berlin: Springer-Verlag, 1985), 62. Also see Richard Harris, *Unplanned Suburbs: Toronto's American Tragedy, 1900–1950* (Baltimore: Johns Hopkins University Press, 1996), 54–64; Stanley Buder, *Pullman: An Experiment in Industrial Order and Community Planning, 1880–1930* (New York: Oxford University Press, 1967); Edward Greer, "Monopoly and Competitive Capital in the Making of Gary, Indiana," *Science and Society* 40 (1976): 465–78; Muller, "Industrial Suburbs"; and "The Pittsburgh Survey and 'Greater Pittsburgh': A Muddled Metropolitan Geography," in *Pittsburgh Surveyed: Social Science and Social Reform in the Early Twentieth Century,* ed. Maurine Greenwald and Margo Anderson (Pittsburgh: University of Pittsburgh Press, 1996), 69–87; Zunz, *Changing Face of Inequality.*

19. Mary Beth Pudup, "Packers and Reapers, Merchants and Manufacturers: Industrial Structure and Location in an Era of Emergent Capitalism" (Master's thesis, University of California, Berkeley, 1983), 47–99; Louise Carroll Wade, *Chicago's Pride: The Stockyards, Packingtown and Environs in the Nineteenth Century* (Urbana: University of Illinois Press, 1987); Thomas Jablonsky, *Pride in the Jungle: Community and Everyday Life in Back of the Yards Chicago* (Baltimore: Johns Hopkins University Press, 1993); Fred Viehe, "Black Gold Suburbs: The Influence of the Extractive Industry on the Suburbanization of Los Angeles, 1890–1930," *Journal of Urban History* 8 (1981): 3–26; Greg Hise, *Magnetic Los Angeles: Planning the Twentieth-Century Metropolis* (Baltimore: Johns Hopkins University Press, 1997), 126–32; Mike Davis, "Sunshine and the

Open Shop: Ford and Darwin in 1920s Los Angeles," *Antipode* 29 (1997): 356–82; Walker, "Industry Builds the City"; Edward Muller, "Industrial Suburbs."

20. Editorial, *Journal of the Town Planning Institute of Canada* (December 1927): 207.

21. Manuel Gottlieb, *Long Swings in Urban Development* (New York: Columbia University Press, 1976); David Harvey, *The Limits to Capital* (Chicago: University of Chicago Press, 1982); and *The Urbanization of Capital* (Baltimore: Johns Hopkins University Press, 1986); Homer Hoyt, *One Hundred Years of Land Values in Chicago, 1830–1933* (Chicago: University of Chicago Press, 1933); Helen Monchow, *Seventy Years of Real Estate Subdividing in the Region of Chicago* (Evanston: Northwestern University Press, 1939); Walker, "Suburban Solution"; and "A Theory of Suburbanization"; Jeremy Whitehand, *The Changing Face of Cities: A Study of Development Cycles and Urban Form* (New York: Blackwell, 1987). On working-class suburbs before World War II, see Harris, *Unplanned Suburbs*; Richard Harris and Matt Sendbuehler, "The Making of a Working-Class Suburb in Hamilton's East End, 1900–1945," *Journal of Urban History* 20 (1994): 486–511; Muller, "Industrial Suburbs"; Golab, *Immigrant Destinations*; Hise, *Magnetic Los Angeles*; Kane and Bell, "Suburbs for a Labor Elite"; McKiven, *Iron and Steel*; Henry Silcox, *A Place to Live and Work: The Henry Disston Saw Works and the Tacony Community of Philadelphia* (University Park: Pennsylvania State University Press, 1994); von Hoffman, *Local Attachments*; Zunz, *Changing Face of Inequality*. Although the factory land development process has been neglected, some studies have looked at the question. See Buder, *Pullman*; Susan Smith, "The Development of Toronto's Crown Reserves as Industrial Areas, 1793–1900" (Master's thesis, University of Toronto, 1998).

22. Michael Ball, "The Built Environment and the Urban Question," *Environment and Planning D* 4 (1986): 447–64; Kevin Cox and Andrew Mair, "Locality and Community in the Politics of Local Economic Development," *Annals of the Association of American Geographers* 78 (1988): 307–25; Harvey, *Limits to Capital*, 419–21; John Logan and Harvey Molotch, *Urban Fortunes: The Political Economy of Place* (Berkeley: University of California Press, 1987).

23. Ball, "Built Environment," 458–61; Logan and Molotch, *Urban Fortunes*, 34–37; Paul-André Linteau, *The Promoter's City: Building the Industrial Town of Maisonneuve, 1883–1918* (Toronto: Lorimer, 1985); Harold Platt, *City Building in the New South: The Growth of Public Services in Houston, Texas, 1830–1910* (Philadelphia: Temple University Press, 1983); and *The Electric City: Energy and the Growth of the Chicago Area, 1880–1930* (Chicago: University of Chicago Press, 1991); Marc Weiss, *The Rise of the Community Builders: The American Real Estate Industry and Urban Land Planning* (New York: Columbia University Press, 1987).

24. John Appleton, *The Iron and Steel Industry of the Calumet District: A Study in Economic Geography* (Urbana: University of Illinois Press, 1927); Alfred Meyer and Paul Miller, "Manufactural Geography of Chicago Heights, Illinois," *Proceedings of the Indiana Academy of Science* 66 (1956): 209–29; Zunz, *Changing Face of Inequality*.

Chapter Two: "Marvellous Rapidity"

1. Canada Railway Advertising Company, *Montreal Business Sketches with a Description of the City of Montreal, Its Public Buildings and Places of Interest* (Montreal, 1864), 1.

2. On the fur trade see R. Cole Harris, "France in North America," in *North America: The Historical Geography of a Changing Continent*, ed. Robert Mitchell and Paul Groves (Totowa, N.J.: Rowman and Littlefield, 1987), 65–92. For the new staples, see Jack Little, *Nationalism, Capitalism and Colonization in Nineteenth Century Quebec: The Upper St. Francis District* (Montreal: McGill-Queen's University Press, 1989); Arthur Lower, *Great Britain's Woodyard: British America and the Timber Trade, 1763–1867* (Montreal: McGill-Queen's University Press, 1973); Thomas McIlwraith, "British North America, 1763–1867," in Mitchell and Groves, *North America*, 220–52; Normand Séquin, *La conquête du sol au 19e siècle* (Montréal: Boréal Express, 1977).

3. Gerald Tulchinsky, *The River Barons: Montreal Businessmen and the Growth of Industry and Transportation, 1837–1853* (Toronto: University of Toronto Press, 1977); Jean Hamelin and Yves Roby, *Histoire économique du Québec, 1851–1896* (Montréal: Fides, 1971); Tom Naylor, *The History of Canadian Business, 1867–1914* (Toronto: Lorimer, 1975), 2 vols.; Ronald Rudin, *Banking en Français: The French Banks in Quebec, 1835–1925* (Toronto: University of Toronto Press, 1985), 3–21; Brian Young and John Dickinson, *A Short History of Quebec: A Socio-Economic Perspective* (Toronto: Copp, Clarke, Pitman, 1988); Louise Dechêne, "Les enterprises de William Price, 1810–1850," *Histoire Sociale / Social History* 1 (1968): 16–52.

4. Serge Courville and Norman Séquin, *Rural Life in Nineteenth Century Quebec* (Ottawa: Canadian Historical Association, 1989); Allan Greer, *Peasant, Lord and Merchant: Rural Society in Three Quebec Parishes, 1740–1840* (Toronto: University of Toronto Press, 1985); R. Cole Harris, *The Seigneurial Regime in Early Canada* (Québec and Madison: Les Presses de l'Université Laval and University of Wisconsin Press, 1966); Jean-Claude Robert, "Activities agricole et urbanisation dans la paroisse de Montréal, 1820–1840," in *Sociétés villageoisies et rapports villes-compagnes aux Québec and dans la France de 21 ouest*, ed. François Lebrun and Norman Séquin (Trois-Rivières: Centre de Recherche en Études Quebecoises, 1987), 101–12.

5. Paul-André Linteau, "Charles-Séraphin Rodier," in *Dictionary of Canadian Biography* (Toronto: University of Toronto Press, 1982), 11:763; John McCallum, *Unequal Beginnings: Agriculture and Economic Development in Quebec and Ontario until 1870* (Toronto: University of Toronto Press, 1980), 83–96; H. Clare Pentland, *Labour and Capital in Canada, 1650–1860* (Toronto: Lorimer, 1981).

6. Hamelin and Roby, *Histoire économique;* David Hanna, "The Importance of Transportation Infrastructure," in *Montreal Metropolis, 1880–1930,* ed. Isabelle Gournay and France Vanlaethem (Montreal: Canadian Centre for Architecture, 1998), 45–57; Pentland, *Labour and Capital;* Young and Dickinson, *Short History of Quebec,* 117–20; Paul Craven and Tom Traves, "Canadian Railways as Manufacturers, 1850–1880," Canadian Historical Association, *Historical Papers* (1983): 254–81; Tulchinsky, *River Barons;* John Willis, *The Process of Hydraulic Industrialization on the Lachine Canal: Origins, Rise and Fall* (Ottawa: Environment Canada, 1987).

7. McCallum, *Unequal Beginnings;* Pentland, *Labour and Capital;* Bruno Ramirez, *On the Move: French-Canadian and Italian Migrants in the North Atlantic Economy, 1860–1914* (Toronto: McClelland and Stewart, 1991); Willis, *Process of Hydraulic Industrialization;* Young and Dickinson, *Short History of Quebec,* 104–14; Bettina Bradbury, *Working Families: Age, Gender, and Daily Survival in Industrializing Montreal* (Toronto: McClelland and Stewart, 1993).

8. "Report of A. H. Blackeby on the State of the Manufacturing Industries of Ontario and Quebec," *Sessional Papers,* no. 37 (1885): 4–5; Naylor, *History of Canadian Business,* 1:260–96; Brian Young, *George-Etienne Cartier. Montréal bourgeois* (Montreal: McGill-Queen's University Press, 1981); and *In Its Corporate Capacity: The Seminary of Montreal as a Business Institution* (Montreal: McGill-Queen's University Press, 1986).

9. Richard Pomfret, *The Economic Development of Canada* (Toronto: Methuen, 1981), 123; Canada, *Census of Canada, 1870–1871* (Ottawa, 1875), vol. 3; Canada, *Census of Canada, 1881* (Ottawa, 1883), vol. 3; Canada, *Census of Canada, 1891* (Ottawa, 1894), vol. 3.

10. Young, *George-Etienne Cartier;* Rudin, *Banking en Français;* Tulchinsky, *River Barons.*

11. Hamelin and Roby, *Histoire économique,* 369–70; Tulchinsky, *River Barons;* Willis, *Process of Hydraulic Industrialization;* C. Warrington and Robert Nicholls, *A History of Chemistry in Canada* (Toronto: Pitman, 1949), 336–37.

12. *Canada Year Book, 1922–1923* (Ottawa: Acland, 1924), 444–46; J. Fountain, "The Growth of the Local Enterprise: From J. M. Schneider Ltd. to the Heritage Group," in *Manufacturing in Kitchener-Waterloo: A Long Term Perspective,* ed. David Walker (Waterloo: Department of Geography Publication Series

No. 26, University of Waterloo, 1987), 87; Leslie Roberts, *From Three Men* (N.p.: Dominion Rubber Co. Ltd., n.d.), 15; Peter Bischoff, "La formation des traditions de solidarité ouvrière chez les mouleurs montréalais. La longue marche vers le syndicalisme," *Labour / Le Travail* 21 (1988): 9–43.

13. For quote, see Canada, House of Parliament, "Report of the Select Committee on the Causes of the Recent Depression of the Manufacturing, Mining, Commercial, Shipping, Lumber and Fishing Interests," *Journals* (Ottawa, 1876), app. 3, 149. Also see John Borthwick, *Montreal, Its History to Which Is Added Biographical Sketches, with Photographs of Many of Its Principal Citizens* (Montreal, 1875), 146.

14. For quote, see Canada, "Report of the Select Committee," app. 3, 130. Also see Craven and Traves, "Canadian Railways as Manufacturers"; Glen Williams, *Not for Export: Toward a Political Economy of Canada's Arrested Industrialization* (Toronto: McClelland and Stewart, 1986).

15. For quotes, see William Wood, *The Days of John Wood, Watchmaker* (Hudson, Quebec: Wood Family Archives, 1996), 225–26; and Canada, "Report of the Select Committee," app. 3, 107.

16. Bradbury, *Working Families*, 118–81; D. Suzanne Cross, "The Neglected Majority: The Changing Role of Women in Nineteenth-Century Montreal," in *The Canadian City*, ed. Gilbert Stelter and Alan Artibise (Toronto: McClelland and Stewart, 1977), 255–81; Jacques Ferland, "'In Search of the Unbound Prometheia': A Comparative View of Women's Activism in Two Quebec Industries, 1869–1908," *Labour / Le Travail* 24 (1989): 11–44; Michelle Payette-Daoust, "The Montreal Garment Industry, 1871–1901" (Master's thesis, McGill University, 1986), 100–102; Gerald Tulchinsky, "Fredrick Warren Harris," in *Dictionary of Canadian Biography* (Toronto: University of Toronto Press, 1976), 9:367; Canada, *Census of Canada, 1880–81*, vol. 3; Canada, "Report of the Select Committee," app. 3, 85; Canada Railway Advertising Co., *Montreal Business Sketches*, 18–19.

17. Bradbury, *Working Families*, 39–43; Ralph Hoskins, "An Analysis of the Payrolls of the Point St. Charles Shops of the Grand Trunk Railway," *Cahiers de Géographie du Québec* 33 (1989): 323–44; William Kilbourn, *The Elements Combined: A History of the Steel Company of Canada* (Toronto: Clarke, Irwin, 1960), 24–25; Sherry Olson, "Ethnic Strategies in the Urban Economy," *Canadian Ethnic Studies* 33 (1991): 39–64; Sherry Olson and Patricia Thornton, "Le Raz de Marée irlandais à Montréal," in *Les chemins de la migration en Belgique et au Québec*, ed. Yves Landry, John Dickinson, Suzy Pasleau and Claude Desama (Beauport, Québec: Academia MNH, 1995), 69–80; Bruno Ramirez and Michael del Balso, "The Italians of Montreal: From Sojourning to Settlement, 1900–1920," in *Little Italies in North America*, ed. Robert Harney and J. Vincenza Scarpaci (Toronto: Multicultural History Society of Ontario, 1981), 67–75; Pa-

tricia Thornton and Sherry Olson, "Family Contexts and Infant Survival in Nineteenth-Century Montreal," *Journal of Family History* 16 (1991): 404.

18. *Montreal: The Metropolis* (Montreal: Gazette Printing, 1907), 15.

19. The history of Montreal's municipality institutions is a rocky one. Montreal's first charter was revoked in 1836 and a second one granted in 1840. This was taken away in 1918 and a third one given in 1921. See Milton Hersey, "Montreal's Many Administrations," *Municipal Review of Canada* 32 (July-August 1932): 13–14; William Lighthall, "City Government," in *Canada and Its Provinces*, ed. Adam Shortt and Arthur Doughty (Toronto: Glasgow, Brooke and Co., 1914), 15:299–320. On the changes to Montreal society, see Young, *George-Etienne Cartier*, 86–118.

20. Michael Gauvin, "The Reformer and the Machine: Montreal Civic Politics from Raymond Préfontaine to Méderic Martin, *Journal of Canadian Studies* 13 (1978): 16–26; Guy Bourassa, "Les élites politiques de Montréal: De l'aristocratie à la démocratie," *Canadian Journal of Economics and Political Science* 31 (1965): 35–51; Annick Germain, *Les mouvements de réform urbaine à Montréal au tournant du siècle* (Montréal: Département de Sociologie, Université de Montréal, 1984); Harold Kaplan, *Reform, Planning, and City Politics: Montreal, Winnipeg and Toronto* (Toronto: University of Toronto Press, 1982), 312–55; Young, *George-Etienne Cartier*, 86–118; Luc Desrochers, "Raymond Préfontaine,"in *Dictionary of Canadian Biography* (Toronto: University of Toronto Press, 1994), 13:842–46.

21. Pierre Brouillard, "La Commission du Havre de Montréal (1850–1896)," in Société Historique de Montréal, *Montréal: Artisans, Histoire, Patrimoine* (Montréal: Fides, 1979), 83–102; Benoît Brouillette, "Le port et les transports," in *Montréal économique*, ed. Edras Minville (Montréal: Fides, 1943), 115–41; Frederick Cowie, "The Great National Port of Canada," *Canadian Engineer* 22 (18 January 1912): 78–83; Hamelin and Roby, *Histoire économique*, 106–10; Jean-Claude Marsan, *Montreal in Evolution* (Montreal: McGill-Queen's University Press, 1981), 170–72; Gerald Tulchinsky and Brian Young, "John Young," in *Dictionary of Canadian Biography* (Toronto: University of Toronto Press, 1972), 10:722–28.

22. Robert Passfield,"Waterways"; and Lettie Anderson,"Water-Supply,"in *Building Canada: A History of Public Works*, ed. Norman Ball (Toronto: University of Toronto Press, 1988), 113–42 and 195–200; A. Doucet, "History of the Montreal Aqueduct," *Journal of the Engineering Institute of Canada* 4 (1921): 601–5; Lighthall, "City Government," 309–10; F. Clifford Smith, *The Montreal Water Works* (Montreal: City of Montreal, 1913), 13–14. The financing of the new reservoirs was the reason for creating the new water tax, from which I draw the rent data for 1861, 1890, and 1929.

23. Paul-André Linteau, *The Promoters' City: Building the Industrial Town*

of Maisonneuve, 1883–1918 (Toronto: Lorimer, 1985), 11–15; Robert Rumilly, *Histoire de Montréal* (Montréal: Fides, 1972), vol. 3; Walter van Nus, "The Role of Suburban Government in the City-Building Process: The Case of Notre Dame de Grâces, Quebec, 1876–1910," *Urban History Review* 13 (1984): 91–103; Paul-André Linteau, Rene Durocher, and Jean-Claude Robert, *Quebec: A History, 1867–1929* (Toronto: Lorimer, 1985), 29; Olson, "Ethnic Strategies in the Urban Economy," 42–44; Olson and Thornton, "Raz de Marée."

24. Marc Choko, *Crises du logement à Montréal (1860–1939)* (Montréal: Albert Saint-Martin, 1980), 5–61; Gilles Lauzon, *Habitat ouvrier et Révolution Industrielle. Le cas du Village St-Augustin* (Montréal: Collection RCHTQ, 1989); Marsan, *Montreal in Evolution*, 256–83; David Hanna, "Montreal, a City Built by Small Builders" (Ph.D. diss., McGill University, 1986); Phyllis Lambert, "Land Tenure and Concepts of Architecture and the City: Milton Park in Montreal," in *Power and Place: Canadian Urban Development in the North American Context*, ed. Gilbert Stelter and Alan Artibise (Vancouver: University of British Columbia Press, 1986), 137; Robert Marshall, "The Development of 'La Côte à Baron'" (MS, School of Urban Planning, McGill University, 1983); David Hanna, "Creation of an Early Victorian Suburb in Montreal," *Urban History Review* 9 (1980): 38–64.

25. Hanna, "Montreal, a City Built by Small Builders," 13–14, 77–85; Linteau, Durocher and Robert, *Quebec*, 158; John Saywell, *Housing Canadians: Essays on the History of Residential Construction in Canada* (Ottawa: Economic Council of Canada, 1975), 41–53; Robert Lewis, "The Segregated City: Class Residential Patterns and the Development of Industrial Districts in Montreal, 1861 and 1901," *Journal of Urban History* 17 (1991): 123–52.

26. The testimony of Costigan, Decrow, Laberge, and Boyd is to be found in Canada, *Report of the Royal Commission on Relations of Labour and Capital in Canada—Quebec Evidence* (Ottawa, 1889), 3:1–4, 732, 309. For home ownership, see Steven Hertzog and Robert Lewis, "A City of Tenants: Homeownership and Social Class in Montreal, 1847–1881," *Canadian Geographer* 30 (1986): 316–23; Marc Choko and Richard Harris, "The Local Culture of Property: A Comparative History of Housing Tenure in Montreal and Toronto," *Annals of the Association of American Geographers* 80 (1990): 73–95.

27. For quote, see Canada, *Report of the Royal Commission on Relations of Labour and Capital*, 69. Also see Hanna, "Montreal, a City Built by Small Builders"; Saywell, *Housing Canadians*, 41–43; Lewis, "Segregated City"; Réjean Legault, "Architecture et forme urbaine. L'exemple du Triplex à Montréal de 1870 à 1914," *Urban History Review* 18 (1989): 1–10.

28. The Gratorex and Muir quotes are from Canada, *Report of the Royal Commission on Relations of Labour and Capital*, 85, 259, 263. Studies of Montreal's working-class housing provide ample evidence of the problem of low

wages and increasing rents. See Choko, *Crises du Logement à Montréal*, 5–61; Lauzon, *Habitat ouvrier et Révolution Industrielle;* Jean de Bonville, *Jean-Baptiste Gangenpetit. Les travailleurs Montréal à la fin du XIXe siècle* (Montréal: L'Aurore, 1975), 85–86; Bradbury, *Working Families*, 80–117.

29. Sherry Olson and David Hanna, "Social Change in Montreal, 1842–1901," in *Historical Atlas of Canada*, ed. Louis Gentilcore (Toronto: University of Toronto Press, 1993), pl. 49; Marshall, "Development of 'La Côte à Baron'"; Cross, "Neglected Majority," 265–68; Mary Davidson, "The Social Adjustment of British Immigrant Families in Verdun and Point St. Charles" (Master's thesis, McGill University, 1933); David Hanna, "The New Town of Montreal: Creation of an Upper Middle Class Suburb on the Slope of Mount Royal in the Mid-Nineteenth Century" (Master's thesis, University of Toronto, 1977); and "Montreal, a City Built by Small Builders," 112–22; Lambert, "Land Tenure"; Lewis, "Segregated City"; Patricia Thornton, Sherry Olson, and Q. Thuy Thach, "Dimensions sociales de la mortalité infantile à Montréal, au milieu du XIXe siècle," *Annales de Démographie Historique 1988* (Paris: Société de Démographie Historique, 1989), 311–12.

Chapter Three: "One Vast Block"

1. David Hanna, "Partage social et partage de l'espace à Montréal, 1847 à 1901" (report presented to Fonds FCAR Québec, Geography Department, McGill University, 30 June 1986). Old Montreal is defined here as the area bounded by the Saint-Lawrence River and McGill, Lagauchetière, and Visitation Streets. The district's share of manufacturing is probably overstated, as most firms combining manufacturing with retailing or importing were located in Old Montreal. The chapter draws on Robert Lewis, "Productive Strategies and Manufacturing Reorganization in Montreal's Central District, 1850–1900," *Urban Geography* 16 (1995): 4–22.

2. *The Commerce of Montreal and Its Manufactures* (Montreal, 1888).

3. Ludger Beauregard, "La rue Saint-Jacques à Montréal. Une géographie des bureaux" (Département de Géographie, Université de Montréal, Notes et Documents, No. 81-02, 1981), 3-14; Raoul Blanchard, *Montréal, esquisse de géographie urbaine* (Grenoble: Allier, 1947); Jean-Claude Marsan, *Montreal in Evolution* (Montreal: McGill-Queen's University Press, 1974). For Toronto, see Gunter Gad and Deryck Holdsworth, "Building for City, Region, and Nation: Office Development in Toronto, 1834–1984," in *Forging a Consensus: Historical Essay on Toronto,* ed. Victor Russell (Toronto: University of Toronto Presss, 1984), 273–319.

4. Allan Stewart, "Settling an 18th-Century Faubourg: Property and Family in the Saint-Laurent Suburb, 1735–1810" (Master's thesis, McGill University,

1988); Bettina Bradbury, *Working Families: Age, Gender, and Daily Survival in Industrializing Montreal* (Toronto: McClelland and Stewart, 1993), 19, 27–28, 34–35.

   5. Nicholas Phelps, "External Economies, Agglomeration and Flexible Accumulation," *Transactions. Institute of British Geographers* 17 (1992): 35–46; Allen Scott, "Production System Dynamics and Metropolitan Development," *Urban Studies* 19 (1982): 111–42; Edward Muller and Paul Groves, "The Emergence of Industrial Districts in Mid-Nineteenth Century Baltimore," *Geographical Review* 54 (1979): 159–78; Richard Stott, *Workers in the Metropolis: Class, Ethnicity and Youth in Antebellum New York City* (Ithaca: Cornell University Press, 1990); David Ward, *Cities and Immigrants* (New York: Oxford University Press, 1971).

   6. Edgar Collard, "Owen McGarvey," in *Dictionary of Canadian Biography* (Toronto: University of Toronto Press, 1990), 12:624–25; Canada, *Report of the Royal Commission on the Relations of Capital and Labour. Evidence—Quebec* (Ottawa, 1889), 3:726–27; Elizabeth Collard, "George Armstrong," in *Dictionary of Canadian Biography* (Toronto: University of Toronto Press, 1982), 11:30–31.

   7. For quotes, see *The Commerce of Montreal and Its Manufactures* (Montreal, 1888), 127, 57; *Commercial Sketch of Montreal and Its Superiority as a Wholesale Market* (Montreal, 1868), 37; *Industries of Canada: City of Montreal* (Montreal, 1886), 147.

   8. For quotes, see "Manufacturing Industry of Montreal," *Montreal Gazette*, 27 July 1864; William Patterson, *Statements Relating to the Home and Foreign Trade of the Dominion of Canada: Also, Annual Report of the Committe of Montreal, 1873* (Montreal, 1874), 40; *Commerce of Montreal*, 168–69; *Canadian Journal of Commerce* (3 May 1889): 754–55. Also see Celebration Committee of the Grand Trunk Railway, *Montreal in 1856* (Montreal, 1856), 45, 49; Joanne Burgess, "Work, Family and Community: Montreal's Leather Craftsmen, 1790–1831" (Ph.D. diss., University of Quebec, Montreal, 1986), 2 vols.; and "L'industrie de la chaussure à Montréal: 1840–1870—le passage de l'artisanat à la fabrique," *Revue d'Histoire de l'Amérique Française* 31 (1977): 187–210; Jacques Ferland, "Evolution des rapports sociaux dans l'industrie canadienne du cuir au tournant du 20e siècle" (Ph.D. diss., McGill University, 1985).

   9. For quotes, see "Bread-Making," *Canadian Journal of Commerce* (23 October 1885): 816–17; and Canada, "Report of the Select Committee to Investigate and Report upon Alleged Combinations in Manufactures, Trade and Insurance in Canada," *Journals of the House of Commons, 1889* (Ottawa, 1888), 132. Also see Ian McKay, "Capital and Labour in the Halifax Baking and Confectionery Industry during the Last Half of the Nineteenth Century," *Labour / Le Travailleur* 3 (1978): 63–108; K. George Huttemeyer, *Les intérêts commerciaux de Montréal et Québec et leurs manufactures—1889* (Montréal, 1889), 126; Paul-André Lin-

teau, "Charles-Théodore Viau," in *Dictionary of Canadian Biography* (Toronto: University of Toronto Press, 1990), 12:1074–1075; Roger Viau, *Une siècle de progrès. Historique de Viau Ltée* (Montréal: Viau, 1967); *Industries of Canada*, 172.

10. For quote, see John Duncan, *Travels through Parts of the United States and Canada in 1818 and 1819*, quoted in Jean-Paul Bernard, Paul-André Linteau, and Jean-Claude Robert, "La société montréalaise dans les années 1820," *Rapport et Travaux, 1973–1975* (Montréal: Groupe de Recherche sur la Société Montréalaise au 19e Siècle, Université of Québec, Montréal, 1975), 2.

11. Charles Glackmeyer, *The Charter of the City of Montreal Together with Miscellaneous Acts of the Legislature Relating to the City* (Montreal, 1865), 245–47; William Lighthall, "City Government," in *Canada and the Provinces*, ed. Adam Shortt and Arthur Doughty (Toronto: Glasgow, Brooke, 1914), 15:306–9.

12. Marsan, *Montreal in Evolution*, 228–42; Ward, *Cities and Immigrants*; Christine Rosen, *The Limits of Power: Great Fires and the Process of City Growth in America* (New York: Cambridge University Press, 1986).

13. G. Goliger, "Le Cours St-Pierre," *Habitat* 25 (1982): 36; Marsan, *Montreal in Evolution*, 72.

14. Christopher Armstrong and H. Viv Nelles, *Monopoly's Moment: The Organization and Regulation of Canadian Utilities, 1830–1930* (Philadelphia: Temple University Press, 1986), 38–49, 88–89; David Hanna, "The Importance of Transportation Infrastructure," in *Montreal Metropolis, 188–1930*, ed. Isabelle Gournay and France Vanlaethem (Montreal: Canadian Centre for Architecture, 1998), 46–53.

15. Yvan Lamonde, *La culture ouvrière à Montréal, 1880–1920* (Québec City: Institute Québécoise de Recherche sur la Culture, 1982). The Outer Core's population is calculated from Saint-Antoine, Saint-Lawrence, Saint-Louis, and Saint-Jacques wards.

16. Ernest Burgess, "The Growth of the City," in *The City*, ed. Robert Park, Ernest Burgess and Roderick McKenzie (Chicago: University of Chicago Press, 1925), 47–62.

17. For quote, see Sam Bass Warner, *The Private City: Philadelphia in Three Periods of Its Growth* (Philadelphia: University of Pennsylvania Press, 1968), 50. For Montreal, see Bradbury, *Working Families*, 39–43; David Hanna and Frank Remiggi, *Montreal Neighbourhoods* (Montreal: Canadian Association of Geographers, 1980); Robert Lewis, "The Segregated City: Class Residential Patterns and the Development of Industrial Districts in Montreal, 1861 and 1901," *Journal of Urban History* 17 (1991): 123–52; Patricia Thornton and Sherry Olson, "Family Contexts and Infant Survival in Nineteenth-Century Montreal," *Journal of Family History* 16 (1991): 401–17; and "Le Raz de Marée irlandais à Montréal," in *Les chemins de la migration en Belgique et au Québec*, ed. Yves Landry,

John Dickinson, Suzy Pasleau and Claude Desama (Beauport, Québec: Academia MNH, 1995), 69–80.

18. *Commercial Sketch*, 4.

19. Jean De Bonville, *La presse québecoise de 1884 à 1914. Genèse d'un média de masse* (Québec: Les Presses de l'Université Laval, 1988); Douglas Fetherling, *The Rise of the Canadian Newspaper* (Toronto: Oxford University Press, 1990), 28–30; Greg Kealey, "Work Control, the Labour Process, and Nineteenth-Century Canadian Printers," in *On the Job: Confronting the Labour Process in Canada*, ed. Craig Heron and Robert Storey (Montreal: McGill-Queen's University Press, 1986), 75–101; Paul Rutherford, *A Victorian Authority: The Daily Press in Late Nineteenth-Century Canada* (Toronto: University of Toronto Press, 1982). For a description of the machines introduced by the *Gazette* during the nineteenth century, see *Montreal: The Metropolis of Canada* (Montreal: Gazette Printing, 1907), iii-xii. The *Montreal Star* introduced fourteen linotype machines in 1895, and *La Presse* had twelve by 1900. For a description of the linotype process, see Fastus, *An Epoch in Printing, by Fastus: Being the First Matter Set on the First Linotype Machine Manufactured in Canada* (Montreal, 1892).

20. For quotes, see *Industries of Canada*, 112, 137. Also see *Dominion Illustrated Christmas Number* (Montreal, 1889).

21. G. Parker, "John Lovell," in *Dictionary of Canadian Biography* (Toronto: University of Toronto Press, 1990), 12:569–74; Eves Martel, "L'industrie à Montréal en 1871" (Master's thesis, Université de Québec, Montréal, 1976), 78–79.

22. Rutherford, *Victorian Authority;* Mona Domosh, *Invented Cities: The Creation of Landscape in Nineteenth-Century New York and Boston* (New Haven: Yale University Press, 1996).

23. Edward Duggan, "Machines, Markets and Labor: The Carriage and Wagon Industry in Late- Nineteenth-Century Cincinnati," *Business History Review* 51 (1977): 308–25; Canada, *Census of Canada, 1871* (Ottawa, 1875), vol. 3; and Canada, *Census of Canada, 1891* (Ottawa, 1894), vol. 3.

24. William Johnson, *Sketches of the Late Depression; Its Cause, Effect and Lessons with a Synoptical Review of Leading Trades during the Past Decade* (Montreal, 1882), 143; Martel, "L'industrie à Montréal," 91–92; Canada, *Census of Canada, 1871* (Ottawa, 1875), vol. 3; and Canada, *Census of Canada, 1891* (Ottawa, 1894), vol. 3. Although Montreal remained the center of carriage making, it did lose firms to other towns. In 1890, for example, Granby lured Guarantee Carriage from Montreal with a fifteen-year tax exemption, a site alongside the Central Vermont tracks, and hydraulic power. Equally important, the town's mayor and several "leading men" subscribed in the company's stock. See *Montreal Gazette*, 17 October 1890, 3. The employment figures given here are taken from the 1891 population manuscripts of the Canadian census, which give the num-

ber of workers an individual employed. *Canadian Engineer* 2 (September 1894): 157; *Canadian Journal of Commerce* (September 1892): 493.

25. For quote, see Celebration Committee, *Montreal in 1856*, 46, 47.

26. *Commerce of Montreal*, 83; Huttemeyer, *Intérêts commerciaux*, 131; *Special Number of the Dominion Illustrated Devoted to Montreal the Commercial Metropolis of Canada* (Montreal, 1891), 42.

27. Steven Fraser, "Combined and Uneven Development in the Men's Clothing Industry," *Business History Review* 57 (1983): 522–47; Peter Hall, "The Location of the Clothing Trades in London, 1861–1951," *Transactions. Institute of British Geographers* 28 (1960): 155–78; Daniel Hiebert, "Discontinuity and the Emergence of Flexible Production: Garment Production in Toronto, 1901–1931," *Economic Geography* 66 (1990): 229–53; Edgar Hoover and Raymond Vernon, *Anatomy of a Metropolis* (New York: Anchor, 1959), 59–69; Edward Muller and Paul Groves, "The Changing Location of the Clothing Industry: A Link to the Social Geography of Baltimore in the Nineteenth Century," *Maryland Historical Magazine* 71 (1976): 403–20. For Montreal, see Michelle Payette-Daoust, "The Montreal Garment Industry, 1871–1901" (Master's thesis, McGill University, 1986); Mary Poutanen, "For the Benefit of the Master: The Montreal Needle Trades during the Transition, 1820–1842" (Master's thesis, McGill University, 1985). On the position of women in local manufacturing, see Bradbury, *Working Families*; D. Cross, "The Neglected Majority: The Changing Role of Women in Nineteenth-Century Montreal," in *The Canadian City*, ed. Gilbert Stelter and Alan Artibise (Toronto: McClelland and Stewart, 1977), 255–81.

28. For quote, see *Commercial Sketch*, 23. Also see Payette-Daoust, "Montreal Garment Industry," 90–102. From an early date Montreal had a sewing machine manufacturing company. In 1861 there were two, while by 1871 it had three producing more than $250,000 worth of machines.

29. Martel, "L'industrie à Montréal," 56; Payette-Daoust, "Montreal Garment Industry"; Celebration Committee, *Montreal in 1856*, 46; *Commercial Sketch*, 18–19.

30. For quotes, see R. Sparks, "The Garment and Clothing Industries," *Manual of the Textile Industry of Canada, 1930* (Montreal: Canadian Textile Journal Publishing Co., 1930), 109; Canada, "Report of the Select Committee on the Manufacturing Interests of the Dominion," *Journals of the House of Commons 1874* (Ottawa, 1874), app. 3, 36; and *Canadian Journal of Commerce* (30 March 1883): 1040–41. Also see Payette-Daoust, "Montreal Garment Industry," 105; Canada, "Report of the Select Committee on the Manufacturing Interests of the Dominion," app. 3, 23; Martel, "L'industrie à Montréal," 57–58. The putting-out system would last well into the end of the nineteenth century, although its importance declined in reverse proportion to that of the contract

system, which by the 1880s was becoming a major organizational form in the clothing industries.

31. For quote, see *Commercial Sketch*, 18. Also see Philip Scranton, "An Exceedingly Irregular Business: Structure and Process in the Paterson Silk Industry, 1885–1910," in *Silk City: Studies on the Paterson Silk Industry, 1860–1940*, ed. Philip Scranton (Newark: New Jersey Historical Society, 1985), 35–72. One effect of the small capital needs was the entry of large numbers of immigrant entrepreneurs. For Montreal, see Sherry Olson, "Ethnic Strategies in the Urban Economy," *Canadian Ethnic Studies* 33 (1991): 39–64.

Chapter Four: "Factories and Industrial Establishments"

1. Merrill Denison, *The Barley and the Stream: The Molson Story* (Toronto: McClelland and Stewart, 1955); Guy Pinard, *Montréal. Son histoire, son architecture* (Montréal: La Presse, 1987), 22–27; Carmen Soucy-Roy, "Le Quartier Ste-Marie" (Master's thesis, Université de Québec, Montréal, 1977); Gerald Tulchinsky, *The River Barons: Montreal Businessmen and the Growth of Industry and Transportation, 1837–1853* (Toronto: University of Toronto Press, 1977), 206–8; City of Montreal, Water Tax Rolls, Saint-Marie ward, 1848–51.

2. Soucy-Roy, "Le Quartier Ste-Marie," 28–34.

3. For quote, see Denison, *Barley and the Stream*, 226. Also see Pinard, *Montréal*, 22–27; Newton Bosworth, *Hochelaga Depicta or the Early History and Present State of the City and Island of Montreal* (Montreal, 1839), 179. By 1856 Converse's Parthenais factory was idle, and a new plant had been built at St. Gabriel Locks on the Lachine Canal. See Celebration Committee of the Grand Trunk Railway, *Montreal in 1856* (Montreal, 1856), 40–41.

4. The Molsons closed their distillery in 1867. Denison, *Barley and the Stream*, chap. 12.

5. J. Douglas Borthwick, *History and Biographical Gazetter of Montreal to the Year 1892* (Montreal, 1892), 409–10; *Commerce of Montreal and Its Manufactures* (Montreal, 1888), 131.

6. Jean Benoit, "Joseph Barsalou," in *Dictionary of Canadian Biography* (Toronto: University of Toronto Press, 1990), 12:62–64. The information on Victoria Straw compiled from Lovell's city directory. Despite the financial backing of the Molson family, the straw company went out of business in 1884 with liabilities of almost $100,000. See *Canadian Journal of Commerce* (8 February 1884): 184–85.

7. For quotes, see *The Dominion Illustrated* (7 December 1889): 359. Also see Leslie Roberts, *From Three Men* (Montreal: Dominion Rubber Co. Ltd., n.d);

Celebration Committee, *Montreal in 1856*, 39; Marcel Bellevance and Jean-Daniel Gronoff, "Les structures de l'espace montréalais l'époque de la confédération," *Cahiers de Géographie de Québec* 24 (1980): 21; *Souvenir Number of the* Montreal Daily Star *Reviewing the Various Financial and Commercial Interests Represented in the City of Montreal* (Montreal, 1890), 45; *Footwear in Canada* 3 (March 1913): 103–4.

    8. J. Douglas Borthwick, *Montreal, Its History to Which Is Added Biographical Sketches, with Photographs of Many of Its Principal Citizens* (Montreal, 1875), 143; Atelier d'Histoire Hochelaga-Maisonneuve, *De fil en Aiguille. Chronique ouvrière d'une filature de coton à Hochelaga en 1880* (Montréal: Atelier d'Histoire Hochelaga-Maisonneuve, 1985); and *L'Industrialisation à Hochelaga-Maisonneuve, 1900–1930* (Montréal: Atelier d'Histoire Hochelaga-Maisonneuve, 1980); Michèle Brassard and Jean Hamelin, "Raymond Préfontaine," in *Dictionary of Canadian Biography* (Toronto: University of Toronto, 1994), 13:842–45; Paul-André Linteau, "Jean-Baptiste Rolland," in *Dictionary of Canadian Biography*, 11:765–66.

    9. Barbara Austin, "Life Cycles and Strategy of a Canadian Company. Dominion Textile: 1873- 1983" (Ph.D. diss., Concordia University, 1985), 126–43; Borthwick, *Montreal, Its History*, 143; Ronald Rudin, "Victor Hudon," in *Dictionary of Canadian Biography* (Toronto: University of Toronto Press, 1990), 12:455–57; Atelier d'Histoire, *De fil en Aiguille*; Luc Desrochers, "Jacques Grenier," in *Dictionary of Canadian Biography*, 13:423–24.

    10. Austin, "Life Cycles and Strategy," 126–43; Borthwick, *Montreal, Its History*, 143; Rudin, "Victor Hudon," 455–57; Atelier d'Histoire, *De fil en Aiguille*; *Annual Financial Review* (Toronto: Houston, 1901); *Le Moniteur de Commerce* (25 February 1881); Jacques Ferland, "Syndicalisme 'parcellaire' et syndicalisme 'collectif'. Une intreprétation socio-technique des conflits ouvriers dans deux industries québécoise, 1880–1914," *Labour / Le Travail* 19 (1987): 49–88; and "'In Search of the Unbound Prometheia': A Comparative View of Women's Activism in Two Quebec Industries, 1869–1908," *Labour / Le Travail* 24 (1989): 11–44.

    11. Austin, "Life Cycles and Strategy," 143–56; and "Managing Marketing in a Commodities Manufacturing Firm: Dominion Textile," *Business and Economic History* 18 (1989): 168–77; *Annual Financial Review* (1901); *Canadian Journal of Commerce* (14 October 1887): 705; Ferland, "'In Search of the Unbound Prometheia,'" 24–29.

    12. Austin, "Life Cycles and Strategy," 143–56; and "Managing Marketing"; *Annual Financial Review* (1901); Raymond Davoud, "The Past History, Present Crisis, and Future Prospects of the Primary Textile Industry in Canada" (Master's thesis, McGill University, 1954), 1–9; Canada, *Report of the Royal Commission on the Textile Industry* (Ottawa: Patenaude, 1938).

13. For quote, see *Canadian Cigar and Tobacco Journal* 23 (July 1917): 25. Also see Pinard, *Montréal,* 284–90; "William Macdonald," *Maclean's* (June 1959): 24–25, 32, 36, 38, 40–41; *Globe and Mail,* 20 April 1877; Henry Small, *The Products and Manufactures of the New Dominion* (Ottawa, 1868), 147; Canada, *Manuscript Census,* Industrial Schedules, 1871.

14. For quote, see *Globe and Mail,* 20 April 1877.

15. *Montreal: The Metropolis of Canada, Illustrated* (Montreal, 1894), 314; *Canadian Journal of Commerce* (3 January 1890): 20.

16. Soucy-Roy, "Le Quartier Ste-Marie," 68, 83.

17. Robert Lewis, "The Segregated City: Class Residential Patterns and the Development of Industrial Districts in Montreal, 1861 and 1910," *Journal of Urban History* 17 (1991): 123–52.

18. For the impact of seasonality and low wages, see the evidence in Canada, *Report of the Royal Commission on Relations of Labour and Capital in Canada—Quebec Evidence* (Ottawa, 1889), 3:21–24, 24–26, 55–60, 84, 250, 529–33; Bettina Bradbury, *Working Families: Age, Gender, and Daily Survival in Industrializing Montreal* (Toronto: McClelland and Stewart, 1993); Atelier d'Histoire, *De fil en Aiguille.*

19. For quote, see *Report of the Royal Commission on Relations,* 545–51. Also see Soucy-Roy, "Le Quartier Ste-Marie," 108–60; Atelier d'Histoire, *L'industrialisation,* 8–14; David Hanna, "Montreal, a City Built by Small Builders, 1867–1880" (Ph.D. diss., McGill University, 1986), 124–27.

20. Hanna, "Montreal, a City Built by Small Builders," 125; Borthwick, *History and Biographical Gazetter,* 409–10; *Commerce of Montreal,* 131; City of Montreal, Water Tax Rolls, Sainte-Marie ward, 1861; Atelier d'Histoire, *De fil en aigulle,* 57–74; Austin, "Life Cycles and Strategy," 133. For company suburbs elsewhere, see Stanley Buder, *Pullman: An Experiment in Industrial Order and Community Planning, 1880–1930* (New York: Oxford University Press, 1967); Henry Silcox, *A Place to Live and Work: The Henry Disston Saw Works and the Tacony Community of Philadelphia* (University Park: Pennsylvania State University Press, 1994).

21. Soucy-Roy, "Le Quartier Ste-Marie," 10–22, 150.

22. For quote, see *Montreal Gazette,* 6 June 1890, 2; Atelier d'Histoire, *De fil en aigulle, l'industrialisation,* 6–29; and *L'histoire du logemont ouvrier à Hochelaga-Maisonneuve* (Montréal: Atelier d'Histoire Hochelaga-Maisonneuve, 1980), 6–12; Brassard and Hamelin, "Raymond Préfontaine," 843.

23. Atelier d'Histoire, *De fil en aigulle, l'industrialisation;* and *l'histoire du logemont ouvrier.*

24. Robert Fishman, *Bourgeois Utopias: The Rise and Fall of Suburbia* (New York: Basic Books, 1987); Kenneth Jackson, *The Crabgrass Frontier: The Suburbanization of the United States* (New York: Oxford University Press, 1985); Sam

Bass Warner, *Streetcar Suburbs: The Process of Growth in Boston, 1870–1900* (Cambridge: Harvard University Press, 1978). On mixed-class suburbs, see John Stilgoe, *Borderland: Origins of the American Suburb, 1820–1939* (New Haven: Yale University Press, 1988), 129–38; Alexander von Hoffman, *Local Attachments: The Making of an American Urban Neighborhood, 1850–1920* (Baltimore: Johns Hopkins University Press, 1994); Henry Binford, *The First Suburbs: Residential Communities on the Boston Peripehry, 1815–1860* (Chicago: University of Chicago Press, 1985). For nineteenth-century industrial and working-class suburban growth, see Buder, *Pullman;* Edward Muller and Paul Groves, "The Emergence of Industrial Districts in Mid-Nineteenth Century Baltimore," *Geographical Review* 69 (1979): 159–78. On industrial suburbanization, see Allen Scott, "Location Patterns and Dynamics of Industrial Activity in the Modern Metropolis," *Urban Geography* 2 (1982): 111–42.

Chapter Five: "The Whirr of Machinery"

1. For quote, see *Montreal Gazette*, 16 July 1831. Also see E. Allan Cureton, "The Lachine Canal" (Master's thesis, McGill University, 1957), 1–90; Larry McNally, *Water Power on the Lachine Canal, 1846–1900* (Ottawa: Parks Canada, 1982); Gerald Tulchinsky, *The River Barons: Montreal Businessmen and the Growth of Industry and Transportation, 1837–1853* (Toronto: University of Toronto Press, 1977); John Willis, *The Process of Hydraulic Industrialization on the Lachine Canal: Origins, Rise and Fall* (Ottawa: Environment Canada, 1987), 168.

2. Tulchinsky, *River Barons*, 228–31; Willis, *Process of Hydraulic Industrialization;* Canada, *Census of Canada, 1901* (Ottawa: Dawson, 1905), 3:233; Celebration Committee of the Grand Trunk Railway, *Montreal in 1856* (Montreal, 1856), 42, 48; Canada Railway Advertising Co., *Montreal Business Sketches with a Description of the City of Montreal, Its Public Buildings and Places of Interest* (Montreal, 1864); McNally, *Water Power on the Lachine Canal;* William Patterson, *Statements Relating to the Home and Foreign Trade of the Dominion of Canada: Also, Annual Report of the Committee of Montreal, 1867* (Montreal, 1868), 120–21; Richard Feltoe, *Redpath: The History of a Sugar House* (Toronto: Natural Heritage–Natural History, 1991), 48; Peter Bischoff, "Des forges du Saint-Maurice aux fonderies de Montréal. Mobilité géographique, solidarité communautaire et action syndicale les mouleurs, 1829–1881," *Revue d'Histoire de l'Amérique Française* 43 (1989): 3–29.

3. Willis, *Process of Hydraulic Industrialization*, 51; McNally, *Water Power on the Lachine Canal*, 77; and "James Shearer," in *Dictionary of Canadian Biography* (Toronto: University of Toronto Press, 1994), 13:947–48; Paul Craven and Tom Traves, "Canadian Railways as Manufacturers, 1850–1880," Canadian His-

torical Association, *Historical Papers* (1983): 255–79 Celebration Committee, *Montreal in 1856*, 42; *Montreal Business Sketches*, 15; Province of Canada, "Report of the Commissioners of Public Works for the Year Ending 31st December, 1855," *Journals of the Legislative Assembly* (Toronto, 1856), app. 31.

4. For quotes, see *Commercial Sketch of Montreal and Its Superiority as a Wholesale Market* (Montreal, 1868), 5; *Montreal Witness*, 7 March 1855, cited in Feltoe, *Redpath*, 46 and app. 4; and Michèle Brassard and Jean Hamelin, "Sir George Alexander Drummond," in *Dictionary of Canadian Biography* (Toronto: University of Toronto Press, 1994), 13:283–84. Also see Tulchinsky, *River Barons*, 228; Canada, *Census of Canada, 1901*, 3:233; Celebration Committee, *Montreal in 1856*, 40, 43–44; Canada, *Manuscript Census*, Industrial schedules, 1871; Canada, "Report of the Select Committee on the Causes of the Recent Depression of the Manufacturing, Mining, Commercial, Shipping, Lumber and Fishing Interests," *Journals of the House of Commons* (Ottawa, 1876), app. 3, 37; Gerald Tulchinsky, "John Redpath," in *Dictionary of Canadian Biography* (Toronto: University of Toronto Press, 1976), 9:654.

5. Celebration Committee, *Montreal in 1856*, 41; Tulchinsky, *River Barons*, 208–10; and "Augustin Cantin," in *Dictionary of Canadian Biography* (Toronto: University of Toronto Press, 1990), 12:158–59.

6. Maurice Milot, "John McDougall," in *Dictionary of Canadian Biography* (Toronto: University of Toronto Press, 1990), 12:620–21; Canada, "Report of the Select Committee on the Causes of the Recent Depression," app. 3, 37; McNally, *Water Power on the Lachine Canal*, 78; and "James Shearer," 947; Celebration Committee, *Montreal in 1856*, 41, 47; Tulchinsky, *River Barons*, 208–10; and "Augustin Cantin," 158–59; Feltoe, *Redpath: Canadian Journal of Commerce* (25 March 1892): 530.

7. For quotes, see Brian Young, *In Its Corporate Capacity: The Seminary of Montreal as a Business Institution, 1816–1876* (Montreal: McGill-Queen's University Press, 1986), 133; and Canada, *Report of the Royal Commission on the Leasing of Water Power, Lachine Canal* (Ottawa, 1887), 7.

8. Marcel Bellavance and Jean-Daniel Gronoff, "Les structures de l'espace montréalais à l'époque de la confédération," *Cahiers de Géographie du Québec* 24 (1980): 380–81; McNally, *Water Power on the Lachine Canal*, Willis, *Process of Hydraulic Industrialization*; Tulchinsky, *River Barons*.

9. William Kilbourn, *The Elements Combined: A History of the Steel Company of Canada* (Toronto: Clarke, Irwin, 1960), 4–10; Celebration Committee, *Montreal in 1856*, 44; Province of Canada, "Report of the Commissioner of Public Works, 1855." There is a discrepency between the number of machines reported by the two latter publications. I have taken the number reported by the report, which is significantly lower than the other publication.

10. William Donald, *The Canadian Iron and Steel Industry* (Boston: Houghton Mifflin, 1915), 60; Willis, *Process of Hydraulic Industrialization*, 368–

69; *Industries of Canada: City of Montreal* (Montreal, 1886), 108; Canada, *Manuscript Census,* Industrial schedules, 1871; Kilbourn, *Elements Combined,* 14.

11. *Canadian Engineer* 8 (January 1901): 187; Peter Bischoff and Robert Tremblay, "James Robertson," in *Dictionary of Canadian Biography* (Toronto: University of Toronto Press, 1990), 12:900–901.

12. *Special Number of the Dominion Illustrated Devoted to Montreal, the Commercial Metropolis of Canada* (Montreal, 1891), 73; *Industries of Canada,* 114; *Commerce of Montreal,* 117, 145.

13. For quote, see Alfred Bray, *Canada under the National Policy: Arts and Manufactures, 1883* (Montreal, 1883), 126. Also see *Commerce of Montreal,* 44–45; *Industries of Canada,* 105.

14. The quote is from Herbert Ames, *The City below the Hill* (1897; rpt., Toronto: University of Toronto Press, 1972), 6). Alan Conter, "The Origins of a Working-Class District: A Portrait of Saint-Ann's Ward in the 1850s" (Undergraduate paper, McGill University, 1976), 4; Mary Davidson, "The Social Adjustment of British Immigrant Families in Verdun and Point St. Charles" (Master's thesis, McGill University, 1933); Young, *In Its Corporate Capacity,* 139; Robert Lewis, "The Segregated City: Class Residential Patterns and the Development of Industrial Districts in Montreal, 1861 and 1901," *Journal of Urban History* 17 (1991): 143; Bischoff, "Des forges du Saint-Maurice aux fonderies de Montréal," 19–24; David Hanna and Frank Remiggi, *Montreal Neighbourhoods,* Canadian Association of Geographers (May 1980), 5–6; Ralph Hoskins, "An Analysis of the Payrolls of the Point St. Charles Shops of the Grand Trunk Railway," *Cahiers de Géographie du Québec* 33 (1989): 323–44; Bettina Bradbury, *Working Families: Age, Gender, and Daily Survival in Industrializing Montreal* (Toronto: McClelland and Stewart, 1993).

15. Conter, "Origins of a Working-Class District"; Claude Larivière, *Petite Bourgogne* (Montréal: Editions Québécoise, 1973), 11–16; Willis, *Process of Hydraulic Industrialization,* 96; Young, *In Its Corporate Capacity.*

16. Conter, "Origins of a Working-Class District," 4–6; Davidson, "Social Adjustment of British Immigrant Families"; David Hanna, "Montreal, a City Built by Small Builders, 1867–1880" (Ph.D. diss., McGill University, 1986), 30–31; Hanna and Remiggi, *Montreal Neighbourhoods,* 2–5; Sherry Olson, "Partage social et partage de l'espace à Montréal, 1847–1901," *Rapport d'Étape* (report presented to the Fonds FCAR, Geography Department, McGill University, 1986), 13, 46; Sherry Olson and Patricia Thornton, "Le Raz de Marée irlandais à Montréal," in *Les chemins de la migration en Belgique et au Québec,* ed. Yves Landry, John Dickinson, Suzy Pasleau, and Claude Desama (Beauport, Québec: Academia MNH, 1995), 69–80; Young, *In Its Corporate Capacity,* 138–41.

17. Young, *In Its Corporate Capacity,* 131–42; McNally, *Water Power on the Lachine Canal,* 17–48; Gerald Tulchinsky and Brian Young, "John Young," in

Dictionary of Canadian Biography (Toronto: University of Toronto Press, 1972), 10:723; Feltoe, *Redpath*, 39. Gerald Tulchinsky, "Frederick Warren Harris," in *Dictionary of Canadian Biography* (Toronto: University of Toronto Press, 1976), 9:367; Willis, *Process of Hydraulic Industrialization*, 173–85.

18. For quote, see Province of Canada, "General Report of Public Works, 1859" (Quebec, 1860), 17. The reports on the Lachine Canal are to be found in several places: "Report of the Commissioners of Public Works," *Journals of the Legislative Assembly of the Province of Canada* for 1841–59; and *Sessional Papers* for 1860–78/79; "The Annual Report of the Minister of Railways and Canals," *Sessional Papers* for 1878/79–1889/90. Cureton, "Lachine Canal," 71–90.

19. For quotes, see *Pilot*, 21 June 1859.

20. Canada, *Census of Canada, 1881* (Ottawa, 1883), vol. 1, table 3.

21. McNally, *Water Power on the Lachine Canal,* 49–50; Willis, *Process of Hydraulic Industrialization;* Gerald Tulchinsky, "John Frothingham," in *Dictionary of Canadian Biography* (Toronto: University of Toronto, 1976), 9:288–89; and "William Workman," in *Dictionary of Canadian Biography* (Toronto: University of Toronto Press, 1982), 11:717–18; Frothingham and Workman, *Price List 1872* (n.p.p., n.d.); John Lovell, *Montreal Directory, 1871–1872* (Montreal, 1871); Michelle Benoit, *La Côte-Saint-Paul* (Montreal: City of Montreal, 1986); *Pilot*, 21 June 1859.

22. For quotes, see J. Douglas Borthwick, *Montreal, Its History to Which Is Added Biographical Sketches, with Photographs of Many of Its Principal Citizens* (Montreal, 1875), 144; and *Pilot*, 21 June 1859. Also see "Report of the Commissioners of Public Works, 1856," app. 3; Celebration Committee, *Montreal in 1856*, 38–39; "General Report of the Commissioners of Public Works, 1866," app.; William Patterson, *Report of the Trade and Commerce of the City of Montreal for 1868* (Montreal, 1869), 121; "General Report of the Commissioner of Public Works, 1864," app. C, 33; McNally, *Water Power on the Lachine Canal,* 64–72, 115–16; *Commerce of Montreal,* 77.

23. For quote, see "General Report of the Commissioners of Public Works, 1864," app. C. Also see Patterson, *Report of the Trade, 1868,* 121; Willis, *Process of Hydraulic Industrialization.*

24. Gilles Lauzon and Lucie Ruellard, *1875/St-Henri* (Montréal: Société Historique de Saint-Henri, 1985), 4, 15; Gilles Lauzon, *Habitat ouvrier et révolution industrielle. Le cas du Village St.-Augustin* (Montréal: Collection RCHTQ, 1989), 20; Willis, *Process of Hydraulic Industrialization,* 97; Joanne Burgess, "Work, Family and Community: Montreal's Leather Craftsmen, 1790–1831" (Ph.D. diss., Université de Québec, Montréal), vol. 2, chaps. 9–10.

25. Lauzon, *Habitat ouvrier,* 15–19, 110–13; Lauzon and Ruellard, *1875/St-Henri,* 11. La Société Historique de Saint-Henri, *Portrait d'une ville. Saint-Henri, 1875–1905* (Montréal: La Société Historique de Saint-Henri, 1987), 4; Claude Ouellet, "Les élites municipales et la municipalisation de Saint-Henri et Sainte-

Cunégonde" (MS, Université de Québec, Montréal, 1982); Hanna and Remiggi, *Montreal Neighbourhoods;* Larivière, *Petite Bourgogne,* 16; Edouard Massicotte, *La cité de Sainte-Cunégonde de Montréal. Notes et souvenirs* (Montréal, 1893), 10; Jean Monet, "Alexandre-Maurice Delisle," in *Dictionary of Canadian Biography* (Toronto: University of Toronto Press, 1972), 10:220.

26. La Société Historique, *Portrait d'une ville,* 6; Ouellet, "Les élites municipales"; *Plan of the Water Distribution of the Town of St. Henry* (Montreal, 1890); *Plan of the Water Distribution of the City of St. Cunegonde* (Montreal, 1890); Jean-Pierre Collin, "La cité sur mesure. Spécialisation sociale de l'espace et autonomie municipale dans la banlieue montréalaise, 1875-1920," *Urban History Review* 13 (1984): 19–34; Massicotte, *La cité de Sainte-Cunégonde,* 21–24; *Monetary Times,* 16 May 1902, 1479; Ernest Small, *The Book of Canada: Illustrating the Great Dominion* (Montreal: Book of Canada Co., 1905), 297–306.

27. For quote, see *Industries of Canada,* 172. Also see Bray, *Canada under the National Policy,* 117; *Commerce of Montreal,* 45–46; Ernest Chambers, *The Book of Montreal: A Souvenir of Canada's Commercial Metropolis* (Montreal: Book of Canada, 1903), 186.

28. Claude Ouellet, "Les industries du cuir à St-Henri, 1871–1891" (MS, Université de Québec, Montréal, 1981), 10–12, 18–22; John Lovell, *Montreal Directory* (Montreal, 1876), 443.

29. For quote, see Thomas Raphael, *Annual Review of the Trade and Commerce of Montreal for 1866* (Montreal, 1867), 9. Also see Kilbourn, *Elements Combined,* 20–21; *Commercial Sketch of Montreal and Its Superiority as a Wholesale Market* (Montreal, 1868), 11; *Montreal Daily Star,* 15 January 1885, 4.

30. *Canadian Journal of Commerce* (14 October 1887): 712; Kilbourn, *Elements Combined,* 23–25; John Lovell, *Lovell's Historic Report of the Census of Montreal, 1891* (Montreal, n.d.), 82; Canada, *Census of Canada, 1891,* vol. 3; Jean Hamelin, Paul Larocque, and Jacques Rouillard, *Répertoire des grèves dans la Province de Québec au XIXe siècle* (Montréal: Les Presses de l'École des Hautes Études Commerciales, 1970), 91. For a discussion of Canada's steel mills in this period, see Craig Heron, *Working in Steel: The Early Years in Canada, 1883–1935* (Toronto: McClelland and Stewart, 1988), 34–42.

31. Henry Small, *The Products and Manufactures of the New Dominion* (Ottawa, 1868), 143. *Commercial Sketch,* 5; Lauzon, *Habitat ouvrier,* 40; Willis, *Process of Hydraulic Industrialization,* 398.

Chapter Six: "One of the Most Magnificent Cities"

1. For quote, see "The Royal Electric Company," *Canadian Engineer* 4 (April 1897): 364. Also see John Dales, *Hydroelectricity and Industrial Development: Quebec, 1898–1940* (Cambridge: Harvard University Press, 1957).

2. Canada, *Census of Canada, 1891* (Ottawa, 1894), vol. 3; Quebec, *Statistical Yearbook of Quebec, 1921* (Quebec: Bureau of Statistics, 1921); Canada, *The Manufacturing Census of Canada, 1929* (Ottawa: Acland, 1931), 70–72.

3. *Annual Financial Review* (Toronto: Houston, 1909), 77; *Canadian Manufacturer* (6 January 1905): 28; Canada, "Report of the Select Committee to Investigate and Report upon Alleged Combinations in Manufactures, Trade and Insurance in Canada," *Journals of the House of Commons, 1888* (Ottawa, 1888), app. 3; Barbara Austin, "Life Cycles and Strategy of a Canadian Company. Dominion Textile: 1873–1983" (Ph.D. diss., Concordia University, 1985); Canada, *Report of the Royal Commission on the Textile Industry* (Ottawa, Patenaude, 1938), 31–39.

4. Paul-André Linteau, René Durocher, and Jean-Claude Robert, *Quebec: A History, 1867–1929* (Toronto: Lorimer, 1983), 336–37; Graham Taylor, "Charles F. Sise, Bell Canada, and the Americans: A Study of Managerial Autonomy, 1880–1915," Canadian Historical Association, *Historical Papers* (1982): 30; Mira Wilkins, *The Emergence of Multinational Enterprise: American Business Abroad from the Colonial Era to 1914* (Cambridge: Harvard University Press, 1970), 144; Earle Gage, "Branch Factories in Canada," *Industrial Management* 64 (1922): 305–8, 318. On Belding Paul, see *Annual Financial Review* (1911): 140; (1913): 150–51; (1920): 156–57; (1923): 165–66; (1926): 139–40.

5. For quote, see *Canadian Engineer* 7 (August 1899): 112. Also see Craig Heron, "The Crisis of the Craftsman: Hamilton's Metal Workers in the Early Twentieth Century," *Labour / Le Travailleur* 6 (1980): 7–48; Wayne Roberts, "Toronto Metal Workers and the Second Industrial Revolution, 1889–1914," *Labour / Le Travailleur* 6 (1980): 49–72.

6. *Canadian Engineer* 11 (January 1904): 27–30; F. Kimball, "Small Electric Motors," *Canadian Manufacturer* (15 September 1905): 33; Dales, *Hydroelectricity*, 103–13; Canada, *Census of Canada, 1901* (Ottawa: Dawson, 1905), 3:100–105; Canada, *Census of Canada, 1911* (Ottawa: Parmalee, 1913), 3:158–63; Canada, *Manufacturing Industries of the Province of Quebec, 1931* (Ottawa: Dominion Bureau of Statistics, 1933), 24; Clarence Hogue, André Bolduc, and Daniel Larouche, *Québec. Un siècle d'Electricité* (Montréal: Libre Expression, 1979), 25–26; Richard Du Boff, "The Introduction of Electric Power in American Manufacturing," *Economic History Review* 20 (1967): 509–18; Thomas Hughes, *Networks of Power: Electrification in Western Society, 1880–1930* (Baltimore: Johns Hopkins University Press, 1983); Harold Platt, *The Electric City: Energy and the Growth of the Chicago Area, 1880–1930* (Chicago: University of Chicago Press, 1991).

7. *Montreal: The Metropolis* (Montreal: Gazette, 1909), 166.

8. C. Warrington and Robert Nicholls, *The History of Chemistry in Canada* (Toronto: Pitman, 1949); Canada, *Directory of the Chemical Industries, 1919* (Ot-

tawa: Labroquerie Tache, 1919), 5–9; Graeme Taylor and Patricia Studnik, *Dupont and the International Chemical Industry* (Boston: Twayne, 1984).

9. This discussion draws on Robert Lewis, "Productive and Spatial Strategies in the Montreal Tobacco Industry, 1850–1918," *Economic Geography* 70 (1994): 370–89. Also see Alfred Chandler, *The Visible Hand: The Managerial Revolution in American Business* (Cambridge: Belknap Press, 1977); Patricia Cooper, *Once a Cigar Maker: Men, Women and Work Culture in American Cigar Factories, 1900–1929* (Urbana: University of Illinois Press, 1987); Mark Prus, "Mechanization and the Gender-Based Division of Labour in the U.S. Cigar Industry," *Cambridge Journal of Economics* 14 (1990): 63–79.

10. Austin, "Life Cycles and Strategy"; Canada, *Report of the Royal Commission on the Textile Industry*; William Kilbourn, *The Elements Combined: A History of the Steel Company of Canada* (Toronto: Clarke, Irwin, 1960). Also see Philip Scranton, *Endless Novelty: Specialty Production and American Industrialization, 1865–1925* (Princeton: Princeton University Press, 1997).

11. Montreal Board of Trade, *The Board of Trade Illustrated Edition* (Montreal: Guertin, 1909), 1; Paul-André Linteau, *Histoire de Montréal depuis Confédération* (Montréal: Boréal, 1992), 187–208; Raymond Tanghe, "La population," in *Montréal économique*, ed. Edras Minville (Montréal: Editions Fides, 1943), 97–114; Walter van Nus, "A Community of Communities: Suburbs in the Development of 'Greater Montreal,'" in *Montreal Metropolis, 1880–1930*, ed. Isabelle Gournay and France Vanlaethem (Montreal: Canadian Centre for Architecture, 1998), 59–67.

12. Terry Copp, *The Anatomy of Poverty: The Condition of the Working Class in Montreal, 1897–1929* (Toronto: McClelland and Stewart, 1974), 43; Marc Choko, *Crises du logement à Montréal* (Montréal: Édition Coopératives Albert Saint Martin, 1980).

13. Harold Ames, *The City below the Hill* (1897; rpt., Toronto: University of Toronto Press, 1972); Percy Robert, "Dufferin District" (Master's thesis, McGill University, 1928).

14. Walter van Nus, "The Role of Suburban Government in the City-Building Process: The Case of Notre Dame de Grâces, Quebec, 1876–1910," *Urban History Review* 13 (1984): 91–103; Larry McCann, "Planning and Building the Corporate Suburb of Mount Royal, 1910–1925," *Planning Perspectives* 11 (1996): 259–301.

15. John Benhart, "Industrial Suburbanization and Residential Segregation in the Upland South: The Lonsdale Suburb of Knoxville, Tennessee—1890–1914," *Historical Geography* 24 (1995): 71–90; Kevin Kane and Thomas Bell, "Suburbs for a Labor Elite," *Geographical Review* 75 (1985): 319–34; Graham Taylor, *Satellite Cities: A Study of Industrial Suburbs* (New York: Appleton,

1915); Richard Harris, *Unplanned Suburbs: Toronto's American Tragedy, 1900–1950* (Baltimore: Johns Hopkins University Press, 1996). For Chicago, see Stanley Buder, *Pullman: An Experiment in Industrial Order and Community Planning, 1880–1930* (New York: Oxford University Press, 1967); Louise Carroll Wade, *Chicago's Pride: The Stockyards, Packingtown and Environs in the Nineteenth Century* (Urbana: University of Illinois Press, 1987), 15–17, 47–60.

16. For quote, see C. Kirkpatrick, "Town Planning in Relation to Industrial Development," *Canadian Engineer* 37 (December 1919): 524–25. Also see James Ford, "Residential and Industrial Decentralization" in *City Planning*, ed. John Nolen (New York: Appleton, 1922), 333–52; Robert Haig and R. McCrea, "Major Economic Factors in Metropolitan Growth and Arrangement," *Regional Survey* (New York: Regional Plan of New York and Its Environs, 1927); Edward Pratt, *Industrial Causes of Congestion of Population in New York City* (New York: Columbia University Press, 1911).

17. Paul-André Linteau, *The Promoters' City: Building the Industrial Town of Maisonneuve, 1883–1918* (Toronto: Lorimer, 1985); Jean-Pierre Collin, "La cité sur mesure. Spécialisation sociale de l'espace et autonomie municipale dans la banlieue montréalaise, 1875–1920," *Urban History Review* 13 (1984): 19–34; David Hanna, "Montreal, a City Built by Small Builders, 1867–1880" (Ph.D. diss., McGill University, 1986), 9; Christopher Boone, "The Politics of Transportation Services in Suburban Montreal: Sorting Out the 'Mile End Muddle,' 1893–1909," *Urban History Review* 24 (1996): 25–39; Gregory Levine, "Class, Ethnicity and Property Transfers in Montreal, 1907–1909," *Journal of Historical Geography* 14 (1988): 360–80; *Canadian Architect and Builder* 21 (March 1908): 15; Choko, *Crises du logement*, 58.

18. For quotes, see Choko, *Crises du logemont*, 25; and Ernest Chambers, *Suburban Montreal as Seen from the Routes of the Park and Island Railway Co.* (Montreal, 1895), 36. Also see Van Nus, "Role of the Suburban Government," 97; Bruno Ramirez, "Montreal's Italians and the Socioeconomy of Settlement: Some Historical Hypotheses," *Urban History Review* 10 (1981): 39–48; John Saywell, *Housing Canadians: Essays on the History of Residential Construction in Canada* (Ottawa: Economic Council of Canada, 1975), 119–22. For Toronto, see Harris, *Unplanned Suburbs*.

19. For quote, see Mary Davidson, "The Social Adjustment of British Immigrant Families in Verdun and Point St. Charles" (Master's thesis, McGill University, 1933), 22. Also see Ames, *City below the Hill*; Copp, *Anatomy of Poverty*; Carl Dawson, "City Planning and Our North American Social Heritage," *Housing and Community Planning* (Montreal: McGill University Press, 1944), 149; Joint Committee of the Montreal Board of Trade and the City Improvement League, *A Report on Housing and Slum Clearance for Montreal* (Montreal: Mon-

treal Board of Trade, 1935); Leo Zakuta, "The Natural Areas of the Montreal Metropolitan Community with Special Reference to the Central Area" (Master's thesis, McGill University, 1948).

20. Rebecca Aiken, *Montreal Chinese Property Ownership and Occupational Change, 1881–1981* (New York: AMS Press, 1989), 79–84; Davidson, "Social Adjustment of British Immigrants"; Ramirez, "Montreal's Italians"; Lloyd Reynolds, *The British Immigrant: His Social and Economic Adjustment in Canada* (Toronto: Oxford University Press, 1935), 114–32; Judith Seidel, "The Development and Social Adjustment of the Jewish Community in Montreal" (Master's thesis, McGill University, 1939); Charles Young, *The Ukrainain Canadians* (Toronto: Thomas Nelson, 1931).

21. For quote, see "Montreal Real Estate Market," *Montreal Gazette* (12 March 1890): 2. Also see Robert Binns, *Montreal's Electric Streetcars: An Illustrated History of the Tramway Era, 1892 to 1959* (Montreal: Railfare, 1973); "Effect of Rapid Electric Transit upon Suburban Real Estate," *Montreal Star,* 10 March 1906, 20; *Canadian Journal of Commerce* (17 January 1890): 118; Linteau, *Promoters' City,* chap. 6. Also see Eugene Erickson and William Yancey, "Work and Residence in Industrial Philadephia," *Journal of Urban History* 5 (1979): 147–82; Caroline Golab, *Immigrant Destinations* (Philadelphia: Temple University Press, 1977); Olivier Zunz, *The Changing Face of Inequality* (Chicago: University of Chicago Press, 1982).

22. For quote, see "The Housing Problem in Canada," *Labour Gazette* 5 (October 1904): 373–74.

23. For quotes, see G. Ferguson, "Decentralization of Industry and Metropolitan Control," *Journal of the Town Planning Institute* 2 (April 1924): 12; and *Canadian Architect and Builder* (5 November 1892); Arsene Lavallé, "City Planning in Montreal," *Proceedings of the Fifth National Conference on City Planning* (Boston: N.p., 1913): 28; and George Stephens, "Montreal Must Plan for Great Future Growth," *Municipal Review of Canada* 25 (May 1929): 194.

24. For quote, see *Contract Record and Engineering Review* (18 June 1919): 568. Also see Zakuta, "Natural Areas of the Montreal Metropolitan Community"; Ferguson, "Decentralization of Industry."

25. For quotes, see Montreal Junior Board of Trade, *Report of Committee on Town Planning and Zoning* (Montreal: N.p., 1935), 4. Also see France Vanlaethem, "Beautification versus Modernization," in Gournay and Vanlaethem, *Montreal Metropolis,* 132–51; Susan Wagg, *Percy Erskine Nobbs: Architect, Artist, Craftsman* (Montreal: McGill-Queen's University Press, 1982), 45–57; Jean-Claude Marsan, *Montreal in Evolution* (Montreal: McGill-Queen's University Press, 1981).

26. Michael Gauvin, "The Reformer and the Machine: Montreal Civic Politics from Raymond Préfontaine to Médéric Martin," *Journal of Canadian Stud-*

ies 13 (1978): 16–26; Annick Germain, *Les mouvements de réform urbaine à Montréal au tournant de siècle* (Montréal: Département de Sociologie, Université de Montréal, 1984); Harold Kaplan, *Reform, Planning, and City Politics: Montreal, Winnipeg and Toronto* (Toronto: University of Toronto Press, 1982); Gregory Levine, "To Tax or Not to Tax? Political Struggle over Personal Property Taxation in Montreal and Toronto, 1870–1920," *International Journal of Urban and Regional Research* 11 (1987): 543–66; D. Russell, "H. B. Ames as Municipal Reformer" (Master's thesis, McGill University, 1972).

27. For quote, see James Ewing, "The Engineer and the Town Plan," *Journal of the Engineering Institute of Canada* 4 (December 1921): 413. Also see Thomas Adams, "Town and Regional Planning in Relation to Industrial Growth in Canada," *Journal of the Town Planning Institute of Canada* 1 (June-August 1921): 13; City of Montreal, *Planning for Montreal* (Montreal: City Planning Department, 1944), 37. For bylaws, see Paul Martineau, "The Civic Administration of Montreal," in *Municipal Government in Canada,* ed. Morley Wickett (Toronto: University of Toronto Studies, 1907), 2:317; *Charter of the City of Montreal* (Montreal, 1889), 41.

28. For quote, see James Ewing, "The Montreal Situation with Reference to Town Planning," *Journal of the Town Planning Institute of Canada* 1 (1921): 4–9. Also see E. Deville, "Chairman's Address," *Journal of the Town Planning Institute of Canada* 1 (1921): 3–4; Copp, *Anatomy of Poverty,* 71–83; Thomas Gunton, "The Ideas and Policies of the Canadian Planning Profession, 1909–1931," in *The Usable Urban Past: Planning and Politics in the Modern Canadian City,* ed. Alan Artibise and Gilbert Stelter (Toronto: Macmillan, 1979), 177–95; Walter Van Nus, "Towards the City Efficient: The Theory and Practice of Zoning, 1919–1939," in Artisbise and Stelter, *Usable Urban Past,* 226–46. Also see Edward Muller, "The Pittsburgh Survey and 'Greater Pittsburgh': A Muddled Metropolitan Geography," in *Pittsburgh Surveyed: Social Science and Social Reform in the Early Twentieth Century,* ed. Maurine Greenwald and Margo Anderson (Pittsburgh: University of Pittsburgh Press, 1996), 69–87.

29. For quotes, see "Montreal, a Hygienic Disgrace to Civilization," *Canadian Engineer* 16 (April 1909): 527; and Percy Nobbs, "Montreal's Town Planning and Housing Problems," *Municipal Review of Canada* 33 (1937): 27–28. Also see Martineau, "Civic Administration of Montreal," 317; City of Montreal, *Planning for Montreal,* 33; Julia Schoenfeld, "A Recreation Survey of the City of Montreal," *Canadian Municipal Journal* 9 (1913): 358; Copp, *Anatomy of Poverty.*

30. Binns, *Montreal's Electric Streetcars;* Christopher Armstrong and H. Viv Nelles, "Suburban Street Railway Strategies in Montreal, Toronto and Vancouver, 1896–1930," in *Power and Place: Canadian Urban Development in the North American Context,* ed. Gilbert Stelter and Alan Artibise (Vancouver: University of British Columbia Press, 1986), 187–218; Dales, *Hydroelectricity;* J. Saint Cyr,

"The Montreal Tramways" (address delivered to the Montreal Branch of the Engineering Institute of Canada, 10 November 1927); and "Le tramways et la circulation" (paper delivered before the Chamber of Commerce, district of Montreal, 15 May 1929).

31. For quote, see Gauvin, "Reformer and the Machine," 23. Also see Armstrong and Nelles, "Suburban Street Railway Strategies"; Linteau, *Promoters' City*, 87–96; Taylor, "Charles F. Sise," 25; Dales, *Hydroelectricity and Industrial Development*; Hogue, Bolduc, and Larouche, *Québec*.

Chapter Seven: "Pierced by Another Giant Skyscraper"

1. "Move Starts Today to New Premises," *Montreal Gazette*, 6 April 1929, 5; "Nineteen-Storey Building Planned" and "Skyscraper to Be 23 Storeys High," *Montreal Gazette*, 9 July 1929, 5; Carol Willis, *Form Follows Finance: Skyscrapers and Skylines in New York and Chicago* (New York: Princeton Architectural Press, 1995).

2. James Ewing, "Plain Words from Montreal," *Journal of the Town Planning Institute of Canada* 4 (August 1925): 3.

3. Ludger Beauregard, "La Rue Saint-Jacques à Montréal: une géographie des bureaux" (Département de Géographie, Université de Montréal, Notes and Documents, No. 81–02, 1981); Jean-Claude Marsan, *Montreal in Evolution* (Montreal: McGill-Queen's University Press, 1981), 228–49; John Weaver, *Shaping the Canadian City: Essays on Urban Politics and Policy, 1890–1920* (Toronto: Institute of Public Administration of Canada, 1979), 12–15. Gunter Gad and Deryck Holdsworth, "Corporate Capitalism and the Emergence of the High-Rise Office Building," *Urban Geography* 8 (1987): 212–31; David Ward, "The Industrial Revolution and the Emergence of Boston's Central Business District," *Economic Geography* 42 (1966): 152–71. Parts of this chapter draw upon my essay "Robert Lewis, "Productive Strategies and Manufacturing Reorganization in Montreal's Central District, 1850–1900," *Urban Geography* 16 (1995): 15–18.

4. Ernest Chambers, *The Book of Montreal: A Souvenir of Canada's Commercial Metropolis* (Montreal: Book of Canada, 1903), 128; *The Commerce of Montreal and Its Manufactures* (Montreal, 1888); 43; *Canadian Enterprise in Manufacturing* (Toronto: Canadian Manufacturers' Association, 1945), 229–30.

5. For quote, see *Footwear* 3 (January 1913): 35.

6. Ewing, "Plain Words from Montreal," *Journal of the Town Planning Institute of Canada* 4 (August 1925): 13. Despite his constant references to downtown and uptown, Ewing never once defined what these were. He probably meant that downtown was Old Montreal and the older sections of the Outer

Core (south of Dorchester Street), while uptown referred to the areas north of Dorchester.

7. For quote, see *Montreal Gazette*, 8 May 1906, 7. Also see *Industries of Canada: City of Montreal* (Montreal, 1886), 112.

8. For quote, see "Montreal—The Commercial Metropolis," *Contract Record and Engineering Review* (23 October 1912): 48–61.

9. "Move Starts Today to New Premises," *Montreal Gazette*, 6 April 1929, 5; Marsan, *Montreal in Evolution*, 242–49; France Vanlaethem, "Montreal Architects and the Challenge of Commissions," in *Montreal Metropolis, 1880–1930*, ed. Isabelle Gournay and France Vanlaethem (Montreal: Canadian Centre for Architecture, 1998), 70–111; Isabelle Gournay, "Prestige and Professionalism: The Contribution of American Architects," in Gournay and Vanlaethem, *Montreal Metropolis*, 112–31.

10. *Industries of Canada;* Chambers, *Book of Montreal*, 184, 208, 215, 217–18; Communauté Urbaine de Montréal, *Architecture industrielle* (Montréal: Communauté de Montréal, 1982).

11. "The Year's Building at Montreal," *Contract Record and Engineering Review* (23 December 1914): 1559.

12. For quote, see "Montreal," *Canadian Engineer* 22 (18 January 1912): 159. Also see Isabelle Gournay, "Gigantism in Downtown Montreal," in Gournay and Vanlaethem, *Montreal Metropolis*, 153–82; "The Blumenthal Building at Montreal," *Contract Record and Engineering Review* (15 March 1911): 32–35; "The Herald Building," *Contract Record and Engineering Review* (8 November 1911): 40–42; "New Office and Factory Building at Montreal," *Contract Record and Engineering Review* (3 January 1912): 45.

13. "Firm's New Premises," *Montreal Gazette*, 6 April 1929, 5.

14. For quote, see Chambers, *Book of Montreal*, 217, 220. Also see City of Montreal, Water Tax Rolls, various years; *Canadian Manufacturer* (6 October 1905): 30; "Herald Building"; "New Office and Factory Building at Montreal," *Contract Record and Engineering Review* (3 January 1912): 45.

15. For quote, see *Centennial Report of the Montreal Board of Trade* (Montreal, 1893), 74. Also see William Atherton, *Montreal, 1534–1914* (Montreal: Clarke, 1914), vol. 2; Marsan, *Montreal in Evolution*, 286; F. Clifford Smith, *The Montreal Water Works* (Montreal: City of Montreal, 1913); *Charter of the City of Montreal* (Montreal, 1889), 75–77.

16. M. Casey, "The Use of the Power for Port Facilities," *Engineering Journal* 7 (1924): 486–88; Frederick Cowie, "The Great National Port of Canada. Features of the Important Extension Work in Progess in Montreal Harbour," *Canadian Engineer* 22 (18 January 1912): 178–83; Paul-André Linteau, "Le développment du port de Montréal du début du 20e siècle," Historical Papers, *Canadian Historical Association* (1972): 181–205.

17. "A Notable Group of Railway Terminals," *Canadian Engineer* 29 (23 September 1915): 403–4; David Hanna, "The Importance of Transportation Infrastructure," in Gournay and Vanlaethem, *Montreal Metropolis*, 46–53.

18. Bell Telephone, *Annual Report*, 1881; Michelle Martin, "Communication and Social Forms: The Development of the Telephone, 1876–1920," *Antipode* 23 (1991): 307–33.

19. For quotes, see *Canadian Journal of Commerce* (17 January 1890): 118; and "Montreal," *Canadian Engineer* 22 (18 January 1912): 159.

20. For quote, see *Canadian Architect and Builder* 5 (December 1892). Also see Joint Committee of the Montreal Board of Montreal and the City Improvement League, *A Report on Housing and Slum Clearance for Montreal* (Montreal: N.p., 1935), 14–21; *Canadian Architect and Builder* 5 (November 1892); Yvan Lamonde, *La culture ouvrière à Montréal, 1880–1920* (Québec City: Institute Québécoise de Recherche sur la Culture, 1982); Canada, *Census of Canada, 1931* (Ottawa: Patenaude, 1933), 2:18–20 and table 33; Canada, *Population of the Municipal Wards of Montreal City, 1931* (Ottawa: Government Printing Office, 1934), 22–29; Bruno Ramirez, "Montreal's Italians and the Socioeconomy of Settlement: Some Historical Hypotheses," *Urban History Review* 10 (1981): 39–48; Judith Seidel, "The Development and Social Adjustment of the Jewish Community in Montreal" (Master's thesis, McGill University, 1939); Charles Young, *The Ukrainain Canadians* (Toronto: Thomas Nelson, 1931).

21. The discussion of Landau and Cormack draws on Robert Lewis, "Productive and Spatial Strategies in the Montreal Tobacco Industry, 1850–1918," *Economic Geography* 70 (1994): 370–89.

22. For quotes, see *Canadian Cigar and Tobacco Journal* 13 (July 1907): 62; and 13 (October 1907): 27–29. Also see Meredith Burrill, "Studies in the Industrial Geography of Montreal" (Ph.D. diss., Clark University, 1930), 183–91.

23. *Canadian Cigar and Tobacco Journal* 13 (July 1907): 62; and 14 (June 1908): 25. The limited range of line contrasts with Imperial, which produced thirty-seven brands of cigarettes, seventy-six brands of cut tobacco, twenty-two brands of chewing tobacco, and ten brands of little cigars.

24. *Canadian Cigar and Tobacco Journal* 13 (July 1907): 62.

25. For quotes, see *Canadian Cigar and Tobacco Journal* 12 (June 1906): 19; and 19 (February 1913): 59. Also see *Canadian Cigar and Tobacco Journal* 7 (September 1901): 437; 12 (February 1906): 43; 12 (March 1906): 53; 22 (May 1916): 29; 22 (August 1916): 77; and 26 (May 1920): 52; Émile Benoist, "L.-O. Grothé," in *Monographies économiques*, ed. Émile Benoist (Montréal: Devoir, 1925), 214–28; Guy Pinard, *Montréal, son histoire, son architecture* (Montréal: La Presse, 1987), 96–101.

26. For quotes, see *Canadian Cigar and Tobacco Journal* 4 (March 1898): 69; 12 (January 1906): 59; and 12 (March 1906): 65. Also see *Canadian Journal of Commerce* (18 January 1884): 76; *Commerce of Montreal*, 104–5; Gerald Tulchin-

sky, "Samuel Davis," in *Dictionary of Canadian Biography* (Toronto: University of Toronto Press, 1990), 12:228–29. The workers in the Cuban department had been such trouble that in 1901 the company threatened to move the factory to Toronto. *Canadian Cigar and Tobacco Journal* 7 (August 1901): 395.

27. For quotes, see *Canadian Journal of Commerce* (1 November 1889): 772; (29 April 1887): 963; *Special Number of the Dominion Illustrated Devoted to Montreal, the Commercial Metropolis of Canada* (Montreal, 1891), 45; and Chambers, *Book of Montreal,* 171. Also see Michelle Payette-Daoust, "The Montreal Garment Industry, 1871–1901" (Master's thesis, McGill University, 1986); Jacques Rouillard, "Les travailleurs juifs de la confection à Montréal," *Labour / Le Travailleur* 8–9 (1981–82): 253–59; Frank Scott and Harry Cassidy, *Labour Conditions in the Men's Clothing Industry* (Toronto: Thomas Nelson, 1935), 1–3; Gregory Teal, "The Organization of Production and the Heterogeneity of the Working Class: Occupation, Gender and Ethnicity among Clothing Workers in Quebec" (Ph.D. diss., McGill University, 1986), 162–98.

28. For quote, see *Canadian Journal of Commerce* (3 August 1888): 212–13. Also see Daniel Hiebert, "Discontinuity and the Emergence of Flexible Production: Garment Production in Toronto, 1901–1931," *Economic Geography* 66 (1990): 229–53; Payette-Daoust, "The Montreal Garment Industry," 105–10; Scott and Cassidy, *Labor Conditions in the Men's Clothing Industry,* 1–2; R. Sparks, "The Garment and Clothing Industries: History and Organization," *Manual of the Textile Industry of Canada, 1930* (Montreal: Canadian Textile Journal Publishing, 1930), 107–30; Teal, "Organization of Production and the Heterogeneity of the Working Class," 136–98. An excellent American example is Edward Muller and Paul Groves, "The Changing Location of the Clothing Industry: A Link to the Social Geography of Baltimore in the Nineteenth Century," *Maryland Historical Magazine* 71 (1976): 403–20.

29. For quotes, see "Montreal's Newest Office and Loft Building," *Contract Record and Engineering Review* (23 April 1924): 393–94. Also see "The Blumenthal Building"; "The 'Amherst Building' in Uptown Montreal," *Contract Record and Engineering Review* (7 October 1925): 977.

30. Burrill, "Studies in the Industrial Geography of Montreal," 156–65; Payette-Daoust, "Montreal Garment Industry," 107; Rouillard, "Les travailleurs juifs de la confection," 255; Canada, *Population of the Municipal Wards.*

31. *Annual Financial Review* (Toronto: Houston, 1911): 107; (1925): 441–42; and (1926): 758–59. For another example, see Chambers, *Book of Montreal,* 177.

Chapter Eight: "Busy Hives of Industry"

1. For quote, see "Metal Shingle and Siding Company," *Canadian Manufacturer* (6 December 1907): 44; *Canadian Manufacturer* (5 July 1907): 27.

2. For quote, see "Dominion Rubber Has New System," *Montreal Gazette*, 15 July 1929, 23. Also see Leslie Roberts, *From Three Men* (Montreal: Dominion Rubber Co., n.d.); *Canadian Engineer* 12 (April 1905): 108; *Annual Financial Review* (Toronto: Houston, 1909): 138–40; and (1930): 473–76; *Canadian Chemical Journal* 1 (September 1917): 118; and 15 (February 1931): 59; "New Montreal Factory," *Contract Record and Engineering Review* (13 September 1911): 53; Atelier d'Histoire Hochelaga-Maisonneuve, *Evolution de l'architecture industrielle. Un quartier-type: Hochelaga-Maisonneuve* (Montréal: Atelier d'Histoire Hochelaga-Maisonneuve, 1982), *Le histoire du logement ouvrier à Hochelaga-Maisonneuve* (Montréal: Atelier d'Histoire Hochelaga-Maisonneuve, 1980); and *L'industrialisation à Hochelaga-Maisonneuve, 1900–1930* (Montréal: Atelier d'Histoire Hochelaga-Maisonneuve, 1980).

3. Ernest Chambers, *The Book of Montreal: A Souvenir of Canada's Commercial Metropolis* (Montreal: Book of Canada, 1903), 192.

4. William Atherton, *Montreal, 1534–1914* (Montreal: Craig, 1914), 466–69; *Commerce of Montreal and Its Manufactures* (Montréal, 1888), 128–29; Huttemeyer, *Les intérêts commerciaux de Montréal et Québec et leurs manufactures* (Montréal, 1889), 170; Chambers, *Book of Montreal*, 226. Charles Goad, *Insurance Plan of the City of Montreal* (Montreal: Goad, 1915): vol. 3, pl. 106.

5. Chambers, *Book of Montreal*, 188; Michael Hinton, "Andrew Frederick Gault," in *Dictionary of Canadian Biography* (Toronto: University of Toronto Press, 1994), 13:372–76.

6. For quote, see Bruno Ramirez, "Brief Encounters: Italian Immigrant Workers and the CPR, 1900–1930," *Labour / Le Travail* 17 (1986): 11–13. Also see Omer Lavallee, *Delorimier and Angus* (Montreal: Canadian Railroad Historical Association, 1962), 7–10; *Canadian Engineer* 11 (March 1904): 73; "A Notable Group of Railway Terminals," *Canadian Engineer* 29 (September 1915): 403–404; R. Naheut, "Une expérience canadienne de taylorisme. Le cas des usines Angus du Canadien Pacifique" (Master's thesis, Université de Québec, Montréal, 1984), 59; John Brown, *The Baldwin Locomotive Works* (Baltimore: Johns Hopkins University Press, 1995).

7. "The Angus Shops of the Canadian Pacific Railway, Montreal," *Canadian Engineer* 28 (January 1915): 174–76; "Locomotive and Car Shops of the Canadian Pacific Railway," *Canadian Engineer* 29 (September 1915): 329; Naheut, "Une expérience canadienne de taylorisme."

8. Paul-André Linteau, *The Promoters' City: The Industrial Town of Maisonneuve, 1883–1918* (Toronto: Lorimer, 1985); Jean-Pierre Gauthier and Pierre Larivière, "La cité de Maisonneuve. Ville modèle du début du siècle," in Société Historique de Montréal, *Montréal. Artisans, histoire, patrimoine* (Montréal: Fides, 1979), 103–19.

9. *Canadian Manufacturer* (March 1904): 25; *Canadian Engineer* 10 (April

1903): 110; "Shops of the Locomotive and Machine Company of Montreal," *Canadian Engineer* 10 (November 1903): 314–16; Herbert Marshall, Frank Southard, and Kenneth Taylor, *Canadian-American Industry: A Study in International Investment* (New Haven: Yale University Press, 1936), 69; *Montreal: The Metropolis of Canada* (Montreal: N.p., 1923), 65–66.

10. "Longue Point Plant of the Canadian Steel Foundries, Ltd," *Canadian Machinery* 8 (November 1912): 325–30; William Donald, *The Canadian Iron and Steel Industry* (Boston: Houghton Mifflin, 1915), 227–28, 269–71; *Annual Financial Review* (1911): 103; (1925): 463; and (1916): 262–63; R. Rogers and A. Turnbull, "Application of Electrical Equipment to Operations of a Cement Mill," *Canadian Chemistry and Metallurgy* 10 (February 1926): 27–29; "The Canada Cement Company's Great Plant at Pointe aux Trembles, Que.," *Contract Record and Engineering Review* (21 January 1914): 87; John Cooper, *Montreal: A Brief History* (Montreal: McGill-Queen's University Press, 1969), 156–57; *Canadian Chemical Journal* 10 (March 1926): 68.

11. Canada, *Census of Canada, 1911* (Ottawa: Parmalee, 1913), 3:302–3; Jacques Ferland, "Évolution des rapports sociaux dans l'industrie canadienne du cuir du tournant du 20e siècle" (Ph.D. diss., McGill University, 1985). The discussion on the shoe industry draws from Robert Lewis, "Restructuring and the Formation of an Industrial District in Montreal's East End, 1850- 1914," *Journal of Historical Geography* 20 (1994): 151–54.

12. *Canadian Journal of Commerce* (16 January 1885): 86–87; *Footwear in Canada* 2 (April 1912): 27. *Montreal Gazette*, 5 February 1890, 8; Canadian Reconstruction Association, *The Boot and Shoe Industry* (Toronto: Canadian Reconstruction Association, 1920); *Canadian Shoe and Leather Journal* 26 (15 June 1912): 50.

13. For quotes, see Victor Clark, *History of Manufactures in the United States* (New York: Peter Smith, 1949), 3:230; and *Canadian Journal of Commerce* (31 August 1883): 47–48. Also see Canadian Reconstruction, *Boot and Shoe Industry*, 7; Jacques Ferland, "'Not for Sale'—American Technology and Canadian Shoe Factories: The United Shoe Machinery Company of Canada, 1899–1912," *American Review of Canadian Studies* 18 (1988): 59–82; Jacques Ferland, "'In Search of the Unbound Prometheia': A Comparative View of Women's Activism in Two Quebec Industries," *Labour / Le Travail* 24 (1989): 23; H. Watson, "Canada's Sixth Industry," *Canadian Magazine of Politics, Science, Art and Literature* (1911): 33–40.

14. For quotes, see *Canadian Shoe and Leather Journal* 27 (1 May 1913): 63; *Montreal Gazette*, 5 February 1890, 8; and "Some Industry Has Now a Wide Range," *Montreal Gazette*, 28 August 1929, 5.

15. Canada, *Leather Boot and Shoe Industry, 1921* (Ottawa: Minister of Trade and Commerce, 1923), 5; "The Progess of the Shoe Manufacturing Industry in

the Province of Quebec," *Footwear in Canada* 1 (February 1912): 42; *Canadian Shoe and Leather Journal* 26 (15 August 1912): 39–40.

16. Ferland, "'In Search of the Unbound Prometheia'"; Jacques Ferland, "Syndicalisme 'parcellaire' et syndicalisme 'collectif'. Une intreprétation sociotechnique des conflits ouvriers dans deux industries québécoise," *Labour / Le Travail* 19 (1987): 55–64; Jean Hamelin, Paul Larocoque and Jacques Rouillard, *Répertoire des grèves dans la province de Québec au XIXe siècle* (Montréal: Les Presses de l'École des Hautes Études Commerciales, 1970).

17. Linteau, *Promoters' City,* 78–79, 82. Joanne Burgess, "Work, Family and Community: Montreal Leather Craftsmen, 1790–1831" (Ph.D. diss., Université de Québec, Montréal, 1986), 2 vols.; Claude Ouellet, "Les industries du cuir à St-Henri" (MS, Université de Québec, Montréal, 1981); *Canadian Shoe and Leather Journal* 27 (15 January 1913): 28.

18. For Dufresne and Locke, see *Footwear in Canada* 2 (July 1912): 41; *Canadian Shoe and Leather Journal* 28 (1 July 1915): 46; Ville de Maisonneuve, *Rôle d'évaluation,* 1904 and 1912. For Dupont and Kirvan-Doig, see *Canadian Shoe and Leather Journal* 24 (April 1910): 61; 24 (October 1910): 101 and 103; 26 (February 1912): 52; 26 (15 October 1912): 55; and 28 (1 July 1915): 46; *Footwear in Canada* 2 (February 1912): 46; 2 (April 1912): 32; and 2 (August 1912): 61. For Rideau, see *Canadian Shoe and Leather Journal* 24 (April 1910): 86.

19. Meredith Burrill, "Studies in the Industrial Geography of Montreal" (Ph.D. diss., Clark University, 1930), 2–4.

20. Burrill, "Studies in the Industrial Geography of Montreal"; C. Warrington and Robert Nicholls, *The History of Chemistry in Canada* (Toronto: Pitman, 1949), 341; Canada, *Directory of the Chemical Indstries in 1919* (Ottawa: Labroquerie Tache, 1919), 5–9; Graeme Taylor and Patricia Sudnik, *DuPont and the International Chemical Industry* (Boston: Twayne, 1984); J. Riley, "Business Outlook in Canada from the Stand-Point of the Paint and Varnish Industry and Kindred Lines," *Canadian Chemical Journal* 8 (May 1924): 115.

21. "Brandram-Henderson Ltd," *Canadian Chemical Journal* 7 (July 1923): 190; Warrington and Nicholls, *History of Chemistry,* 336; *Annual Financial Review* (1913): 153–54; (1920): 161–63; and (1925): 181–82.

22. Burrill, "Studies in the Industrial Geography," 193–98; *Annual Financial Review* (1929): 615.

23. For quote, see *Canadian Cigar and Tobacco Journal* 27 (March 1921): 52. Also see *Annual Financial Review* (1925): 308; and (1928): 877; *Canadian Cigar and Tobacco Journal* 27 (June 1921): 27.

24. For quote, see *Canadian Shoe and Leather Journal* 28 (April 1914): 63. For Cimon and McFarlane, also see *Canadian Shoe and Leather Journal* 26 (February 1912): 52; 26 (May 1912): 65; 26 (November 1912): 54; 27 (June 1913): 47;

Footwear 2 (February 1912): 62; 2 (August 1912): 39; and 3 (March 1913): 119. Also see *Annual Financial Review* (1928): 803; and (1929): 854; *Canadian Chemical Journal* 10 (July 1926): 170; "David Yuile," in *Dictionary of Canadian Biography* (Toronto: University of Toronto Press, 1994), 13:1126–27; Tom King, "History of the Canadian Glass Industry," *Journal of the Canadian Ceramic Society* 34 (1965): 86–91; George Miller and Antony Pacey, "Impact of Mechanization in the Glass Container Industry: The Dominion Glass Company of Montreal—a Case Study," *Historical Archaeology* 19 (1985): 38–50.

25. For quote, see *Footwear* 2 (July 1912): 50. Also see *Canadian Shoe and Leather Journal* 25 (February 1911): 51; 25 (April 1911): 51; and 26 (September 1912): 65; *Footwear* 2 (November 1912): 64; and 13 (September 1923): 64.

26. *La Patrie*, 26 June 1909, 13.

27. Frederick Cowie, "The Great National Port of Canada," *Canadian Engineer* 22 (18 January 1912): 182; Benoît Brouillette, "Le port et les transports," in *Montréal économique*, ed. Edras Minville (Montréal: Fides, 1943), 115–82; M. Casey, "The Use of Power for Port Facilities," *Engineering Journal* 7 (1924): 468–88; "Montreal Harbour Development," *Canadian Engineer* 19 (1 December 1910): 684; Jean-Pierre Collin, "Histoire de l'urbanisation de la paroisse de Montréal, 1851–1941" (Montréal: INRS-Urbanisation, 1984), 83–85; Paul-André Linteau, "Le développement du port de Montréal au début du 20e siècle," in Canadian Historical Association, *Historical Papers* (1972): 181–205; Graham Taylor, "A Merchant of Death in the Peaceable Kingdom: Canadian Vickers, 1911–1927," in *Canadian Papers in Business History*, ed. Peter Baskerville (Victoria: University of Victoria Press, 1989), 213–44; "Constructional Features of the Five-Million-Dollar Plant of Canadian Vickers, Ltd.," *Contract Record and Engineering Review* (9 December 1914): 1482–92.

28. For quote, see *Montreal Gazette*, 31 August 1891, 3. Also see *Montreal Star*, 10 March 1906, 18; Lorenzo Prince, *Montreal Old and New* (Montreal: International Press Syndicate, 1915): 191; Michèle Brassard and Jean Hamelin, "Joseph-Octave Villeneuve," in *Dictionary of Canadian Biography* (Toronto: University of Toronto Press, 1994), 13:1058–60; Marc Choko, *Crises du logemont à Montréal (1860–1939)* (Montréal: Éditions Coopératives Albert Saint-Martin, 1980), 58; Michelle Benoit and Roger Gratton, *La "Cité du Nord"* (Montreal: City of Montreal, 1985); Ernest Chambers, *Suburban Montreal as Seen from the Routes of the Park and Island Railway Co.* (Montreal, 1895); Collin, "Histoire de l'urbanisation"; Robert Rumilly, *Histoire de Montréal* (Montréal: Fides, 1970), 152–226; Christopher Boone, "The Politics of Transportation Services in Suburban Montreal: Sorting Out the 'Mile End Muddle,' 1893–1909," *Urban History Review* 24 (1996): 25–39.

29. For quotes, see *Montreal Star*, 10 March 1906, 19, 21. Also see Canada,

Census of Canada, 1901 (Ottawa: N.p., 1903), vol. 2; Canada, Bureau of Statistics, *Population of the Municipal Wards of Montreal City, 1931* (Ottawa: Dominion Bureau of Statistics, 1934).

30. For quotes, see "New Plant of National Bridge Company," *Contract Record and Engineering Review* (2 August 1911): 51; and "Activities North and East," *Montreal Star*, 10 March 1906, 20. Also see Ernest Chambers, *The Book of Canada* (Montreal: Book of Canada Co., 1905), 108; Bruno Ramirez, "Montreal's Italians and the Socioeconomy of Settlement: Some Historical Hypotheses," *Urban History Review* 10 (1981): 39–48; and "Brief Encounters."

31. "Outremont, P.Q.," *Canadian Municipal Journal* 11 (February 1915): 66; Lloyd Reynolds, *The British Immigrant: His Social and Economic Adjustment in Canada* (Toronto: Oxford University Press, 1935); Carl Dawson, "City Planning and Our North American Social Heritage," in *Housing and Community Planning* (Montreal: McGill University Press, 1944); Réjean Legault, "Architecture et forme urbaine. L'exemple du triplex à Montréal de 1870 à 1914," *Urban History Review* 18 (1989): 1–10; Leo Zakuta, "The Natural Areas of the Montreal Metropolitan Community with Special Reference to the Central Area" (Master's thesis, McGill University, 1948); Atelier d'Histoire, *Evolution de l'architecture industrielle; Le histoire du logement ouvrier;* and *L'industrialisation à Hochelaga-Maisonneuve.*

32. For quote, see W. Bowman Tucker, "Eliminating the Slum: How the Montreal City Mission Is Attempting It," *Canadian Municipal Journal* 9 (August 1913): 315. Also see Linteau, *Histoire de Montréal*, 361–64; Reynolds, *British Immigrant*; Ramirez, "Montreal's Italians and the Socioeconomy of Settlement," 41–44; "Brief Encounters"; Bruno Ramirez and Michael Del Balso, "The Italians of Montreal: From Sojourning to Settlement, 1900–1920," in *Little Italies in North America*, ed. Robert Harney and J. Vincenza Scarpaci (Toronto: Multicultural Society of Ontario, 1981): 67–75; Judith Seidel, "The Development and Social Adjustment of the Jewish Community in Montreal" (Master's thesis, McGill University, 1939): 24–29; Charles Young, *The Ukrainian Canadians* (Toronto: Nelson, 1931): 107–25.

33. For quotes, see "Pointe-Aux-Trembles, P.Q.," *Canadian Municipal Journal* 15 (April 1919): 165. Also see Fredrick Wright, "A Garden City: The Housing Problem in the Vicinity of Montreal," *Canadian Municipal Journal* 14 (August 1918): 238–40; Abbé Ovila Fournier, *Joseph Versailles. Le fondateur de Montréal-est* (Saint-Etienne de Bolton: Les Éditions de la Libellule, 1974); Collin, "Histoire de l'urbanisation," 85–88.

34. Linteau, *Promoters' City*, 55–57, 65–104; *La Patrie* (26 June 1909): 5; Jean-Pierre Collin, "La cité sur mesure: spécialisation sociale de l'espace et autonomie municipale dans la banlieue montréalaise," *Urban History Review* 13 (1984): 19–34; Gauthier and Larivière, "La cité de Maisonneuve."

35. Linteau, *Promoters' City*; Herbert Ames, *The City below the Hill* (1897; rpt., Toronto: University of Toronto Press, 1972); Collin, "La cité sur mesure"; Terry Copp, *Anatomy of Poverty: The Condition of the Working Class in Montreal, 1897–1929* (Toronto: McClelland and Stewart, 1974), 70–105.

Chapter Nine: "Expanded in All Directions"

1. For quote, see Gilles Lauzon, "Le développement industriel de St-Henri, 1896–1913" (MS, Université de Québec, Montréal, 1982), 29. Also see La Société Historique de Saint-Henri, *Portrait d'une ville. Saint-Henri, 1875–1905* (Montréal: La Société Historique de Saint-Henri, 1987), 14.

2. *Canadian Journal of Commerce* (1889): 772b.

3. For quote, see "Canadian Industries. I. The Milling Industry," *Dominion Illustrated* (6 July 1889): 6–7. Also see Meredith Burrill, "Studies in the Industrial Geography of Montreal" (Ph.D. diss., Clark University, 1930), 135; John Willis, *The Process of Hydraulic Industrialization on the Lachine Canal: Origns, Rise and Fall* (Ottawa: Environment Canada, 1987), 342–61; Larry McNally, *Water Power on the Lachine Canal, 1846–1900* (Ottawa: Parks Canada, 1982), 19; A. Levine, "William Watson Ogilvie," in *Dictionary of Canadian Biography* (Toronto: University of Toronto Press, 1990), 12:800–801; Michèle Brassard and Jean Hamelin, "Alexander Walker Ogilvie," in *Dictionary of Canadian Biography* (Toronto: University of Toronto Press, 1994), 13:782–84; George Stevens, *Ogilvie in Canada, Pioneer Miller, 1801–1951* (Toronto: Ashton Potter, 1951), 10.

4. Brassard and Hamelin, "Alexander Ogilvie"; McNally, *Water Power on the Lachine Canal*, 17–32, 57–66. *Canada Year Book, 1922–1923* (Ottawa: Acland, 1924), 444–46; *Canadian Journal of Commerce* (6 June 1884): 791; Burrill, "Studies in the Industrial Geography of Montreal," 135–43; *Souvenir Number of the Montreal Daily Star: Reviewing the Various Financial and Commercial Interests Represented in the City of Montreal* (Montreal, 1890), 41; Stevens, *Ogilvie in Canada*; Willis, *Process of Hydraulic Industrialization*; *Canadian Engineer* 12 (21 April 1905): 24; *Annual Financial Review* (Toronto: Houston, 1910): 94–96; and (1911): 104–6; "Canadian Industries. I. The Milling Industry," 6–7.

5. *Canadian Engineer* 5 (March 1897): 338–39; 8 (January 1901): 195; 8 (August 1901): 380; and 38 (8 April 1920): 353; *Annual Financial Review* (1901): 84–85; Herbert Marshall, Frank Southard and Kenneth Taylor, *Canadian-American Industry: A Study in International Investment* (New Haven: Yale University Press, 1936), 72–75; Tom Naylor, *The History of Canadian Business, 1867–1914* (Toronto: Lorimer, 1975), 2:59–61; "Description of the Manufacture of Electric Wire," *Canadian Machinery* (February 1908): 50; Communauté Urbaine de Montréal, *Architecture industrielle* (Montréal: Communauté Urbaine de Montréal, 1982), 166–67; Canada, *Combines Investigation Act: Electrical Wire and Ca-*

ble Products (Ottawa: Department of Justice, 1953), 2–3; *Canadian Manufacturer* (19 May 1907): 27; and (17 January 1908): 20–21; "Phillips' Electrical Works," *Municipal Review of Canada* 25 (January 1929): 24–25. It should be noted that even early twentieth-century greenfield sites were not immune from capital flight in this period. In 1921 Phillips built a new plant in Brockville to produce new lines and eventually to take over the entire operations of the Mile End factory. By 1929 the wire-drawing and two of the insulating departments had been moved from Montreal to Brockville.

6. *La Patrie*, 26 June 1910, 13; Paul-André Linteau, *The Promoters' City: Building the Industrial Town of Maisonneuve, 1883–1918* (Toronto: Lorimer, 1985), 82–83; *The Commerce of Montreal and Its Manufactures* (Montreal, 1888), 46–47, 110; *Canadian Enterprise in Manufacturing* (Toronto: Canadian Manufacturers' Association, 1945), 232; *Canadian Journal of Commerce* (19 August 1892): 293; *Montreal Gazette*, 16 August 1892, 3; *Canadian Engineer* 2 (June 1894): 60; Peter Bischoff and Robert Tremblay, "William Clendinneng," in *Dictionary of Canadian Biography* (Toronto: University of Toronto Press, 1994), 13:203–5; *Industries of Canada* (Montreal, 1886), 125

7. *Annual Financial Review* (1913): 100; (1926): 363–64; and (1927): 395–96; "David Yuile," in *Dictionary of Canadian Biography* (Toronto: University of Toronto Press, 1994), 13:1126–27; *Canadian Chemical Journal* 4 (December 1920): 360; 13 (January 1929): 29; and 14 (December 1930): 360; *Canadian Engineer* 2 (May 1894): 27; 2 (December 1894): 245; and 6 (November 1898): 203.

8. *Canadian Enterprise in Manufacturing*, 212; *Canadian Engineer* 3 (September 1895): 126; "Factory of the Northern Electric and Manufacturing Co., Montreal," *Canadian Municipal Journal* 8 (1912): 355; "Features of the Northern Electric Company's New Plant at Point St. Charles, Que," *Contract Record and Engineering Review* 28 (May 1914): 684–87; *Canadian Chemical Journal* 10 (May 1926): 126; Canada, *Combines Investigation Act*, 2.

9. Paul Craven and Tom Traves, "Canadian Railways as Manufacturers, 1850–1880," Canadian Historical Association, *Historical Papers* (1983): 254–81; Mary Davidson, "The Social Adjustment of British Immigrant Families in Verdun and Point St. Charles" (Master's thesis, McGill University, 1933); Ralph Hoskins, "An Analysis of the Payrolls of the Point St. Charles Shops of the Grand Trunk Railway," *Cahiers de Géographie du Québec* 33 (1989): 323–44; *Canadian Manufacturer* (6 January 1905): 20–23; "'Old Timers' Move Working Quarters," *Montreal Gazette*, 13 July 1929, 7; William Atherton, *Montreal, 1534–1914* (Montreal: Craig, 1915), 3:440–41; "Montreal Branch," *Journal of the Engineering Institute of Canada* 13 (June 1930): 391; *Canadian Engineer* 7 (August 1899): 116; *Contract Record and Engineeering Review* (27 November 1918): 941; *Iron and Steel of Canada* (June 1918): 207; *Canadian Enterprise in Manufacturing*, 214–15; *Canadian Chemical Journal* 12 (January 1928): 29.

10. For quote, see *Canadian Engineer* 10 (July 1903): 179. Also see *Annual Financial Review* (1911): 110–11; *Canadian Manufacturers' Products List, 1929* (Toronto: Manufacturers' Products List, 1929).

11. "An Important Montreal Industry: Marconi Radio," *Montreal Year Book, 1930* (Montreal: Canadian Record of National Development, 1930), 95–97.

12. Edward Cureton, "The Lachine Canal" (Master's thesis, McGill University, 1957); La Société Historique, *Portrait d'une ville;* Normand Mousette, *En ces lieux que l'on nomma "La Chine"* (Lachine: Cité de Lachine, 1978).

13. "The Iron and Steel Works of Canada," *Canadian Manufacturer* (6 January 1905): 21; William Kilbourn, *The Elements Combined: A History of the Steel Company of Canada* (Toronto: Clarke, Irwin, 1960), 19–32; *Annual Financial Review* (1910): 102; (1917): 325–28; (1922): 422–25; and (1927): 819–22; *Canadian Chemical Journal* 3 (November 1919): 414; Willis, *Process of Hydraulic Industrialization;* Bernard Elbaum and Frank Wilkinson, "Industrial Relations and Uneven Development: A Comparative Study of the American and British Steel Industries," *Cambridge Journal of Economics* 3 (1979): 275–303.

14. For Canadian Car, see *Annual Financial Review* (1911): 74–77; and (1913): 82–87; *Canadian Enterprise in Manufacturing*, 213–14; "Iron and Steel Works of Canada," 21; "The Most Self-Contained Car-Building Company in the World Is in Canada," *Financial Post Survey* (Toronto: Macleans, 1925), 40. For other firms, see *Annual Financial Review* (1916): 262–63; (1920): 232–33; (1928): 708; and (1930): 460–63; Communauté Urbaine, *Architecture industrielle*, 292–99; "The St. Lawrence Bridge Company's Shops," *Canadian Engineer* 26 (22 January 1914): 207–11; and 7 (August 1899): 116; Doug Fetherling and Steve Dunwell, *Vision in Steel, 1888–1992: One Hundred Years of Growth, Dominion Steel Bridge to AMCA International* (Toronto: Key Porter, 1992); Cureton, "Lachine Canal," 31; Ernest Chambers, *The Book of Montreal: A Souvenir of Canada's Commercial Metropolis* (Montreal: Book of Canada, 1903), 215–18; Alfred Bray, *Canada under the National Policy: Arts and Manufactures, 1883* (Montreal, 1883), 105–7; *Industries of Canada; Canadian Manufacturer* (6 May 1904): 21–28; and (5 May 1905): 22–24; *Dominion Illustrated* (1891): 133.

15. For quote, see Chambers, *Book of Montreal*, 186. Also see City of Montreal, Water Tax Rolls, St. Louis ward, 1906, and Saint-Henri ward, 1912; *Annual Financial Review* (1911): 110; Communauté Urbaine, *Architecture industrielle*, 178–81.

16. For quote, see Burrill, "Studies in the Industrial Geography of Montreal," 129. Also see Cureton, "Lachine Canal," 115–17; *Annual Financial Review* (1911): 107; (1926): 758–59; (1927): 441–42; and (1930): 461; Communauté Urbaine, *Architecture industrielle*, 152–53; Chambers, *Book of Montreal*, 170, 197; William Donald, *The Canadian Iron and Steel Industry* (Boston: Houghton Mifflin, 1915), 221–95; Barbara Austin, "Life Cycles and Strategy of a Canadian

Company: Dominion Textile, 1873–1983" (Ph.D. diss., Concordia University, 1985); Canada, *Report of the Royal Commission on the Textile Industry* (Ottawa: Patenaude, 1938).

17. For quote, see *Canadian Cigar and Tobacco Journal* 21 (May 1915): 9. For details, see Robert Lewis, "Productive and Spatial Strategies in the Montreal Tobacco Industry, 1850–1918," *Economic Geography* 70 (1994): 370–89.

18. For quotes, see *Canadian Cigar and Tobacco Journal* 13 (July 1907): 57; and 13 (September 1907): 53.

19. For quotes, see "Expansion of Suburbs," "Westward from St. Henri," and "Plateau Lots Going Fast," *Montreal Gazette*, 10 March 1906, 20. Also see Jean-Pierre Collin, "La cité sur mesure: spécialisation sociale de l'espace et autonomie municipale dans la banlieue montréalaise, 1875–1920," *Urban History Review* 13 (1984): 19–34; and *Histoire de l'urbanisation de la Paroisse de Montréal, 1851–1941* (Montréal: INRS-Urbanisation, 1984), 72–76; Carl Dawson, "City Planning and Our North American Social Heritage," *Housing and Community Planning* (Montreal: McGill University Press, 1944), 148–55; Annick Germain, *Les mouvements de réforme urbaine à Montréal au tournant du siècle* (Montréal: Université de Montréal, 1984), 281–92; Paul-André Linteau, *Histoire de Montréal depuis la confédération* (Montréal: Boréal, 1992), 187–208, 351–72; Leo Zakuta, "The Natural Areas of the Montreal Metropolitan Community with Special Reference to the Central Area" (Master's thesis, McGill University, 1948).

20. Dawson, "City Planning"; Harold Ames, *The City below the Hill* (1897; rpt., Toronto: University of Toronto Press, 1972).

21. The data compiled from a one-in-ten sample of Lachine and the complete listing of Ville Saint-Pierre taken from *Lovell's Montreal Directory, 1921–1922* (Montreal: Lovell, 1921).

22. E. Gauthier, "Why You Should Come to Verdun," *Montreal, the Metropolis of Canada* (Montreal, 1932), 103; Davidson, "Social Adjustment of British Immigrants," 19; Montreal Junior Board of Trade, *Town Planning* (Montreal: N.p., 1915), 8; Christopher Boone, "Private Initiatives to Make Flood Control Public: The St. Gabriel Levee and the Railway Company in Montreal, 1886–1890," *Historical Geography* 25 (1997): 100–112; Henry Hadley, "Recent Improvements in the Town of Verdun," *Canadian Municipal Journal* 5 (1909): 245–46; Lloyd Reynolds, *The British Immigrant: His Social and Economic Adjustment in Canada* (Toronto: Oxford University Press, 1935), 123–24, 133–40.

23. For quotes, see Davidson, "Social Adjustment of British Immigrant Families," 18, 22; and "Ville Saint-Pierre, P.Q.," *Canadian Municipal Journal* 11 (January 1915): 27. Also see Dawson, "City Planning"; Terry Copp, *The Anatomy of Poverty: The Conditions of the Working Class in Montreal, 1897–1929* (Toronto: McClelland and Stewart, 1974); Marc Choko, *Crises du logement à Montréal (1860–1939)* (Montréal: Albert Saint-Martin, 1980); Greg Levine, "Class, Eth-

nicity and Property Transfers in Montreal, 1907–1909," *Journal of Historical Geography* 14 (1988): 360–80; Reynolds, *British Immigrant*, 258–59; John Cooper, *Montreal: A Brief History* (Montreal: McGill University Press, 1969), 128; Montreal Junior Board of Trade, *Town Planning*, 6; Bray, *Canada under the National Policy*, 106–7; David Hanna, "Montreal, a City Built by Small Builders" (Ph.D. diss., McGill University, 1986).

24. For quotes, see "Ville St. Pierre," 31; and Reynolds, *British Immigrant*, 194, 200. Also see Hoskins, "An Analysis of the Payrolls"; *Canadian Engineer* 4 (March 1896): 315; Christopher Armstrong and Viv Nelles, "Suburban Street Railway Strategies in Montreal, Toronto and Vancouver, 1896–1930," in *Power and Place: Canadian Urban Development in the North American Context*, ed. Gilbert Stelter and Alan Artibise (Vancouver: University of British Columbia Press, 1986), 187–218.

25. For quotes, see Davidson, "Social Adjustment of British Immigrant Families," 23, 13, 14; and "Ville Saint-Pierre," 29. Also see "'Old Timers' Move Working Quarters"; David Hanna and Frank Remiggi, *Montreal Neighbourhoods* (Montreal: Canadian Association of Geographers, 1980), 2–6.

26. For quote, see Davidson, "Social Adjustment of British Immigrant Families," 21–22. Also see Collin, "La cité sur mesure"; Lauzon, "Le développement industriel de St-Henri"; Claude Ouellet, "Rapport de recherche sur les implantations industrielles et commerciales dans la Ville de Saint- Henri, 1876–1905" (MS, Université de Québec, Montréal, 1982).

27. For quote, see *Canadian Journal of Commerce* (20 May 1892): 863.

28. *Canadian Journal of Commerce* (21 April 1899): 564. Also see *Canadian Journal of Commerce* (14 October 1892): 602; (13 January 1893): 50; and (August 1898): 114; *Canadian Manufacturer* (2 June 1905); Chambers, *Book of Montreal*, 212; Austin, "Life Cycles and Strategy," 196–97.

29. For quotes, see Montreal Junior Board of Trade, *Town Planning*, 10, 5; Gabrielle Roy, *The Tin Flute* (1945; rpt., Toronto: McClelland and Stewart, 1980), 34. Also see Copp, *Anatomy of Poverty*.

30. "Progress of Westmount," *Montreal Gazette*, 10 March 1906, 20; Willam Lighthall, "Westmount: A Municipal Illustration," in *Municipal Government in Canada*, ed. S. Morley Wickett (Toronto: University of Toronto Studies in History and Economics, 1907), 2:27–34; William Lighthall, "City Government," in *Canada and Its Provinces*, ed. Adam Shortt and Arthur Doughty (Toronto: Glasgow Brooke, 1914), 15:308; Walter van Nus, "The Role of Suburban Government in the City-Building Process: The Case of Notre Dame de Grâces, Quebec, 1876- 1910," *Urban History Review* 13 (1984): 91–103.

Conclusion

1. Leo Schnore, "Metropolitan Growth and Decentralization," *American Journal of Sociology* 63 (1957): 171–80; Robert Fishman, *Bourgeois Utopias: The Rise and Fall of Suburbia* (New York: Basic Books, 1987); Brian Berry and Yehoshua Cohen, "Decentralization of Commerce and Industry: The Restructuring of Metropolitan America," in *The Urbanization of the Suburbs*, ed. Louis Masotti and Jeffrey Haddon (Beverly Hills: Sage, 1973), 431–55; Allen Scott, "Production System Dynamics and Metropolitan Development," *Annals of the Association of American Geographers* 72 (1982): 185–200; Kenneth Jackson, *Crabgrass Frontier: The Suburbanization of the United States* (New York: Oxford University Press, 1985); Margaret Marsh, *Suburban Lives* (New Brunswick: Rutgers University Press, 1990); Gérald Martin, "Études des facteurs qui ont déterminé la localisation de l'industrie à Montréal et des les banlieues," *Revue Trimestrielle Canadienne* 34 (1934): 297–335; Jean Delage, "L'industrie manufacturière," in *Montréal économique*, ed. Edras Minville (Montréal: Fides, 1943), 183–241; David Slater, "Decentralization of Urban Peoples and Manufacturing Activity in Canada," *Canadian Journal of Economics and Political Science* 27 (1961): 72–84. But see Richard Harris and Robert Lewis, "Constructing a Fault(y) Zone: Misrepresentations of American Cities and Suburbs, 1900–1950," *Annals of the Association of American Geographers* 88 (1998): 622–39.

2. For a similar argument, see Richard Walker and Robert Lewis, "Beyond the Crabgrass Frontier: Industry and the Spread of the City, 1850–1950," *Journal of Historical Geography* (in press); Robert Lewis, "Running Rings around the City: North American Industrial Suburbs, 1850-1950," in *Changing Suburbs*, ed. Richard Harris and Peter Larkham (London: E & FN Spon, 1999), 146–67.

3. David Harvey, *The Limits to Capital* (Oxford: Basil Blackwell, 1982); Doreen Massey, *Spatial Divisions of Labour: Social Structures and the Geography of Production* (London: Macmillan, 1984); David Meyer, "Midwestern Industrialisation and the American Manufacturing Belt in the Nineteenth Century," *Journal of Economic History* 49 (1989): 921–37; Gordon Winder, "The North American Manufacturing Belt in 1880: A Cluster of Regional Industrial Systems or One Large Industrial District?" *Economic Geography* 75 (1999): 71–92.

4. Delage, "L'industrie manufacturière," 187; A. Ross, "Electrication of the Canadian Copper Refiners Plant, Montreal East, Que.," *Journal of the Engineering Institute of Canada* 14 (December 1931): 593–99.

5. Fishman, *Bourgeois Utopias;* Jackson, *Crabgrass Frontier;* Delage, "L'industrie manufacturière"; Scott, "Production System Dynamics"; Stanley Buder, *Pullman: An Experiment in Industrial and Community Planning, 1880–1930* (New York: Oxford University Press, 1967); Edward Greer, "Monopoly and

Competitive Capital in the Making of Gary, Indiana," *Science and Society* 40 (1976): 465–78.

6. On annexation, see Jackson, *Crabgrass Frontier*, 138–56; John Teaford, *City and Suburb: The Political Fragmentation of Metropolitan America, 1850–1970* (Baltimore: Johns Hopkins University Press, 1979). For a discussion of the problems with the definition of *industrial suburbanization*, see Lewis, "Running Rings around the City."

7. Richard Stott, *Workers in the Metropolis: Class, Ethnicity and Youth in Antebellum New York City* (Ithaca: Cornell University Press, 1990); Edward Muller and Paul Groves, "The Emergence of Industrial Districts in Mid-Nineteenth Century Baltimore," *Geographical Review* 69 (1979): 159–78; Philip Scranton, *Proprietary Capitalism: The Textile Manufacture at Philadelphia, 1800–1885* (New York: Cambridge University Press, 1983); John Appleton, *The Iron and Steel Industry of the Calumet District: A Study in Economic Geography.* (Urbana: University of Illinois Press, 1927); Edward Muller, "Industrial Suburbs and the Growth of Metropolitan Pittsburgh, 1870–1920," *Journal of Historical Geography* (in press); Olivier Zunz, *The Changing Face of Inequality: Urbanization, Industrial Development, and Immigrants in Detroit, 1880–1920* (Chicago: University of Chicago Press, 1982); Lewis Thomas, *The Localization of Business Activities in Metropolitan St. Louis* (St. Louis: Washington University Studies–New Series, 1927); Richard Walker, "Industry Builds the City: The Suburbanization of Manufacturing in the San Francisco Bay Area, 1850–1945," *Journal of Historical Geography* (in press).

8. Glen Williams, *Not for Export: Toward a Political Economy of Canada's Arrested Industrialization* (Toronto: McClelland and Stewart, 1986); Michael Goldberg and John Mercer, *The Myth of the North American City: Continentalism Challenged* (Vancouver: University of British Columbia Press, 1986); Richard Harris, *Unplanned Suburbs: Toronto's American Tragedy, 1900 to 1950* (Baltimore: Johns Hopkins University Press, 1996), 275–84.

9. *Canadian Engineer* 12 (January 1905): 2.

10. Fishman, *Bourgeois Utopias;* Jackson, *Crabgrass Frontier;* David Ward, *Cities and Immigrants* (New York: Oxford University Press, 1972). On the complex social geography of Montreal's inner city, see David Hanna and Sherry Olson, "Métiers, loyers et bouts de rue. L'armature de la société montréalaise de 1881 à 1901," *Cahiers de Géographie du Québec* 27 (1983): 255–75.

11. Harold Mayer, "Localization of Railway Facilities in Metropolitan Centers as Typified by Chicago," *Journal of Land and Public Utility Economics* 20 (1944): 299–315; William Cronon, *Nature's Metropolis: Chicago and the Great West* (New York: Norton, 1991); John Marshall, "Railroads and Urban Growth," in *Growing Metropolis: Aspects of Development in Nashville*, ed. James Blum-

stein and Benjamin Walter (Nashville: Vanderbilt University Press, 1975), 65–80; Robert Fogelson, *The Fragmented Metropolis: Los Angeles, 1850–1930* (Cambridge: Harvard University Press, 1967); Buder, *Pullman;* Greer, "Monopoly and Competitive Capital"; Anne Mosher, "'Something Better than the Best': Industrial Restructuring, George McMurty and the Creation of the Model Industrial Town of Vandergrift, Pennsylvania, 1883–1901," *Annals of the Association of American Geographers* 85 (1995): 84–107; Harris, *Unplanned Suburbs;* Anthony Orum, *City-Building in America* (Boulder: Westview Press, 19995), 79–81; Graham Taylor, *Satellite Cities: A Study of Industrial Suburbs* (New York: Appleton, 1915).

12. David Gordon, "Capitalist Development and the History of American Cities," in *Marxism and the Metropolis,* ed. William Tabb and Larry Sawers (New York: Oxford University Press, 1978), 25–63. But see Annick Germain, *Les mouvements de réform urbaine à Montréal au tournant de siècle* (Montréal: Départment de Sociologie, Université de Montréal, 1984).

13. Buder, *Pullman;* Mosher, "'Something Better than the Best.'"

14. For quote, see Roderick McKenzie, "The Neighborhood: A Study of Local Life in the City of Columbus, Ohio," *American Journal of Sociology* 27 (1921): 151. Also see Margeret Byington, *Homestead: The Households of a Mill Town* (1910; rpt., Pittsburgh: University of Pittsburgh, 1996); Carl Dawson, "City Planning and Our North American Social Heritage," *Housing and Community Planning* (Montreal: McGill University Press, 1944), 150–51; Caroline Golab, *Immigrant Destinations* (Philadelphia: Temple University Press, 1977); Dominic Pacyga, *Polish Immigrants and Industrial Change: Workers on the South Side, 1880–1920* (Columbus: Ohio State University Press, 1991).

15. For quote, see Bettina Bradbury, *Working Families: Age, Gender, and Daily Survival in Industrializing Montreal* (Toronto: McClelland and Stewart, 1993), 47. Also see Terry Copp, *The Anatomy of Poverty: The Condition of the Working Class in Montreal, 1897–1929* (Toronto: McClelland and Stewart, 1974); Daniel Hiebert, "Discontinuity and the Emergence of Flexible Production: Garment Production in Toronto, 1901–1931," *Economic Geography* 66 (1990): 229–53; Richard Harris and Victoria Bloomfield, "The Impact of Industrial Decentralization on the Gendered Journey to Work, 1900–1940," *Economic Geography* 73 (1997): 94–117.

16. A good summary of transportation-led suburbanization is Peter Muller, *Contemporary Suburban America* (Englewood Cliffs, N.J.: Prentice-Hall, 1981).

17. For quote, see Ralph Hoskins, "An Analysis of the Payrolls of the Point St. Charles Shops of the Grand Trunk Railway," *Cahiers de Géographie du Québec* 33 (1989): 343. Also see Theodore Hershberg, Dale Light, Harold Cox, and Richard Greenfield, "The 'Journey to Work': An Empirical Investigation of

Work, Residence and Transportation, Philadelphia, 1850 and 1880," in *Philadelphia: Work, Space, Family and Group Experience in the Nineteenth Century*, ed. Theodore Hershberg (New York: Oxford University Press, 1981), 128–73; Eugene Erickson and William Yancy, "Work and Residence in Industrial Philadelphia," *Journal of Urban History* 5 (1979): 147–82; McClellan and Junkersfeld Inc., *Report on Transportation in the Milwaukee Metropolitan District* (New York: McClellan and Junkersfeld, 1928), 2 vols.; Harris and Bloomfield, "Impact of Industrial Decentralization." There is also evidence that women, because of the small number of work niches open to them and that residential decisions were primarily determined by the location of the male head of household's workplace, worked farther from their place of work than men.

Appendix

1. For quotes, see William Hutton, "First Report," in *Census of the Canadas, 1851–52* (Quebec, 1853), ix; Canada, *Census of Canada, 1870–1871* (Ottawa, 1875), 3:x; Canada, *Census of Canada, 1901* (Ottawa: Dawson, 1902), 1:ix. The ambiguity of the phrases "any importance" and "separate establishments or workshops" could lead to problems in the collection of industrial establishments. It is not known how census enumerators interpreted this ambiguity. Canada has a one-hundred-year confidentiality rule for all census material. Therefore, the 1901 census manuscripts (which have been released early) are the last that can be consulted.

2. The water tax rolls have also been used to examine the city's social geography. See David Hanna and Sherry Olson, "Métier, loyers et bouts de rues. L'Armature de la société montréalaise 1881 à 1901," *Cahiers de Géographie de Québec* 27 (1983): 255–75; Stephen Hertzog and Robert Lewis, "A City of Tenants: Homeownership and Social Class in Montreal, 1847–1881," *Canadian Geographer* 30 (1986): 316–23; Robert Lewis, "The Segregated City: Class Residential Patterns and the Development of Industrial Districts in Montreal, 1861 and 1901," *Journal of Urban History* 17 (1991): 123–52; Sherry Olson, "Occupations and Residential Space in Nineteenth-Century Montreal," *Historical Methods* 22 (1989): 81–96. For a discussion of the problems with the rolls, see Gregory Levine, "Criticizing the Assessment: Views of the Property Evaluation Process in Montreal, 1870 to 1920, and Their Implications for Human Geography," *Canadian Geographer* 28 (1984): 276–84. The 1871 capital and employment figures are from MS Census, Industrial Schedules, 1871, Saint-Ann ward. The 1907 capital figures were taken from the "investments in manufacturing industries" found in *The Manufacturers' List Buyers' Guide of Canada, 1907* (Toronto: Manufacturers' List Co., 1907).

3. For the three sample years I collected all of the manufacturing estab-

lishments in the city of Montreal water tax rolls. Along with these, the following suburban municipalities' water tax rolls were collected: Cité de Saint-Henri (1881, 1890), Cité de Sainte-Cunégonde (1881, 1890), Côte Saint-Paul (1895), Ville de Lachine (1890, 1912, 1929), Ville Saint-Pierre (1912, 1929), Maisonneuve (1890, 1904, 1918), Outremont (1929), Westmount (1929), Verdun (1929), and Montréal Est (1929).

4. Andrew Sayer, *Method in Social Science: A Realist Approach* (London: Hutchinson, 1984).

5. For a discussion of defining suburban, see Robert Lewis, "Running Rings around the City: North American Industrial Suburbs, 1850–1950," in *Changing Suburbs*, ed. Richard Harris and Peter Larkham (London: E & FN Spon, 1999), 146–67.

6. The map is "Ville de Montréal, service de l'urbanisme," *Utilisation du sol* (1983), at 1:1,000. Each cell covers 1,332 square meters.

7. Richard Walker, "Technological Determination and Determinism: Industrial Growth and Location," in *High Technology, Space and Society*, ed. Manual Castells (Beverly Hills: Sage, 1985), 228–31.

Index

Page references to figures, maps, and tables are printed in italic type.

Acton Vale, 160
Agriculture, 27–28, 30, 106, 116–17, 145
Ahuntsic, 144
Alaska Feather and Down, 241–42, 245
Aldred Building, 157, *158*
Alexandra Park, 213
Allen, William, 102
Allis-Chalmers, 12, 136, 241
American Locomotive Company, 136, 138, 195
Ames, Herbert, 111–13, 145, 246
Ames, Holden & McCready, 160, 199, 210–11
Amherst Park, 213, *214*
Annexation, 41, 143–44, 151, 194, 260; reasons for, 98, 123
Automobile, 269
Automobile industry, 4, 16, 206, 210; suburbanization of, 2–3, 17

Baking industry, 55–59; Griffin, James, 55; Hall and Scott, 55; Lang Manufacturing, 58, 245; Luttrell, James, 123; Viau and Frère, 58
Baltimore, 4, 53, 71, 109, 262
Banking, 26–28, 31; financial district, 49, *51*, *52*, *164*; and manufacturing, 88, 125

Barsalou Soap, 83–85
Belding, Paul and Co., 136, *137*, 143, 165
Bell Telephone, 163, 171
Berry, Brian, 2
Bigelow Nail, 107–9
Boston, 3, 53
Bradbury, Bettina, 269
Brandram-Henderson, 208, *209*
Brewing industry, 81; Frontenac Brewery, 208–9; National Breweries, 167
British-American Oil, 198
Browne, William, 131, 138
Brownell, Blaine, 3
Building cycles: and East End, 96, 212–13; and harbor, 155–56; and manufacturing districts, 6–9, 257–58; and urban growth, 20, 162–63, 264; and West End, 115–16, 244–45
Burgess, Ernest, 63
Bylaws: housing, 146, 154, 250, 253; nuisance, 153, 261; street widening, 169

Cambridge, Massachusetts, 98
Canada Malting, 245
Canada Paint, 235, *236*
Canada Sugar. *See* Redpath Sugar
Canada Wire and Cable Co., 258–59

Canadian Car and Foundry, 143, 241, 242, 245
Canadian Copper Refineries, 1, 258–59
Canadian National Railway. *See* Grand Trunk Railway
Canadian Pacific Railway: Angus shops, 150, 191–95; and city growth, 29; and electricity, 138; and factories, 185, 205, 213, 219, 264; freight yards, 88, 170, 212; Hochelaga shops, 63, 83, 193; and housing, 149–50; and North End, 181, 208, 213, 215; terminals, 63, 70, 170
Canadian Rubber, 80, 81, 85–87, *86*, 92, 136, 165, 187–89, *188*
Canadian Steel Foundries, 195–96, *198*
Canadian Vickers Ltd., 212
Canal: capital disinvestment in, 227–30; factory sites, 267; flour industry, 223–27; growth of, 106–7, 221–37; industry, 7, 100–111, 222–23, 261; manufacturing district, 47–48, 132, 135, 256; manufacturing structure, *107, 227*; residential development, 111–16
Cantin Shipyards. *See* Montreal Marine Works
Capital investment: foreign, 136, 195, 205, 208, 227–28, 262–63; in Montreal's industry, 31–32, 100–111, 117, 135–43; place bound and mobile, 14, 17, 20–21, 256–63; in urban fabric, 20–22
Carriage making industry, 69–71, *70*; Gravelle, Martin, 70; Heney, E. N., 69; Larivière's Canada Coach and Sleigh, 69; Mercier, Felix, 71
Case study: lack of, 5; reason for, 275–76
Cement industry, 259; Canada Cement, 136, 194, 196–97; National Cement, 196
Central manufacturing district: chains of production in, 50–58, 65–76, 159–60, 257; combining manufacturing and retailing, 54, 59, 161; ethnic and occupational composition, 63–64, 172; functions, 161–65; industry, 49, 90, 262; land use, 157–84, 230, 268; locational assets, 14–15, 58–64, 203; manufacturing structure, *50, 159;* office employment, 158–59, 161–63; property markets, 165–67; redevelopment, 58–63, 166–69; residential areas, 265; transportation, 60–63, 169–70
Chambers, Ernest, 179, 185
Charrune and Daoust, 213
Chicago: annexation, 260; industrial suburbs, 4, 17, 20, 22, 146, 267; industry, 139; infrastructures, 21; manufacturing geography, 262; office building, 157, 167; Pullman, 3, 96, 259, 268; social inequalities, 154; stockyards, 18, 97, 146; working-class areas, 269
Chicago School of Sociology, 3, 63, 255, 268
Cincinnati, 4, 69
City Passenger Railway, 123
Clendinneng, W. and Sons, 165, 230, *231*
Clothing industry, 65, 71–76, *75, 162*, 177–83; Aitken, John, 72; Black, J., 167; Cassils and Cameron, 65, 72, 74; McFarlane and Baird, 72; Montreal Whitewear, 178; Muir, William, 74
Cochenthaler Co., 161
Cohen, Yehoshua, 2
Colonialism, 25–30, 262–63
Columbus, Ohio, 269
Columbus Rubber, 187
Contract system, 177–80, 181–82
Costigan, William, 44
Côte à Baron, 42
Côte Saint-Louis, 41, 44, *144*, 148, 266
Côte Saint-Paul, *124;* Frothingham and Workman, 117–21; housing, 244–47; industrial policy, 250–51; industry,

116–21, 237–44, 246–47, 256; population, 117, *118*, 246–47
Coursol, Charles-Joseph, 122
Crawford, John, 250

Dandurand, U., 213–14, 250
Darling Bros., 232–35
Davidson, Mary, 149, 249
Davis, S. and Sons, 176–77, *178*, 209
Davis and Lawrence Co., 57
Dawson, Charles, 246
Decrow, Douglass, 44
Delage, Jean, 1, 2, 255, 258, 259
Delisle, Alexandre, 122
Delorimier, 216
Deneau, L., 245, 250, 252
Detroit, 4, 17, 22, 262
Deville, E., 154
DeWitt, Jacob, 115
Dominion Glass Co., 230–31
Dominion Oil Cloth Co., 191, *192*
Dufferin, 145
Dufresne and Locke, Ltd., 203, *204*
Duncan, John, 58
Dupont Frère, 203

Eagle Foundry, 101, 105
East End, 122, 248; class and ethnic structure, 92–95, 214–16; industry, 47, 79–92, 183, 185–204; manufacturing district, 10, 256; population, *93*, 213; residential growth, 92–99, 211–20
Economies: agglomeration, 111, 160, 189, 244, 257, 265; external, 53–54; noncentral agglomeration, 22, 90, 240, 242, 258; scale and scope, 11, 14, 243, 256
Electrical industry, 131, 227–37; Black and Decker, 232; Victor Talking Machine, 232
Electricity: industrial, 138–39, 206; street lighting, 218; telephone, 138, 170–71

Emard, *144*, 245
Ethnicity, and suburbanization, 93, 113, 122, 149, 215–16, 246, 269
Evans and Sons, 160
Ewing, James, 153–54, 157–58, 159, 160, 165

Factories: and competition, 109; in situ growth of, 125, 191, 210–11, 231–35, 240–41, 256, 267–68; and technology, 81
Factory layout: and central manufacturing districts, 55, 163–65, 176; and East End, 81, 83, 85–86, 190–94; and locational options, 16–18, 235; and North End, 210–11; role in suburbanization, 91–92, 123, 241–44; and West End, 123–27, 222, 229–230
Fairmount, 213
Family economy, 95–96
Ferguson, G., 15, 151
Fishman, Robert, 2, 3, 98, 259
Flour industry, 223–27; Dominion Flour, 242; Gould Flour, 106; Mount Royal Flour, 119; St. Lawrence Flour, 242
Frosst Co., 160
Frothingham, John, 31, 117–21
Frothingham and Workman Hardware, 31, 117–21, *120*

Gantt, Henry, 193
Garth Foundry, 56
Gary, Indiana, 259
General Electric Co., 136, 227–28, 232
Gilbert and Bartley Foundry, 106
Glass industry, 126–27, 230–31; Perfection Glass, 210
Goldfield, David, 3
Goose Village, 45, 149
Gordon, David, 267–68
Gould, Ira, 115
Grand Trunk Railway: and bonuses, 248; and city growth, 28–29; and

Grand Trunk Railway (*continued*) ethnic division of labor, 36; and factories, 105, 230, 264; and housing, 149, 248, 270; locomotive shops, 63, 102–3, 193, 223, 232; terminals, 63, 70, 170

Gratorex, Thomas, 44–45

Greenfield, 235; features of, 17, 22, 46–47, 83, 136, 192, 208, 237; and manufacturing districts, 5, 55, 78, 132, 241, 256, 266; and mechanization, 15–16; and propulsive firms, 99, 172, 183, 191–95, 258; search for, 18, 222, 230, 232

Grey Nun, 59–60, *61*, 265

Griffintown, *108;* additions, 222, 231; capital disinvestment, 227–30; commodity production, 101–2, 106, 114; industry, 100–111, 165, 208, 221–37; manufacturing district, *47*, 48, 135, 256; manufacturing structure, *107*, 227; residential development, 45, 113–16

Grothé, L. O., Cigar Factory, 174–76, *177*, 209

Halifax, 29, 183

Hamilton, 20, 136, 146, 208, 231

Handicraft production: and fur industry, 26; in Griffintown, 101–2; in Old Montreal, 49; in Sainte-Marie, 79, 81; transition to modern industry, 6, 30, 50; workshops, 16, 31, 87, 121

Harbor: in central manufacturing districts, 60–63, 169–70; development, 39–40, 155–56, 211–12; in East End, 216, 217-18; locational assets of, 7, 19–20, 29, 264–65; views of, *62*

Hastings Park, 213

Haugh, E., 245, 250, 252

Henderson, James, 96

Hochelaga: annexation, 260; early growth, 87; harbor, 264; industry, 7, 81, 87–92, 153, 185–94, 261; manufacturing district, 1, 10, *47, 80,* 135, 250, 256; manufacturing structure, *80, 186;* population, *41, 93, 144;* residential growth, 92–99, *144,* 147, 149, 211–20; working-class, 41, 44, *94*

Hôtel Dieu, 59–60, *61,* 265

Housing: central manufacturing district, 145, 265–66; company, 96–97, 247; conditions, 7, 20, 43–45, 95–96, 145, 219, 247; cost, 43–45, 145, 216; demolition, 145, 166–67, 235, 265; market, 40–45, 150; owner-built, 148, 247; slum, 148–49, 165, 171–72

Houston, 21

Hudon, Victor, 97

Hudon Cotton Mill, 87–90, *89,* 136, 189, *190,* 191

Hutton, William, 273

Hygienic conditions, 154–55

Immigrants: American, 113; British, 29, 113, *214,* 215, 248; Chinese, 149; as entrepreneurs, 31–32; Irish, 29, 41, 113; Italians, 148–49, 172, *207, 214,* 215, 269; Jewish, 149, *162,* 172, 179, 181–82, *207,* 215, 269; in Montreal, 29, 145, 154, 265; in Montreal suburbs, 149, 215–16; in Philadelphia suburbs, 4–5; Southern and Eastern European, 64, 172, 215, 246; Syrian, 149; in tobacco industry, 173, 177; Ukrainian, 149, 216, 269

Imperial Oil, 194, 198

Imperial Tobacco, 172, 173, 183, 243–44

Industrial bonuses, 217–18, 230, 250–52, 261

Industrial linkages: absence of, 189; and agriculture, 30; and central districts, 53–58, 65–77, 159–60, 174–82; and factory districts, 81–83, 105, 185–86, 253, 256–59, 261–62; and geographic scales, 242–43; interfirm, 14, 125–27, 208, 242–43; intrafirm, 167, 176, 196, 210–11, 235, 241–43, 258–59; and rail-

Index 331

road, 87, 109–11, 185; views on, 221–22
Industrial suburbanization, 37, 45–48, 131–35, 156, 266–68; and clothing firms, 182–83; and East End, 78–92, 185–204; and North End, 205–11; views of, 1–5, 11–18; and West End, 100–111, 116–28, 237–44
Infrastructures, 7; roads, 58–59, 97, 122, 155, 169; sewers, 97, 122, 155–56, 169, 256, 265; water, 50, 156
In-house production, 72, 179–80
Ives and Co., 111, *112*

Jackson, Kenneth, 98, 259
Jamaica Plains (Boston), 98
Jamieson, R. C., Paint, 232, *234*
Jersey City, 117
Journey to work: and electric streetcar, 171, 248, 269–70; and horsecars, 63; and housing, 147, 219, 245; and office workers, 171; working-class, 95, 113, 213–14, 245, 248, 269–70

Kennedy, William, 94
Kerry, Watson Co., 160
Kirvan Doig Shoe, 203
Knoxville, 4, 146

Laberge, Louis, 44
Labor, child, 29, 36, 71–74, 88, 91
Labor, division of, 15, 36–37, 137–38, 141–43; in central districts, 65–76; ethnic, 26, 64, 113, 173–74, 176–82, 191; and occupations, 63–64, 94–95
Labor: clothing, 71–76, 179, 182; female, 29, 36, 65, 88, 91; tobacco, 141–42, 244
Labor supply, 146–47, 269–71; and central districts, 14, 55, 63–65, 160–62; and conflict, 193–94, 200–201, 267–68, 173–84; and East End, 85, 219; and North End, 209–10; and West End, 102, 116–17, 236–37, 244
Lachine, City of: housing, 246–48; industrial policy, 260; industry, 132, 147, 183, 221, 237–44, 264; manufacturing district, *47*, 256, 267; occupations and ethnicity, 246; population, *41, 144*, 147, 246; residential areas, 269
Lachine Canal: building 28, 223, 259, 264; corridor, 10, 221, 237, 245–47, 267; industry, 34, 103, 106, 121, 205–6, 223–32, 241–42; locational assets, 29, 40, 223–27; views of, *224–26;* water power, 106, 115–16
Landau and Cormack Co., 173–74, *175*
Langlois, Mendoza, 211
La Parisienne Shoe, 200
LaSalle, 1
Laurier (ward), 144
Lauzon, Gilles, 221
Lavallé, Arsene, 151
Leather industry, 121–22, 125; Barrington, George, 57, 252; Everleigh Trunk, 242. *See also* Shoe industry
Lefebvre, Michel, Vinegar Works, 83, *84, 96*
Linkages. *See* Industrial linkages
Local alliances, 38–40, 145–48, 151–56; and East End, 87, 92–99, 198–203, 211–14, 216–20; and manufacturing districts, 7, 20–22, 31–32, 263–71; uncoordinated activities, 252–53, 263–64; and West End, 115–16, 117–23, 240, 244–45, 249–50
Locational assets: 37–40, 45–48, 155–56, 261, 264–71; and central manufacturing districts, 58–64, 159–60; and East End industry, 87–92, 191–94, 211–20; and manufacturing districts, 7, 19–22; and North End industry, 205–211; and West End industry, 117, 123–28, 230, 240–44
Loft Buildings, 145, 166–68, 180–81, 265, 269; Amherst, 180, *182;* Blumenthal, 166, 180; Caron, 180, *181;* Gillette Safety Razor, 167; Herald, 167, *168*

Loid, Thomas, 79
Longue Pointe. *See* Mercier
Longueuil, *144*
Los Angeles, 4, 10, 18, 20, 21, 267
Lyman, Clare Co., 103

Macdonald, William, 95, 96
Macdonald Tobacco, 90–92, 172, 173, 183, 189, *190*
Maisonneuve: annexation, 260; harbor, 264; industrial policy, 147, 216–19, 230, 260; industry, 194; local alliances, 198–99, 203, 216–19, 267; manufacturing district, 1, 21, 132, 147, 256; manufacturing structure, *195*; residential growth, *144*, 148, 211–20, 269; and shoe industry, 198–204
Manufacturing pathways: development of, 30–37, 131–43; and firm strategy, 6–7, 11–18, 99, 111, 240–44, 256–59; and manufacturing districts, 87; and spatial strategy, 11–18, 172–84, 194, 257, 261–62
Manufacturing scale, 16–17, 19–20, 32–34, 135–36, 277–78
Marcil Trust Building, 157
Marconi Radio, 235–37
Markets, 27–28, 34–35, 102, 131, 232, 257, 262–63
Martin, Gérald, 1, 2, 255
McArthur, Colin, Wallpaper, 189–91, 230
McArthur, John, 32
McColl-Frontenac Oil, 198
McCord, John, 113
McDougall Foundry, 105
McGarvey, Owen, Furniture, 53–55, *54*, 59
McGibbon, D. Lorne, 211
McGill, Peter, 26–27
McKenzie, Roderick, 269
McLaughlin, Glenn, 4
Medicine Hat, 208

Merchant Manufacturing Co., 123, 136, 250, *251*
Mercier: annexation, 260; harbor, 264; industrial policy, 260; industrial structure, 194–98, 222; manufacturing district, 1, 133, 256; population, *41*, *144*; residential growth, 211–20
Mergers, 135–36, 226–27, 235, 240–41
Metal Shingle and Siding Co., 185–86
Metalworking industry, 124–26, 232–35, 240–44; and American firms, 136, 165; Canada Axe and Harvest Tool (J. Higgins), 119, 250; Canada Engine Works, 125; Dominion Brass, 109–11; Dominion Bridge, 241–43, 247; Dominion Type Foundry, 36; Dominion Wire, 165; Dominion Wire and Cable, 165; Dunn's Nail, 119; Gardner & Son, 34; National Bridge, 213; Peck Nail, 106, 223; Simplex Railway Appliance, 245, 250; Singer Manufacturing Co., 123, 136; St. Lawrence Bridge, 241–42; Vulcan Iron, 109; Vulcan Works, 119; Williams Manufacturing, 123; Wire and Cable Co., 167
Middle class: residential patterns, 41–45, 92–94, 161, 253; suburbanization, 3, 20, 145–46, 214–15, 255
Mile End: bonuses, 176; factory sites, 267; industry, 183, 205–11, 229, 257, 261, 266; manufacturing district, *133*, 256; manufacturing structure, *206*; and paint industry, 205–8, 222, 235, 258; population, *41*, *144*; residential growth, 148, 149, 181, 211–16, 219–20, 266
Mill Street: factory sites, 115–16, 226–227, *228*; industry, 106, 109
Milwaukee, 4, 267, 270
Mitchell, Robert Foundry, 34, *56*, 123, 241
Moffatt, George, 26–27

Molson Brewery, 79–83
Molson Family, 79, 96
Montreal: built environment, 37–40; commercial city, 26–30, 37; industrial development, 25, 30–37, 135–43; industrial structure, 32–34, *33*, 139–41, *140;* land development, 41–44, 145–50; locational assets, 37–40, 45–48, 155–56; manufacturing districts, 1, 5–22, 46–48, *47*, 131–35, *133*, *134;* manufacturing pathways, 30–37, 141–43; parallels with other cities, 10; political institutions, 38–40, 150–56; population, 29, 40–41, *41*, 143–45, *144;* social geography, 40–45, 143–50
Montréal Est: harbor, 155, 264; industrial policy, 260; industry, 1, 7, 194–98, 222, 257–59; manufacturing district, 133, 147, 256; manufacturing structure, *195;* residential, 211–20
Montreal Junior Board of Trade, 152, 252
Montreal Locomotive and Machine Co., 194–95, *197*
Montreal Marine Works, 103–5, 114
Montreal Park and Island Railway Co., 148, 248
Montreal Rolling Mills, 124–26, *127*, 143, 240–44
Montreal Rubber, 106
Montreal Street Railway Co., 155, 156
Montreal West, 253
Mont Royal, 45
Morgan, Henry, 94
Morland, Thomas, 125
Muir, George, 44

Nashville, 4, 267
Nazerth fief, 113
New York: annexation, 260; harbor, 39; industry, 53, 71, 161; manufacturing geography, 3, 4, 10, 262; office building, 157, 167; social inequalities, 154

New York Insurance Co. Building, 158
Nobbs, Percy, 152, 154
North End, 10, 132, 183, 205–20, 266
Northern Electric Co., 232, *233*
Notre Dame de Grâce, *41*, *144*, 146, 148, 253
Nye, G., 35

Ogilvie Flour, 165, 226–27
Old Montreal: head offices, 136, 259; industry, 7, 50–52, 85, 131, 158–59, 172–84, 241; locational assets, 52, 159–60; manufacturing decline, 157–59, 262; manufacturing district, 1, 10, 46, *47*, 180, 209; manufacturing structure, *50*, *159;* occupations, *64;* population, 63; redevelopment, 59–60, 265; residential change, 45
Ostell, John, 115
Ostell Sawmill, 106
Outer Core: industry, 52–53, 131, 160–61, 262; manufacturing district, 10, 46, 205; manufacturing structure, *50*, *159;* occupations, *64;* population, 63, 172; redevelopment, 265, 267
Outremont, *144*, 145, 146, 205, *207*, 212, 214–15
Outwork, 72–73, 177–80

Paint industry, 32, 205–8, 235; Fergusson, Alexander, 222; McCaskill Varnish, 123; Sherwin-Williams, 242
Park Extension, 148, 149, 212
Park Realty Co., 148
Paterson, New Jersey, 75
Paul, Walter, 57
Paxton Barrel, 119
Peck, John Clothing, 183, *184*
Pelletier, E., 148
Petite Bourgogne, 114
Petty commodity production. *See* Handicraft production

Philadelphia: annexation, 260; industry, 12; journey to work, 270; manufacturing districts, 262; suburbs, 3–5, 20, 98; urban structure, 64; working-class areas, 269
Phillips Electrical Co., 229–30
Pillow & Hersey Co., 107–9, 240
Pittsburgh: factory districts, 262, 269; growth of, 10; industry, 12, 139; mill towns, 4, 17, 18, 98, 267, 268; social inequalities, 154
Planning, 19–22, 122–23, 150–56, 211–20, 263–71; ideas about, 150–51, 165–66; lack of planning controls, 152, 153, 169; middle-class suburbs, 145–46, 252–53; uncoordinated suburban growth, 150–52, 211–16, 252–53, 263–66; zoning, 151, 153–54, 216, 250
Plateau, 180, 205, *206*, 266
Pointe-aux-Trembles, 213, 216, *217*
Point Saint-Charles, *108;* community, 248–49, *249;* housing, 115, 247, 270; industry, 109, 230–32
Political autonomy, 259–61
Portland, Maine, 29
Préfontaine, Raymond, 87, 92
Prieur, Rosaire, 216
Printing industry, 65–69, *68, 162;* Burland Lithographic, 66; Canada Bank Note Engraving, 66; Lovell Printing, 67; Sabiston Lithographic, 66
Property industry, 20–22, 41–44, 145–50; and building cycles, 20; in central manufacturing districts, 58–60, 171–72; in East End, 92–99, 211–20; and locational assets, 45–46; in West End, 111–16, 122–23, 244–47, 265–66
Propriéte Préfontaine, 213
Propulsive firms: character of, 17, 257; and manufacturing districts, 100, 205, 220, 258; in Montreal, 139; and suburbanization, 22, 47, 133–35, 156, 257–59
Putting out system, 121–22, 125

Québec City, 27, 39, 200
Quebec Housing Act, 216
Quesnel, Frédéric Auguste, 122

Railroad, 136; and firm location, 2–3, 7, 19–20, 63, 264; and intercity links, 28–29; as locational assets, 170, 242, 264–65; spur lines, 20, 195–96, 209, 229, *233*, 256
Ramsey, Alexander, 32
Raphael, Thomas, 126
Ravenhill and Molson Straw Works, 85
Redmond Foundry, 106
Redpath, Peter, 115
Redpath Sugar, 103, *104*, 105, 115, 165, 223
Reform movement, 152, 154, 264
Retail districts: central, *51*, 52, 59, 69, 161, *164;* suburban, 248–49, *249*
Reynolds, Lloyd, 248
Rideau Shoe, 203
Robertson Lead Works, 109, *110*
Rolland Family, 87, 92, 94
Rosemont: ethnicity of, 215–16; housing, 219, 269; population, 144, 213; residential growth, 148, 213–14
Rosemont Land Improvement Co., 148
Roy, Gabrielle, 252
Royal Electric Co., 131, *132*, 138, 227
Rural migrants, 29, 41, 93, 145
Rutherford Door and Sash, 123

Saint-Anne Ward: harbor, 155; industry, 101, 102, 105, 189, 264; occupations, *114;* population, 111, *113*, 115, 143; working-class, 44, 113, 244–46
Saint-Antoine, 48, 163; cigar firms, 176–77; lofts, 166–67; manufacturing, 7, 52–53, 58, 71, 160–61, 183–84; manufacturing district, 46–47; occupation and ethnicity, *64*, 265; railroad terminal, 63
Saint-Antoine Ward, 45, 113
Saint-Augustin. *See* Saint-Henri
Saint-Denis, 144

Sainte-Cunégonde, *124;* annexation, 260; ethnic structure, 122; industrial policy, 250–53, 260; industry, 121–28; manufacturing district, 31, 48, 256; politics, 122–23, 147; population, *41, 118,* 122, *144,* 147; residential growth, 122–23, 244–46, 266

Sainte-Marie: industry, 79–87, 185, 194, 210, 261; manufacturing district, 1, 10, 47, *80,* 132, 135, 256, 259, 270; manufacturing structure, *80, 186;* occupations, *94;* population, *93;* residential growth, 92–99, 144, 211–20; working-class, 44, 266

Saint-Gabriel Domain, *108,* 114, 115

Saint-Gabriel Ward, *41,* 111, *113, 114,* 115, *124, 144,* 245

Saint-Henri, *124;* annexation, 260; ethnic structure, 122; factory sites, 267; industrial policy, 230, 250–53, 260; industry, 121–28, 172, 183, 201, 221, 230, 237–44, 261; manufacturing district, 31, *47,* 48, 88, 153, 256; politics, 122–23, 147; population, *41, 118,* 121, 122, *144,* 147; residential growth, 44, 122–23, 244–46, 266

Saint-Henri des Tanneries. *See* Saint-Henri

Saint-Jacques, 49; lofts, 166–67; manufacturing, 7, 52–53, 57–58, 71, 160–61; manufacturing district, 46–47; occupation and ethnicity, *64,* 205; railroad terminal, 63

Saint-Jacques Ward, 203

Saint-Jean-Baptiste, *41,* 42, 44, *144,* 148, 266

Saint-Lambert, *144*

Saint-Laurent, town, 1, *144*

Saint-Lawrence, 49; cigar firms, 176–77; clothing industry, 180–81; lofts, 166–67; manufacturing, 7, 52–53, 69, 161, 183–84; manufacturing district, 46–47; occupation and ethnicity, *64,* 265

Saint-Lawrence Ward, 107, 172

Saint-Louis de Mile End. *See* Mile End

Saint-Paul. *See* Côte Saint-Paul

Saint-Pierre (town): housing, 247–49; industry, 237–44; population, *144,* 147, 221, 256, 269

San Francisco, 4, 18, 262

Satellite Cities, 4

Schnore, Leo, 2

Schoenfeld, Julia, 155

Scientific managerialism, 16, 138, 141

Scott, Allen, 2, 3, 259

Scranton, Philip, 12

Shearer, James Sash and Door, 102, 105

Shoe industry, 35–36, 50–51, 57, 198–204; Ames, Millard, 57; Boston Shoe, 210; Cimon, A., 210; Daoust, Lalonde, 160; McFarlane Shoe, 210

Shorey, H. and Co., *73,* 74

Skyscrapers, 157–58

Slater, David, 2, 255, 259

Sources: city directory, 6, 274; government reports, 6; industrial journals, 6; manufacturing census, 273–74; newspapers, 6; water tax assessments (rôle d'évaluation), 6, 274–75

South Shore, 111, 151, 171

Spatial scale, range of, 276–77

St. Ann Spinning, 78, 90, 191

St. Gabriel Locks, 102, 103, 106, 115, 222

St. Hyacinthe, 176, 211

St. Lawrence Glass, 126–27

St. Lawrence Sugar, 111, 194, 217–18, *218*

St. Louis, 262

Standard Shirt, 183, 189

Staples, 26–28

Steel Company. *See* Montreal Rolling Mills

Stephans, George, 151

Suburbanization, population, 40–41, 143–45, 150–51, 255–56. *See also* Industrial suburbanization; Working-class suburbanization

Sulpicians: manufacturing sites, 115, 267; urban growth, 101, 113, 114, 121, 214

Tacony (Philadelphia), 96
Tarte, Israel, 169, 211–12
Taylor, Graham, 4
Technology, 6, 15–16, 34, 136–38; and clothing industry, 72; and flour industry, 34, 226–27; and metal industry, 240–41; and shoe industry, 199–200; and tobacco industry, 243–44
Telegraph, 28–29
Tellier, James, 160
Textile industry, 87–90, 191; Colonial Bleaching and Printing, 242, 245; Converse Rope, 81; Dominion Cotton, 136, 138, 242; Dominion Textile, 242; Dominion Wadding, 123; Gault Excelsoir Woolen Mills, 191; Harris, Fred, 36, 102, 115; Hochelaga Manufacturing Company, 35, 90, *190;* Mount Royal Spinning Wool, 243; Woods Cotton, 35, 88
Tobacco industry, 90–92, 141–42, 172–77, 243–44; American Tobacco, 11, 136, 142, 176; Benson and Hedges, 166; Ritchie, D., 243; Simon, H., 209–10
Tooke Shirt, 183, 245, 250
Toronto: foreign investment, 136; fringe employment, 17, 20, 146, 267; growth, 10; housing, 148; industry, 69, 71, 231, 235, 262; journey to work, 270
Town of Mount Royal, 146, *207*
Transit: commuting, 145, 147, 215; development of trolley, 149–50, 212, 247–48; horsecar, 63; and manufacturing, 7, 155, 171, 218; and manufacturing districts, 19–20, 269–70
Turcot family, 122

Union Abattoir, 123
United Shoe Machinery Co., 136, 201–3, *202*
Urban America, 3
Urban geography, interpretations of, 1–5, 11–18, 98–99, 133–35, 146, 240, 255–56, 265–69

Vancouver, 29, 136, 183, 208
Verdun: growth of, 149, 247; housing, 248–49; planning of, 250; population, *144,* 269
Versailles, Joseph, 216
Victoriatown, *108,* 115
Villeneuve, Joseph-Octave Villeneuve, 212
Villeray, 216

Wallpaper industry, 230
Warner, Sam Bass, 2, 64, 98
Watson, Charles, 125
Watson, J. C., Wallpaper, 189, 230
Webster, Arthur, 122
West End, 100–128, 221–53; industry, 100, 105, 221, 238–41, 257, 258; locational assets, 101–11, 237–44; manufacturing district, 10–11, 47–48, 256; residential growth, 111–16, 244–53, 269
Westmount, *41, 144, 145,* 146, 253
Wheat economy, 226
Winnipeg, 136, 235
Withall and Hood, 57
Woodlands Park, 245
Working-class suburbanization, 20–22, 44–46, 138, 146–50, 265–71; and East End, 92–99, 211–20; views of, 3–5; and West End, 111–17, 122–23, 244–53
Workman, William, 31, 117–21, 122

Young, John, 39–40, 60, 115

ABOUT THE AUTHOR

Robert David Lewis was born in Malta and was raised in England and Australia. He received his B.A. degree in geography and history, with honors, from the University of Toronto and completed his M.A. and Ph.D. degrees in geography at McGill University in Montreal. Dr. Lewis has published articles on the historical geography of urban economic change in North America in the *Annals of the Association of American Geographers, Canadian Geographer, Economic Geography, Journal of Historical Geography, Journal of Urban Affairs, Journal of Urban History, Urban Geography,* and *Urban History Review*. He has also edited and organized special issues of the *Journal of Historical Geography* and *Urban History Review,* and he has received various research awards from Canadian organizations and universities. Dr. Lewis is an associate professor of geography at the University of Toronto.

RELATED BOOKS IN THE SERIES

Boston's "Changeful Times": Origins of Preservation and Planning in America
   Michael Holleran

Cities and Buildings: Skyscrapers, Skid Rows, and Suburbs
   Larry R. Ford

The City Beautiful Movement
   William H. Wilson

The Cotton Plantation South since the Civil War
   Charles S. Aiken

From Aztec to High Tech: Architecture and Landscape across the Mexico–United States Border
   Lawrence A. Herzog

Invisible New York: The Hidden Infrastructure of the City
   Stanley Greenberg, with an introductory essay by Thomas H. Garver

Local Attachments: The Making of an American Urban Neighborhood, 1850 to 1920
   Alexander von Hoffman

Magnetic Los Angeles: Planning the Twentieth-Century Metropolis
   Greg Hise

The North American Railroad: Its Origin, Evolution, and Geography
   James E. Vance Jr.

Redevelopment and Race: Planning a Finer City in Postwar Detroit
   June Manning Thomas

The Rough Road to Renaissance: Urban Revitalization in America, 1940–1985
   Jon C. Teaford

Unplanned Suburbs: Toronto's American Tragedy, 1900 to 1950
   Richard Harris

Library of Congress Cataloging-in-Publication Data
Lewis, Robert, 1954–
 Manufacturing Montreal: the making of an industrial landscape, 1850 to 1930 / Robert Lewis.
   p.   cm. — (Creating the North American landscape)
 Includes bibliographical references and index.
   ISBN 0-8018-6349-X (hc : alk. paper)
   1. Industries—Québec (Province)—Montréal. 2. Manufacturing industries—Québec (Province)—Montréal. 3. Industrial productivity—Québec (Province)—Montréal. 4. Montréal (Québec)—Economic conditions. I. Title. II. Series.
 HC118.M6 L49 2000
 338.4′767′0971428—dc21
 99-050708